DC Microgrids

Scrivener Publishing
100 Cummings Center, Suite 541J
Beverly, MA 01915-6106

Publishers at Scrivener
Martin Scrivener (martin@scrivenerpublishing.com)
Phillip Carmical (pcarmical@scrivenerpublishing.com)

DC Microgrids

Advances, Challenges, and Applications

Edited by
Nikita Gupta
Mahajan Sagar Bhaskar
Sanjeevikumar Padmanaban
and
Dhafer Almakhles

Scrivener Publishing

WILEY

This edition first published 2022 by John Wiley & Sons, Inc., 111 River Street, Hoboken, NJ 07030, USA and Scrivener Publishing LLC, 100 Cummings Center, Suite 541J, Beverly, MA 01915, USA

For more information about Scrivener publications please visit www.scrivenerpublishing.com.

Wiley Global Headquarters
111 River Street, Hoboken, NJ 07030, USA

For details of our global editorial offices, customer services, and more information about Wiley products visit us at www.wiley.com.

Library of Congress Cataloging-in-Publication Data

ISBN 978-1-119-77716-8

Cover image: Electric Power Lines, Tifonimages | www.dreamstime.com
Cover design by Kris Hackerott

Set in size of 11pt and Minion Pro by Manila Typesetting Company, Makati, Philippines

Printed in the USA

10 9 8 7 6 5 4 3 2 1

Contents

Preface xv

1 On the DC Microgrids Protection Challenges, Schemes, and Devices – A Review 1
Mohammed H. Ibrahim, Ebrahim A. Badran and Mansour H. Abdel-Rahman
1.1 Introduction 2
1.2 Fault Characteristics and Analysis in DC Microgrid 4
1.3 DC Microgrid Protection Challenges 7
1.3.1 Low Inductance of DC System 7
1.3.2 Fast Rise Rate of DC Fault Current 7
1.3.3 Difficulties of Overcurrent (O/C) Relays Coordination 7
1.3.4 Fault Detection and Location 8
1.3.5 Arcing Fault Detection and Clearing 10
1.3.6 Short-Circuit (SC) Analysis and Change of Its Level 13
1.3.7 Non-Suitability of AC Circuit Breakers (ACCBs) 16
1.3.8 Inverters Low Fault Current Capacity 17
1.3.9 Constant Power Load (CPL) Impact 17
1.3.10 Grounding 18
1.4 DC Microgrid Protection Schemes 21
1.4.1 The Differential Protection-Based Strategies 25
1.4.2 The Voltage-Based Protection Strategies 27
1.4.3 The Adaptive Overcurrent Protection Schemes 28
1.4.4 Impedance-Based Protection Strategy (Distance Protection) 29
1.4.5 Non-Conventional Protection Schemes (Data-Based Protection Scheme) 32
1.5 DC Microgrid Protective Devices (PDs) 34
1.5.1 Z-Source DC Circuit Breakers (ZSB) 35
1.5.2 Hybrid DC Circuit Breakers (HCB) 38
1.5.3 Solid State Circuit Breakers (SSCBs) 42

1.5.4 Arc Fault Current Interrupter (AFCI) 45
1.5.5 Fuses 47
1.6 Conclusions 48
References 50

2 Control Strategies for DC Microgrids 63
Bhabani Kumari Choudhury and Premalata Jena
2.1 Introduction: The Concept of Microgrids 63
2.1.1 DC Microgrids 64
2.2 Introduction: The Concept of Control Strategies 65
2.2.1 Basic Control Schemes for DC MGs 66
2.2.1.1 Centralized Control Strategy 66
2.2.1.2 Decentralized Controller 67
2.2.1.3 Distributed Control 68
2.2.2 Multilevel Control 68
2.2.2.1 Primary Control 69
2.2.2.2 Secondary Control 73
2.2.2.3 Tertiary Control 74
2.2.2.4 Current Sharing Loop 74
2.2.2.5 Microgrid Central Controller (MGCC) 74
2.3 Control Strategies for DGs in DC MGs 76
2.3.1 Control Strategy for Solar Cell in DC MGs 76
2.3.1.1 Control Strategy for Wind Energy in DC MGs 77
2.3.1.2 Control Strategy for Fuel Cell in DC MGs 77
2.3.1.3 Control Strategy for Energy Storage System in DC MGs 78
2.4 Conclusions and Future Scopes 79
References 80

3 Protection Issues in DC Microgrids 83
Bhabani Kumari Choudhury and Premalata Jena
3.1 Introduction 83
3.1.1 Protection Challenge 84
3.1.1.1 Arcing and Fault Clearing Time 84
3.1.1.2 Stability 85
3.1.1.3 Multiterminal Protections 85
3.1.1.4 Ground Fault Challenges 85
3.1.1.5 Communication Challenges 86
3.1.2 Effect of Constant Power Loads (CPLs) 86
3.2 Fault Detection in DC MGs 87
3.2.1 Principles and Methods of Fault Detection 87

3.2.1.1 Voltage Magnitude-Based Detection 87
3.2.1.2 Current Magnitude-Based Detection 88
3.2.1.3 Impedance Estimation Method 88
3.2.1.4 Power Probe Unit (PPU) Method 88
3.3 Fault Location 92
3.3.1 Passive Approach 92
3.3.1.1 Traveling Wave-Based Scheme 92
3.3.1.2 Differential Fault Location 93
3.3.1.3 Local Measurement-Based Fault Location 93
3.3.2 Active Approach for Fault Location 94
3.3.2.1 Injection-Based Fault Location 94
3.4 Islanding Detection (ID) 94
3.4.1 Types of IDSs 95
3.4.2 Passive Detection Schemes (PDSs) for DC MGs 96
3.4.3 Active Detection Schemes (ADS) for DC MGs 96
3.5 Protection Coordination Strategy 97
3.6 Conclusion and Future Research Scopes 97
References 97

4 Dynamic Energy Management System of Microgrid Using AI Techniques: A Comprehensive & Comparative Study 101
Priyadarshini Balasubramanyam and Vijay K. Sood
Nomenclature 102
4.1 Introduction 103
4.1.1 Background and Motivation 103
4.1.2 Prior Work 103
4.1.3 Contributions 104
4.1.4 Layout of the Chapter 104
4.2 Problem Statement 104
4.3 Mathematical Modelling of Microgrid 105
4.3.1 Cost Functions 106
4.3.1.1 Diesel Generator 106
4.3.1.2 Solar Generation 106
4.3.1.3 Wind Generation Unit 106
4.3.1.4 Energy Storage System (ESS) 107
4.3.1.5 Transaction with Utility 108
4.3.2 Objective Function 109
4.3.3 Constraints 109
4.4 Optimization Algorithm 110
4.4.1 Heuristic-Based Genetic Algorithm (GA) 110
4.4.2 Pattern Search Algorithm (PSA) 111

4.5 Results 113
4.6 Conclusion 118
References 118

5 Energy Management Strategies Involving Energy Storage in DC Microgrid 121
S. K. Rai, H. D. Mathur and Sanjeevikumar Padmanaban
5.1 Introduction 121
5.2 Literature Review 123
5.2.1 Classic Approaches of EMS 124
5.2.2 Meta-Heuristic Approach of EMS 129
5.2.3 Artificial Intelligence Approach of EMS 134
5.2.4 Model Predictive, Stochastic and Robust Programming Approach of EMS 139
5.3 Case Study 142
5.3.1 Energy Management System 144
5.3.2 Objective Functions 144
5.3.3 Result and Discussion 145
5.4 Conclusion 151
References 151

6 A Systematic Approach for Solar and Hydro Resource Assessment for DC Microgrid Applications 159
Sanjay Kumar, Nikita Gupta, Vineet Kumar and Tarlochan Kaur
6.1 Introduction 160
6.1.1 Micro Hydro and Solar PV 162
6.1.2 Renewable Energy for Rural Electrification in Indian Perspective 162
6.1.3 Solar Resource Assessment 163
6.1.4 Hydro Resource Assessment 166
6.1.5 Demand Assessment 167
6.2 Methodology 168
6.2.1 Data Collection 168
6.2.1.1 Meteorological and Geographical Data 168
6.2.1.2 Discharge Data for Hydro Potential Estimation 168
6.3 Result and Discussion 172
6.3.1 ANN Architecture 172
6.3.2 Hydro Resource Estimation 176
6.4 Conclusion 178
References 179

7 **Secondary Control Based on the Droop Technique for Power Sharing** **183**
Waner W.A.G. Silva, Thiago R. de Oliveira, Rhonei P. Santos and Danilo I. Brandao
7.1 Introduction 184
7.2 Voltage Deviation and Power Sharing Issues in Droop Technique 186
7.2.1 Approaches for Correcting Power and Current Sharing 190
7.2.2 Hybrid Secondary Control: Distributed Power Sharing and Decentralized Voltage Restoration 197
7.2.2.1 Dynamics and Convergence of the Power Sharing Correction 200
7.2.2.2 Communication Delays in Consensus-Based Algorithm 203
7.2.2.3 Secondary Control Modeling 204
7.2.2.4 Computational and Experimental Validation 208
7.2.3 Secondary Level Control Based on Unique Voltage-Shifting (*vs*) 215
7.2.3.1 Power Sharing and Average Voltage Convergence Analysis 218
7.2.3.2 Secondary Control Level Modeling 223
7.2.3.3 Computational and Experimental Validation 226
7.3 Design and Implementation of the Communication System 230
7.4 Conclusions 234
References 235

8 **Dynamic Analysis and Reduced-Order Modeling Techniques for Power Converters in DC Microgrid** **241**
Divya Navamani J., Lavanya A., Jagabar Sathik, M.S. Bhaskar and Vijayakumar K.
8.1 Introduction 242
8.2 Need of Dynamic Analysis for Power Converters 243
8.3 Various Modeling Techniques 245
8.3.1 Analysis from Modeling Method 249
8.4 Reduce-Order Modeling 253
8.4.1 Faddeev Leverrier Algorithm 253
8.4.1.1 Procedure for Faddeev Leverrier Algorithm 253

8.4.1.2 Illustrative Example with Switched-Inductor-Based Quadratic Boost Converter 254
8.4.2 Order Reduction of Transfer Function 257
8.4.3 Techniques for Model Order Reduction 257
8.4.4 Pole Clustering Method 258
8.4.5 Procedure for Improved Pole Clustering Technique 258
8.4.5.1 Computation of Denominator Polynomial of Lower-Dimensional Model 259
8.4.5.2 Computation of Numerator Polynomial of Lower-Dimensional Model 261
8.4.5.3 Design of Controller 261
8.5 Illustrative Example with the Power Converter 262
8.5.1 Derivation of the Denominator 263
8.5.2 Derivation of the Numerator 264
8.6 Controllers for Power Converter 265
8.6.1 Need of Controller 265
8.6.2 Types of Controller 265
8.7 Conclusion 267
References 267

9 Matrix Converter and Its Probable Applications 273
Khaliqur Rahman
9.1 Introduction 274
9.2 Classification of Matrix Converter 275
9.2.1 Classical Matrix Converter 277
9.2.2 Sparse Matrix Converter 277
9.2.3 Very Sparse Matrix Converter 277
9.2.4 Ultra-Sparse Matrix Converter 278
9.3 Problems Associated with the MC and the Drives 280
9.3.1 Commutation Issues 280
9.3.2 Modulation Issues 280
9.3.3 Common-Mode Voltage and Common-Mode Current Issues 280
9.3.4 Protection Issues 281
9.4 Control Techniques 282
9.5 Basic Components of the Matrix Converter Fed Drive System 283
9.6 Industrial Applications of Matrix Converter 289
9.7 Summary 294
References 294

10 Multilevel Converters and Applications **299**
P. Prem, Jagabar Sathik and K.T. Maheswari
10.1 Introduction 300
10.2 Multilevel Inverters 301
10.2.1 Multilevel Inverters vs. Two-Level Inverters 301
10.2.2 Advantages of Multilevel Converters Based on Waveforms 303
10.2.3 Advantages of Multilevel Converters Based on Topology 304
10.3 Traditional Multilevel Inverter Topologies 305
10.3.1 Diode Clamped Multilevel Inverter 305
10.3.1.1 Features of DCMLI 308
10.3.1.2 Advantages of DCMLI 308
10.3.1.3 Disadvantages of DCMLI 308
10.3.1.4 Applications of DCMLI 309
10.3.2 Flying Capacitor Multilevel Inverter 309
10.3.2.1 Features of FCMLI 312
10.3.2.2 Advantages of FCMLI 312
10.3.2.3 Disadvantages of FCMLI 312
10.3.2.4 Applications of FCMLI 313
10.3.3 Cascaded H Bridge Multilevel Inverter 313
10.3.3.1 Features of CHBMLI 315
10.3.3.2 Advantages of CHBMLI 315
10.3.3.3 Disadvantages of CHBMLI 316
10.3.3.4 Applications of CHBMLI 316
10.4 Advent of Active Neutral Point Clamped Converter 316
10.4.1 Comparison with Traditional Topologies 319
10.4.2 Advantages of ANPC MLI 320
10.4.3 Disadvantages of ANPC MLI 320
10.5 Conclusion 322
References 322

11 A Quasi Z-Source (QZS) Network-Based Quadratic Boost Converter Suitable for Photovoltaic-Based DC Microgrids **325**
Amir Ghorbani Esfahlan and Kazem Varesi
11.1 Introduction 326
11.2 Proposed Converter 328
11.3 Steady-State Analyses 331
11.4 Comparison with Other Structures 335
11.5 Converter Analyzes in Discontinuous Conduction Mode (DCM) 335

11.6 Simulation Results 342
11.7 Real Voltage Gain and Losses Analyzes 346
11.8 Dynamic Behavior of the Proposed Converter 352
11.9 The Maximum Power Point Tracking (MPPT) 354
11.10 Conclusions 356
11.11 Appendix 357
References 358

12 Research on Protection Strategy Utilizing Full-Scale Transient Fault Information for DC Microgrid Based on Integrated Control and Protection Platform 361
Shi Bonian and Sun Gang
12.1 Introduction 362
12.2 Topological Structure and Grounding Model of Studied Microgrid 363
12.2.1 Proposed DC Distribution Network Topology 363
12.2.2 Neutral Grounding Model 366
12.2.2.1 Grounding Position Selection 366
12.2.2.2 Grounding Mode Selection 366
12.3 Fault Characteristics of DC Microgrid 367
12.3.1 DC Unipolar Fault Characteristics 368
12.3.2 DC Bipolar Fault Characteristics 370
12.4 DC Microgrid Protection Strategy 373
12.4.1 Protection Zone Division and Protection Configuration 373
12.4.1.1 Protection Zone Division 373
12.4.1.2 Protection Configuration 375
12.4.2 Integrated Control and Protection Platform 376
12.4.3 Fault Isolation and Recovery Strategy Utilizing Full-Scale Transient Fault Information 378
12.4.3.1 Unipolar Fault Isolation and Recovery of DC Line/Bus 378
12.4.3.2 Bipolar Fault Isolation and Recovery of DC Line/Bus 380
12.5 Simulation Verification 384
12.5.1 Verification under DC Unipolar Fault 386
12.5.1.1 Metal Short Circuit Fault of DC Line 386
12.5.1.2 Unipolar Fault with High Transition Resistance 386
12.5.1.3 High Resistance Unipolar Fault with Parallel Resistance Switching Strategy 386

12.5.2 Verification under DC Bipolar Fault 390
12.6 Conclusion 394
References 395

13 A Decision Tree-Based Algorithm for Fault Detection and Section Identification of DC Microgrid 397
Shankarshan Prasad Tiwari and Ebha Koley
Acronyms 398
Symbols 398
13.1 Introduction 398
13.2 DC Test Microgrid System 400
13.3 Overview of Decision Tree-Based Proposed Scheme 401
13.4 DC Microgrid Protection Using Decision Tree Classifier 403
13.5 Performance Evaluation 404
13.5.1 Mode Detection Module 408
13.5.2 Fault Detection/Classification 409
13.5.3 Section Identification 409
13.5.4 Comparative Analysis of the Proposed Scheme with other DC Microgrid Protection Techniques 412
13.6 Conclusion 416
References 417

14 Passive Islanding Detection Method Using Static Transfer Switch for Multi-DGs Microgrid 421
Rahul S. Somalwar and S. G. Kadwane
14.1 Introduction 422
14.1.1 Technical Challenges of Microgrid and Benefits 424
14.1.2 System with Multi-DGs 425
14.1.3 Power Sharing Methods 426
14.1.3.1 Conventional Droop Control Method 426
14.2 Islanding 427
14.2.1 Challenges with Islanding 427
14.2.2 Different Standards for Microgrid 428
14.2.3 Islanding Detection Methods 428
14.3 Static Transfer Switch (STS) 431
14.3.1 Simulation Results of STS 432
14.4 Proposed Scheme of Islanding 435
14.4.1 Proposed PV System 435
14.4.2 Mathematical Analysis of Harmonic Extraction 436
14.5 Flow Chart 437
14.6 Simulation Results 438

14.7 Experimental Results 441
14.8 Conclusion 445
References 446

Index **449**

Preface

The electric grid is on the threshold of a significant change. In the past few years, the picture of the grid has changed dramatically due to the introduction of renewable energy sources, advancements in power electronics, digitalization, and other factors. All these mega-trends are pointing toward a new electrical system based on Direct Current (DC). DC power systems have inherent advantages of no harmonics, no reactive power, and high efficiency, over conventional AC power systems. Hence, DC power systems have become an emerging and promising alternative in various emerging applications, which include distributed energy sources like wind, solar, and Energy Storage Systems (ESS); distribution networks; smart buildings, remote telecom systems; and transport electrification like electric vehicles (EVs) and shipboard. All these applications are designed at different voltages to meet their specific requirements individually because of the lack of standardization. Thus, the factors influencing the DC voltages and system operation needed to be surveyed and analyzed, which include voltage standards, architecture for existing and emerging applications, topologies and control strategies of power electronic interfaces, fault diagnosis and design of the protection system, optimal economical operation, and system reliability.

This groundbreaking new volume presents these topics and trends of DC microgrids, bridging the research gap on DC microgrid architectures, control, and protection challenges to enable wide-scale implementation of energy-efficient DC microgrids. Whether for the veteran engineer or the student, this is a must-have for any library. This book also presents a detailed analysis of DC electrical architectures and power management methods with their technological aspects. Since the late twentieth century, power electronics technology has progressively gained traction. This technology is now practically allowing DC to reclaim ground from AC power. Of course, standard inertia is a huge roadblock, and it may be a long time before DC overtakes AC, but it will happen eventually. DC power distribution is expected to play a vital role in the optimal use of renewable resources as well as the enhancement of electrical system resilience. DC will cause higher-performing smart grids that are more stable and efficient throughout their life cycle while conserving energy and raw materials. In this book, at various stages of power conversion, the authors give a variety of energy models, management tactics, and control rules. They concentrate on renewable energy conversion like solar, energy storage systems, backup generators, and, of course, smart microgrids. Furthermore, the numerous experimental results and associated simulations contribute significantly to the book's high quality.

Chapter 1 discusses DC microgrid protection challenges, fault detection methods, and design criteria for an efficient protective system. DC microgrid protection strategies and both line-to-ground and line-to-line faults, besides their impacts, are reviewed. Also, DC fault current interrupting devices are presented. Chapter 2 provides a comprehensive overview of different existing control schemes for DC microgrids, along with the motivation and challenges behind them. It covers the basic and multi-level control schemes in detail, along with the converter control scheme for solar, wind, battery, and fuel cell-based systems. Chapter 3 describes different basic fault detection, location, and islanding detection methods for DC microgrids, along with the advantages and disadvantages of the schemes. Chapter 4 proposes an optimized energy management system for a microgrid consisting of solar and wind generation units, an energy storage system, and a diesel generator. An optimization model for the energy management system was developed to minimize the overall operating and maintenance costs for the real power flow in the microgrid. The optimization was done using genetic algorithms and pattern search algorithms and a comparison is made based on the total operating and maintenance costs for real power flow. Chapter 5 presents the literature review of various energy management strategies involving energy storage in a microgrid, and a case study is presented by optimizing the involved objective functions using the linear programming method for a residential microgrid. The optimal operating mode is decided hourly to get the optimal cost. Chapter 6 presents the design of a hybrid renewable energy system using parameters such as available solar radiation, available hydro potential, demand assessment, etc., for DC microgrid application. The present study developed a systematic approach for solar and hydro potential assessment for DC microgrid applications. Chapter 7 details the influence of line equivalent resistances on power-sharing and on average DC bus voltage errors caused by droop techniques, reviewing the secondary control methods proposed in the literature to correct such errors. Moreover, two secondary control techniques based on distributed control, proposed by the authors, are highlighted, as well as the design of the communication system. Chapter 8 discusses the improvement of power electronic interface performance using dynamic analysis. An improved pole clustering method is proposed for dynamic analysis of quadratic boost converters with switched inductor cells only to unravel the resolvent matrix and prove the methodology's easiness. Additionally, a view of various controllers applied to DC-DC converters is given for researchers to have as a guide for their analytical study of mathematical modelling. Chapter 9 presents a detailed study of the matrix converter. The control of a multiphase matrix converter for multiphase drive applications is discussed to make the system more reliable. A brief description of each part of the drive system fed from the matrix converter is presented. Moreover, several issues associated with the matrix converter, such as modulation, control, and mitigating effects of non-idealities, are discussed. The different applications, especially in more electric aircraft applications, wind power conversion, and applications in electrical drives, are also discussed. Chapter 10 proposes an Active Neutral-Point Clamped Multilevel Inverter that has the advantages of both flying capacitors and neutral point-clamped multilevel inverters, such as flexibility of switching redundancy and robustness to produce multilevel voltages. The operation of the proposed Active Neutral-Point clamped multilevel inverter is explained in detail. Chapter 11 presents a quasi-Z-source network-based DC-DC converter with a quadratic voltage conversion ratio that applies to photovoltaic systems and DC microgrids. In Chapter 12, a typical multi-terminal VSC DC microgrid for the study is proposed, with careful consideration of topology design and grounding mode selection. Then the DC fault characteristics of a multi-terminal DC microgrid are analyzed, including DC unipolar fault and DC bipolar fault, and are used as the theoretical basis for protection configuration. Integrated control and protection platform are introduced, on which a layered setting of protection action delays and some supplementary control measures, such as de-blocking the VSCs and re-closing the DC breakers in the non-fault area, can be implemented. In Chapter 13, a decision tree-based protection scheme is designed for fault detection/classification and faulty section identification of DC microgrids during a faulty condition. The simple structure of the decision tree allows for achieving high accuracy in executing the protection task. The proposed algorithm is based on the acquisition of the voltage and current signals from the relaying buses. In Chapter 14, various passive and active islanding methods are discussed along with their merits and demerits, and individual lower-order harmonics are investigated along with the static transfer switch, which, as a new effective passive islanding method, is proposed for the inverter centered utility connected system.

Nikita Gupta, Mahajan Sagar Bhaskar,
Sanjeevikumar Padmanaban and Dhafer Almakhles
Editors

1

On the DC Microgrids Protection Challenges, Schemes, and Devices – A Review

Mohammed H. Ibrahim*, Ebrahim A. Badran and Mansour H. Abdel-Rahman

Electrical Engineering Department, Faculty of Engineering, Mansoura University, Mansoura, Egypt

Abstract

Expansion in using renewable energy sources has demonstrated a promising and growing penetration of distributed generations into utility networks, causing safety, reliability, and quality challenges which have led to a renewed interest in DC power systems. DC microgrids are an enticing solution for these challenges owing to the use of power electronic devices. Also, it is a relevant approach to conjugate distributed energy systems with power grids, furthermore, its excellent performance and operational capability in grid-connected and islanded modes. Protecting DC microgrids from various faults is a major challenge because of the essence of DC power networks, like enormous DC capacitors, small impedance of DC cables, lack of natural zero-crossing points, high transient current, and voltage. Converters limited fault current short lines and convenient grounding scheme are other protection issues of DC microgrids. This paper discusses DC microgrid protection challenges, fault detection methods, and design criteria for an efficient protective system. DC microgrid protection strategies and both line to ground and line to line faults besides their impacts are reviewed. Also, DC fault current interrupting devices are presented.

***Keywords*:** DC microgrids, protection challenges, protection schemes, DC fault, grounding, and protection devices

**Corresponding author*: eng.mhamed1@gmail.com; mhibrahim@std.mans.edu.eg

Nikita Gupta, Mahajan Sagar Bhaskar, Sanjeevikumar Padmanaban and Dhafer Almakhles (eds.) DC Microgrids: Advances, Challenges, and Applications, (1–62) © 2022 Scrivener Publishing LLC

1.1 Introduction

The enlargement of using Renewable Energy Sources (RESs) and the speed of power electronics technology evolution have led to an increase in the number of Distributed Generators (DGs). DGs also have a few negative effects on utility grids, partially because of the periodic presence of most RESs. The critical challenges include the rise of voltage, protection coordination, power quality and stability of grid, and these must be solved. Hence, dependable, configurable, and smart energy distribution systems are required. Thus, microgrids have arisen and been an enticing arrangement for the implementation of renewable-based DGs [1, 2]. The number of DC-powered devices and applications has grown, including fuel cells, photovoltaic panels, storage batteries, electrical vehicles (EVs), and DC loads which have made use of DC microgrids inevitable, in addition to increasing the efficiency due to a decrease in the stages of power transfer [3]. DC microgrids have several connections linked to the DC bus that can be interfaced by individual systems such as loads, sources, and bi-directional devices using power converters as shown in Figure 1.1. Power converters are sometimes Voltage-Source Converters (VSCs) or Line Commutated Converters (LCCs). The thyristors in LCCs have high power capacity and voltage, but the control performance is lower compared to Isolated Gate Bipolar Transistors (IGBTs) in VSCs. One of several inherent problems with VSC-based DC microgrid protection is the quick observation of fault current and rapidly interrupted it because the current fault that withstands VSC ratings is twice the converter load capacity current [4]. DC microgrids are categorized into two voltage-based levels as follows [5]:

1. Medium voltage DC (MVDC) microgrid and,
2. Low voltage DC (LVDC) microgrid.

LVDC microgrids are connected with AC network through converters, and in this scenario the power stream is bi-directional, so various protection strategies are obligatory for the DC microgrids [6]. Also, as a result of the variances between the protection approaches of AC and DC microgrids, the fault position, classifying and detecting are protection concerns. Furthermore, DC microgrids normally have many load converters, which act as Constant Power Loads (CPLs) [7]. CPLs have an unfavorable impact on the protection system after a voltage decrease, and this is due to their negative incremental impedance [8]. In DC microgrids, there are challenges regarding the continuity of their operation efficiently considering

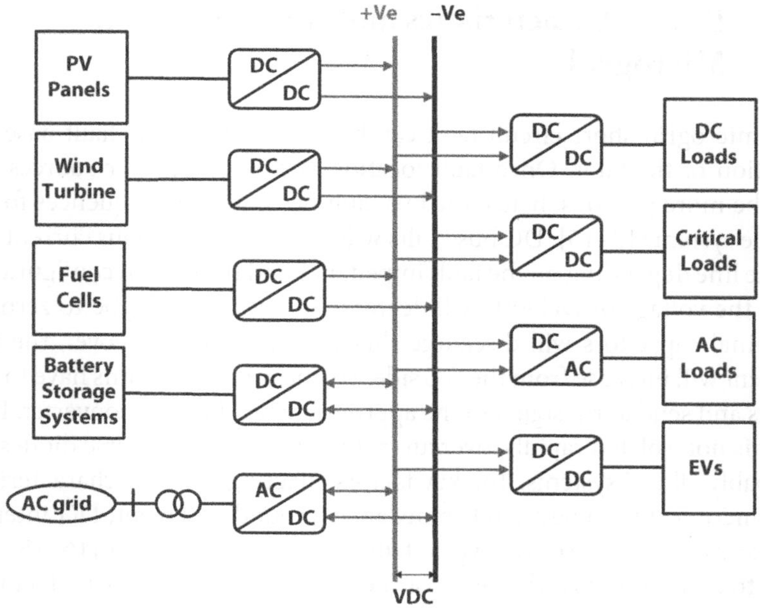

Figure 1.1 Unipolar DC microgrid architecture diagram.

some unusual conditions, such as current of short circuit in the DC bus and instability due to transient; therefore a compatible protection system is necessary for more reliability and efficacy of operation. There are two sorts of faults in DC systems:

1. Line to Line (L-L) fault is known as the path between positive and negative poles formed in the line section. DC voltage across the fault location is reduced to zero, leading to high, fast-rising fault current [4, 9].
2. Line to Ground (L-G) fault, which is the most common type of fault in industrialized distribution systems [10, 11] and is induced by a path connecting either the positive or negative poles to ground. Relying on the grounding layout and the type of grounding, the fault impedance can be either high or low [12].

In this paper, the DC microgrids protection challenges are introduced. Also, the protection strategies and schemes are discussed. Furthermore, the protective devices in DC microgrids are explained.

1.2 Fault Characteristics and Analysis in DC Microgrid

DC microgrid short-circuit fault can be either bus or line fault based on location of the fault. Once fault location is closer to power sources, this will be more perilous, hence the bus fault has major consequences for the whole system [13–15]. DC bus faults will lead to an enormous current flow in the line depending on the fault impedance and the system configuration, thus the voltage of DC link will decrease to low level, maybe to zero, the terminal capacitors will discharge through the fault. Moreover, the fault current will increase from the AC side. So, the protection units have to spot faults and send a trip signal to the appropriate breaker and converter. If the fault is not isolated rapidly, overcurrent might spoil the reverse diodes and probably all DC systems. The key factors affecting the fault characteristics are microgrid topology, fault impedance, grounding layout, DG interfacing converters, DG source types, fault type and its location [16–18]. The pole to pole fault in DC bus is illustrated in Figure 1.2 [4]. The fault current consists of two parts [4, 14]:

1. Transient part represented by (I_c) from bus capacitors and cable discharge of converters
2. Steady-state part represented by (I_s) from power resources

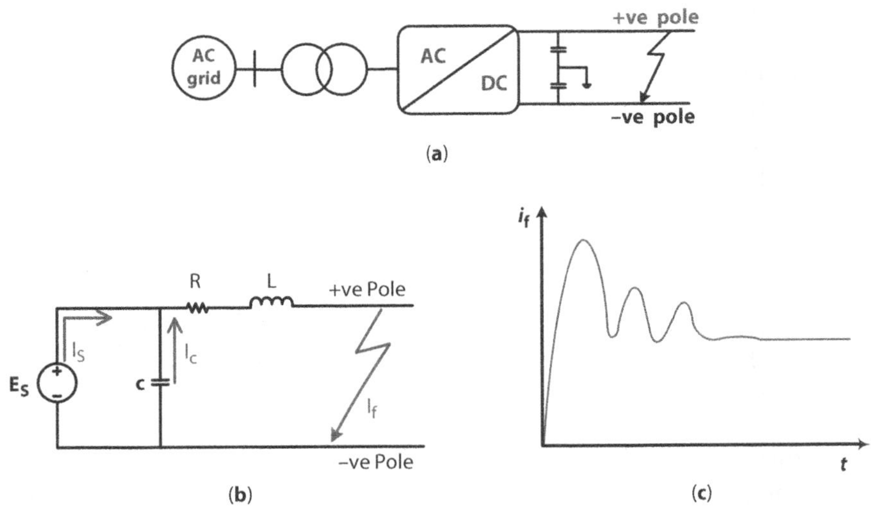

Figure 1.2 (a) Pole - Pole fault in a bipolar DC bus, (b) an equivalent circuit model, (c) Typical fault current waveform.

Equation (1.1) shows the mathematical expression of fault current whenever the equivalent resistance $\left(R < 2\sqrt{L/C}\right)$ in the fault circuit is minimal enough to trigger an oscillation [4]:

$$I_f(t) = \frac{E_S}{R}\left(1 - e^{\frac{-R}{L}t}\right) + \frac{E_s}{\omega_\circ\sqrt{1-\xi^2 L}} e^{-\alpha t}\sin\left(\omega_\circ\sqrt{1-\xi^2 t}\right) \quad (1.1)$$

Where $\omega_\circ = \frac{1}{\sqrt{LC}}$ is natural frequency and $\xi = \frac{\alpha}{\omega_\circ}$ is the damping factor of fault current and $\alpha = \frac{R}{2L}$ [14]. Es is the line voltage, R & L are the equivalent system resistance and inductance, C is the equivalent capacitance in the fault path.

If the fault resistance or bus capacitance is too high, as in a line-to-ground fault with slightly higher ground resistance, the fault current can be modelled as first-order damping with no oscillation [4]. Short-circuit currents calculation in a DC network is very important to design proper protection strategy. So, it is reasonable and interesting to study short-circuit current in a DC network and how it can be calculated. It should also be observed that when calculating the current of short-circuit, the current load is not taken into account. Figure 1.3 illustrates the International Electrotechnical Commission (IEC) standard approximation function for short-circuit fault current that covers the variations of a short-circuit fault current according to different sources, and the equations controlling the function are [4, 14, 19]:

$$i_f(t) = i_1(t) + i_2(t) \quad (1.2)$$

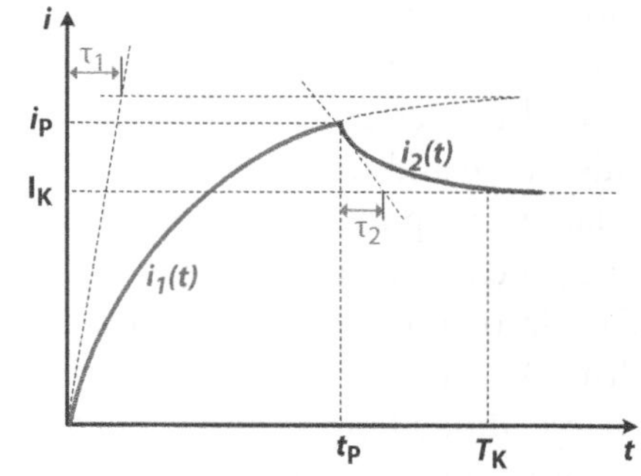

Figure 1.3 Short-circuit fault current in DC system standard approximation.

Where,

$$i_1(t)=i_p\left(\frac{1-e^{-\frac{t}{\tau_1}}}{1-e^{-\frac{t_p}{\tau_1}}}\right) \quad for\ 0\leq t\leq t_p \qquad (1.3),$$

$$and,\ i_2(t)=i_p\left[(1-\alpha)e^{\frac{-(t-t_p)}{\tau_2}}+\alpha\right] \quad for\ t\geq t_p\ and\ \alpha=\frac{I_k}{i_p} \qquad (1.4)$$

where $i_f\ (t)$ is the total fault current, $i_1\ (t)$ and $i_2\ (t)$ are the partial fault currents; i_p is the peak short-circuit current; I_k is the quasi steady-state short-circuit current and it is the value of the short-circuit current 1 second after the beginning of short circuit, t_p is the time to peak τ_1 is the rise-time constant, τ_2 is the decay-time constant and T_K is short circuit duration time [19, 20]. Both currents $i_1\ (t)$ and $i_2\ (t)$ determine the maximum short-circuit current that rules the rating of electrical equipment and the minimum short-circuit current that can be taken as the basis for fuse and protection device ratings, respectively [4, 12].

In DC microgrids, the DC fault might take place either in the DC bus or cables that connect the microgrid apparatuses. DC bus, and DC inter-connections are proposed to perform as a single node of power interface between DG units, Energy Storage Systems (ESSs) and loads. On the viewpoint of protection, the drawback is that a fault on a DC bus or DC cable link affects DG units, ESSs, and loads, both of which may add and participate to the fault current. Hence, if the protection system design is insufficient, a single fault anywhere within the system may have unrecoverable impacts. VSCs play a major role in DC microgrids, where the crossing point to the AC side is via the inductor and to the DC side via the capacitor. During fault condition and due to the VSC design, the DC side capacitor discharges through the DC network, then the fault current involvement from the interfaced converter sources forms the second part of the response capacitor discharge and would result in high current amplitude that could affect the faulty components and other components. If the current fault is part of the protection plan, the excess maximum value of fault current must be taken into consideration on all system and component design procedures. The DC faulted current has three phases during a VSC fault: capacitor discharge, diode freewheeling, and grid-side current feeding phase. The phases were analyzed and characterized mathematically in [14].

1.3 DC Microgrid Protection Challenges

Protecting the DC microgrids with RES from different faults is a major obstacle. Therefore, accurate diagnosis (detect and locate) of the fault leads to fast restoration and recovery of the system and moreover decreases interval time of power outage [21]. The main obstacles associated with DC microgrid protection challenges are presented in the following subsections.

1.3.1 Low Inductance of DC System

During faults in AC power systems, the fault current is limited by fault impedance which has two components: reactance and resistance. The reactance value of fault impedance is generally larger than resistance value. Unlike DC systems, the reactance value is insignificant relative to the resistance of system. Subsequently, not only is the maximum value of DC fault currents greater than the maximum value of fault currents in equivalents AC networks but also, the lower value of the fault impedance results in a higher rate of DC fault current change [22]. The fault impedance is primarily resistive in nature which contributes to a high peak in fault currents [1, 23].

1.3.2 Fast Rise Rate of DC Fault Current

As DC faults generate high-rise currents, faults in VSC-based DC systems evolve faster than AC systems. In contrast, the withstand rate of semiconductor devices used in VSCs is less than that of AC power generators. As a result, the protection systems in DC systems must run comparatively faster to avoid any damage to the converter's semiconductor. Fast rate of rise in DC fault current such AC protection devices can't handle with it [23].

1.3.3 Difficulties of Overcurrent (O/C) Relays Coordination

Owing to the line reactance small value and the high increasing rate of DC fault currents, harmony and discrimination between O/C relays in DC networks is a difficult issue. The DC fault currents rate of increase is quite higher than AC fault currents. So, post-fault, all in series O/C relays simultaneously measure a high ratio of fault current and setting values. This can result in a lack of protection coordination. Additionally, O/C relays must be capable of detecting both positive pole to negative pole (P-P) and pole to ground (P-G) faults efficiently; however, there is a substantial difference in the current magnitudes of these two forms of fault [24]. O/C relays face the challenges of securing and protecting DC systems in a coordinated method which may reduce selectivity. Even though definite time O/C relays may be used as a choice to obtain selectivity in DC networks, coordination of certain time relays can lead to a long time to clear the fault in bulky DC systems with many radial lines [25].

1.3.4 Fault Detection and Location

The key criteria for effective locating and detecting faults in a DC microgrid are reliability, accuracy, functionality during system configuration modifications, and certainly the cost [26]. Absence of frequency and phasor data makes it hard to detect and locate faults in DC systems. Moreover, the short lines of DC microgrids, and small line resistance. Hence, the fault locating error in a DC microgrid is greater than in AC power networks. In [27], the scheme of protection for a DC Ring microgrid was suggested for detecting and locating the fault by using a number of line segments and the Intelligent Electronic Device (IED) at both ends of the line for controlling Solid-State Circuit Breakers (SSCBs). The fault location was recognized by approximating the fault inductance on the onset of fault by the least-square estimation technique. The proposal did not study pole to pole fault. In [14], a DC bus microgrid fault protection method including backup protection was proposed for detecting and isolating the fault without de-energizing the entire system based on use of a ring bus with overlapping nodes and links controlled by IEDs.

In [28], the protection method for fault detection and location configurations for loop type LVDC microgrids systems was proposed based on Multi-Criterion System (MCS) and Neural Network (NN).

In [29], during short-circuit fault, the transient characteristics of a VSC-based DC microgrid must be accurately determined in order to achieve fault detection, parameter and protection configuration. So, a transient modeling approach based on the VSC- DC microgrid was proposed by analyzing both DC bus voltage and active power control mode. Then, the faulted DC microgrid was modeled containing the VSC model and DC line model, and the calculation terms of state variables in the DC microgrid were provided. Following, the accuracy of the suggested modeling methodology was proved by comparing it with the electromagnetic transient (EMT) simulation by MATLAB/Simulink.

In [30], a protection strategy for a loop-type DC bus microgrid was implemented in order to locate the fault and isolate the faulted section so that the healthy part of the system could continue to operate functionally. One master controller, two slave controllers, and freewheeling branches between each line and ground formed the protection scheme. Slave controllers measure the current at each end of the bus section that connects two components and send it to the master controller.

In [31], a protection technique for the smart DC microgrid with ring configuration was introduced based on a parameter estimation. The method used the least squares-based method (LS) for predicting of observed inductance at each IED during fault from which forward and reverse faults were distinguished. By using the fault data path of both ends of the line section in a loop network, internal and external faults were identified to protect the network. It was mentioned that using directional information for local faults, every IED recognized any internal faults accurately.

In [32], a centralized protection system was employed to cover DC bus faults, short circuit, open circuit line faults and overload conditions based on impedance slope for 96 V DC ring microgrid with solar PV arrays, energy storage system and DC loads. The protection systems were developed by accumulative sum average techniques. The slopes were observed and noted at both ends of every line during the fault. The forward and reverse faults were classified and if the relevant slopes were observed at both ends of the faulted line. The slopes were either negative or positive based on bus fault and line fault section, respectively. SSCBs were used to isolate the faulty part. The research didn't take into consideration pole to pole fault.

In [33], a ring configuration DC microgrid protection scheme was introduced to calculate the fault distance and resistance. The proposal used two localized Protection Detection (PD) at one place, which used fault current deviation equations based on the nature of DC current to determine the location of the fault across two stages. Firstly, the fault resistance value was determined by an equation based on the fault current and voltage at the PDs place. Secondly, the fault resistance value was obtained, the fault distance was computed also by using voltage and current equations based on the PDs measured values.

In [34], a non-unit-protection scheme was proposed for mesh-type bipolar LVDC microgrid with different RES. The proposed approach detects a fault by comparing the calculated values of fault current and P-P voltage to pre-defined thresholds and then identifies between internal and external faults relied line parameter estimation.

In [35], the impedance-based fault location technique, based on a digital Lock-In Amplifier (LIA) for DC microgrids was presented which allowed the detection of the impedance by extracting the distances relied on the wiring characteristics of the system. The LIA detected both the magnitude and phase of the impedance (resistive and inductive components) and allowed two distances to be estimated: one based on resistance and the other based on inductance.

In [36], in MVDC microgrids, an ultra-high-speed technique was proposed for detecting, classifying, and locating various fault types based on the polarity and wave shape properties of Traveling Waves (TWs) without any communication. The fault launched voltage and current TWs which propagated in both directions from the fault location towards the line terminals. The current TW polarity was used to recognize the fault direction.

In [37], in MVDC shipboard power systems, a graph traversal-based approach was introduced for automating system-wide programming of differential fault protection. The Percentage Differential Protection (PDP) scheme based on KCL was used for fault detection and identification. It was mentioned that if the PDP rules detected a fault, a minimum set of devices capable of isolating the fault was determined. As disconnect switches (DSs) only open when there is no or insignificant current flowing through them, the current flow through all DSs in the isolation set had to stop.

1.3.5 Arcing Fault Detection and Clearing

The periodic zero-crossing is implicit in AC systems as the polarity of voltage and current alternates many times per second depending on operating frequency. However, in DC networks, the lack of natural zero-crossing point is making self-sustaining arcs during normal switching or fault clearing. The interruption of DC current is a main problem which not only causes a serious risk for personnel safety, but also results in the contact corrosion of CBs. To successfully extinguish an arc in DC systems, the electrical contacts must move not only further away from each other but also faster than in AC systems [38]. It is necessary to know the properties and nature of the arc to handle it well and know how to extinguish it quickly. The channel of arc is related with high impedance and consequently reasons lower fault current than nominal current and increases considerable problems to conventional arc fault interrupter and large number of the DC protection schemes [39].

Arc faults are classified into two categories, series and parallel. Series arc faults frequently happen due to loose electrical connections. The parallel arc faults across transmission lines or within loads oftentimes incite a rapid increase in current and activate overcurrent circuit breakers. Two models were introduced in [40] to describe the dynamics of the evaluated DC network. The first model was large-signal analysis which was tested for arc fault boundary conditions and also for system response. The second model was small-signal analysis which calculated the noise propagation within the network. Series arc fault was characterized as an accidental power discharge between two conductors in series with the circuit [41]. When the air in the gap becomes ionized by the great relative charge on either side of the conductors, it permits the current to stream by producing heat and can ultimately cause fires. Therefore, in a series arc fault, a resistance was added in series with the line or circuit. Series arc faults detection and localization in DC microgrid were proposed using Kalman Filter (KF) and Adaptive Kalman Filter (AKF) algorithms-based admittance estimation in [41]. The KF/AKF algorithms were tested via three cases: nominal operation, series arc fault on specified line followed by another series arc fault on another line and disconnection of third line with various times, and series arc fault on identified line with change in load tracked by another series arc fault on third line.

In [42], a model of a series fault for transient simulations of low-voltage DC microgrids was proposed for electromagnetic transient simulations of series DC arc faults. Three types of series faults were investigated:

1. Constant-speed gap (conductors tearing longitudinally at a constant speed),
2. Fixed-distance gap (the electrodes still stationary after reaching a preset distance), and
3. Accelerated gap (the separation of two conductor sections at accelerated rate).

The DC microgrid under study didn't contain complex parameters like constant power load (CPL), renewable energy sources (RES) or parallel inverters. In [43], a method of series fault detection algorithm in LVDC microgrid was proposed by detecting the occasional discharges at the beginning of the fault due to the high input capacity of the interface. The open-circuit performances and high current spikes were used for proposed identification algorithm by calculating the cross-correlation of the current signal with a pre-determined signal to identify the sparks. Experiments were planned and performed on a DC arc test platform with a buck converter based on constant power load. The outstanding features of spark due to DC arc fault are mentioned as following:

1. Sparks are precedent during open circuit conditions.
2. A current spike was usually observed at the beginning, or sometimes in the middle, of the discharging process.
3. The sparks were typically repetitive. The interval between discharges corresponds to the time constant of the input capacitor of the power electronics interface.

In [44], a statistical analysis approach was suggested for a fault detection technique that was suited for suppressing the effect of interfacing device operation, power circuits for system elements. Normal and arc signals were evaluated using Fast Fourier Transform (FFT). The presence of arc fault leads to a transition in the frequency spectrum and the arcing signal simply raises the spectral power density of the current. The probability curves of all operating conditions as a function of threshold values have been plotted. There was a straightforward trade-off between false-positive and false-negative. False-positive was identified as a detector malfunction that decides the presence of arc in non-arcing state, whereas false-negative was defined as a failure of arc detection. Thus, the probability of false-positive increases as one attempts to set a lower threshold to obtain a greater probability of effective identification.

In [45], detection of load voltage drop during series arc beginning was proposed as unified arc fault detection in low voltage DC microgrid. The reduction in the load voltage associated with the series arc activation was directly based on the electrode material and was observed by the load side power electronic converter, which was then acted similarly to disconnect the load from the grid by reducing the load current to zero and thereby the arc is extinguished easily and quickly. DC systems require the use of DC-DC converters in the distribution system and for this the high-frequency noise from DC-DC converter switching and other electromagnetic interference obscure the arc signature, allowing an arc to initiate and proceed unnoticed. During arc fault, switching harmonics are present, signal-to-noise ratio is low, and arc signal is not a periodic signal, so the time domain or frequency domain analysis using a Fourier transform (FT) do not work well. Therefore, in [46], Wavelet Transformation (WT) was proposed to analyze and detect arc fault and arc flash in DC microgrid systems by simulation synthetic waveforms generated in MATLAB/Simulink and then experimentally use arc waveforms measured from photovoltaic (PV) modules and an experimental arc generator.

In [47], a steady-state model of DC arc was proposed for studying series the impact of arc fault on a ring type DC microgrid. The arc impedance was modeled using macroscopic V-I equation to make the decision on DC arc resistance with a given gap length (a non-linear resistor) and arc noise was modelled by using the Gaussian distribution fitting of filtered arc current using a 1.5 - 45 kHz bandpass filter. The DC arc fault was characterized as a 10 Ω resistor in parallel with a controlled current source generating Gaussian distribution noise. The research did not take into consideration the transients at the initial stage of arc development; also, the impact of DC arc fault was studied only with resistive loads.

1.3.6 Short-Circuit (SC) Analysis and Change of Its Level

A DC microgrid can be defined as a small-scale distributed power system consisting of DG sources and loads. It can be effectively combined with RES [48, 49]. A DC microgrid operates in various operating modes including stand-alone with PV energize power to the load, stand-alone with PV charging the battery banks and battery system only feeding the load, and finally grid-tied operation [50, 51]. So, the operating modes of DC microgrids influence the SC level and direction also, needed various results at different fault levels in the microgrid operation and produced a requirement for separate protection settings. In the grid-tied mode of operation, the fault rate of the microgrid is controlled by the short-circuit fault level of the grid. In the other hand, in stand-alone mode of operation, either PV or battery-powered mode of operation, the fault level is enormously low limited by inverter short circuit and battery current rating [52]. So, the fault level in grid-connected mode is higher than the islanded mode, and the direction of this current may be different in each operating mode. Therefore, for beating this challenge, the relays settings must fine-tune according to the variations of fault current levels, or an adaptive protection scheme must be applied in the DC microgrids [53]. When fault takes place in DC microgrid lines, a huge current of short circuit would cause the decline of DC bus voltage, and make power electronics devices overheat and become damaged, which has a bad effect on the safety and reliability of the DC microgrid. To eradicate or minimize short-circuit effects, it is important to evaluate and calculate short-circuit current [38]. Short-circuit current calculation is important to study the protection and safety concerns for DC microgrids.

In [54], energy storage systems response characteristics were studied with different fault stages under constant voltage control and droop control. The DC energy storage system topology under study was linked to AC power network through the IGBT-based VSC and the transformer. The energy storage system was connected to the DC bus through the Buck/Boost bidirectional converter. The load was connected to the DC bus by the

DC/DC or DC/AC converter. Constant Voltage and droop Control were used to study Buck/Boost converter fault characteristics. The fault transient response process was generally divided into three stages as follows:

1. Capacitor discharge stage,
2. Diode freewheeling stage, and
3. Short circuit stable control stage.

The constant voltage and droop control modes have little impact on the fault response at a short-circuit fault in an energy storage system. However, once the short circuit had stabilized, the short-circuit current and voltage will be reduced to a certain range. It is mentioned that a current limiting device was installed at the converter's outlet to ensure the converter's operating safety. This fault must be removed before the fault current peak value by using dependable and effective protection device (SSCBs) [54].

In [55], a microgrid fault simulation was presented and implemented to provide a protection model for multiple energy sources. The wind turbine, solar system, and fuel cell categories are all modeled in the microgrid system. On the same DC bus that feeds the load through a three-phase operated inverter, all energy sources are connected with rectifier models. Every energy source's output was protected by a circuit breaker. Short-circuit fault scenarios had been developed on designed microgrid. Control signals of switching devices had been generated by PI controller. The PI controller was controlled the output voltage to fix voltage with pulse width modulation (PWM). This situation was created for each of energy sources, which were wind, solar and fuel cell.

In [56], a model of equivalent short circuit was recommended to study the short circuit to assess the involvement from DC and utility side of microgrid. A DC microgrid system model consisting of PV array, booster converter, battery, buck-boost converter, inverter, dynamic load and grid network was proposed for fault analysis. The cumulative short circuit of DC systems and the AC system was accomplished by superposition of two side fault current contribution. It has been pointed out that weak short-circuit current for microgrid autonomous operation has raised challenges for protection planners to advance proper protection configurations for the autonomous microgrid.

In [57], the peak amplitude of short-circuit current in the LVDC system was relatively higher compared to LVAC system as a result of DC bus capacitors at the power converter output. The PV and energy storage system (ESS) share in the SC current during DC bus fault. Oppositely, converters in DC networks have the capability to reduce short-circuit current and can stop high SC current in DC systems. IEC 61660-1 is the most used standard for DC short-circuit characterization, the DC short-circuit current source can be from rectifier, battery, bus-capacitor and DC motor with independent excitation. SC current is simulated for multiple LVDC grid situations. It was noted that the high number of parallel strings in the case of PV improves the overall SC capacity of the PV unit. Similarly, the SC current from the MV link was basically constant where the rectifier and the single DC-DC converter were used. If the back-to-back configuration of the converter was used, the fault current from the MV relation was greatly decreased because the LVDC side converter voltage source was not a constant voltage source but was based on the capacitive MVDC side converter DC output voltage bus.

In [58], the transient characteristics of ring type wind turbine-based DC microgrid system under DC line fault were studied on the basis of the strategy of current differential protection as the main protection and under-voltage protection as backup protection was proposed. During the transmission line bipolar short-circuit fault, DC side capacitor discharges rapidly through the failure point and the AC system current was clamped to the three-phase short-circuit current. The equivalent circuit of Voltage Source Converter (VSC) under short-circuit fault of DC cable was illustrated as a complex nonlinear circuit categorized into three stages:

1. Discharge phase of the capacitor,
2. Diode continued flow phase, and excitation phase of AC system.

It was mentioned that during low voltage wind power DC microgrid short-circuit fault of, the current differential protection technique required only to detect the DC transmission lines differential current for locating and isolating the fault lines rapidly and precisely.

In [59], the DC short-circuit current technique of IEC 61660 Std (which implemented the calculation approach of short-circuit current for different sources in the DC radial network) has been improved and applied to the DC mesh network. Sources contain rectifier with ideal AC source, battery, capacitor, and DC motor enhanced and expanded to mesh DC networks. A rapid DC short-circuit current analysis method has been developed for DC mesh systems with sources like rectifiers, batteries, and capacitors. It was reported that the established method accurately estimated the fault currents for batteries and that the error was less than 5%, and that when the fault was close to the capacitors, the capacitor peak currents were precisely estimated with about 7% error.

In [60], during a transmission line short-circuit occurs, DC solid-state switches start to interrupt fault current. Under high current and voltage conditions, the cut-off progress of solid-state switches is hard switching, which may cause permanent damage to the solid-state switches. The short-circuit current impact factors in DC lines (including solid-state switches with snubber circuits, an equivalent circuit of fault current freewheeling direction, and short-circuit fault point positions in line) were investigated. In addition, a suppression approach was developed to highlight the negative effects of short-circuit current.

In [61], a DC microgrid with a large short-circuit current might lead to a decrease in DC bus voltage, which produces overheating and even failure to power electronics equipment, compromising the DC microgrid's security, reliability, and stability. To eliminate or mitigate short-circuit impacts, it was essential to analyze and measure short-circuit current; hence, a short-circuit current measurement approach for low-voltage DC microgrid was implemented, which can be used for solid-state switch selection, power resistance, and protection running time. An equivalent circuit of the transmission line was suggested using π type equivalent circuit, solid-state bidirectional switches with RCD snubber circuits, and freewheeling branches.

1.3.7 Non-Suitability of AC Circuit Breakers (ACCBs)

The current increasing time in DC systems is extremely short, leading to quickly rising transient fault current. As a result, a fault in a DC microgrid must be observed and removed as soon as possible before it approaches critical fault current levels [62]. ACCBs are classic fault current interrupters in AC power systems that can be used to obtain the required DC protection [63]. It uses a mechanical mechanism to clear the fault in many tens of milliseconds and during that period the interfaced converter should be able to withstand the fault current until the ACCB has been disconnected. Because of the lack of a cross-zero point in DC microgrids, traditional ACCBs cannot be completely used in these schemes. Furthermore, suitable DC breakers must interrupt short-circuit currents quickly and dissipate the massive amount of energy stored in the system inductances [62, 64]. In contrast, faster protection schemes are needed for DC networks that can operate in milliseconds or less to prevent damage to the freewheeling diodes of VSCs [65, 66]. So, DCCBs are used to interrupt the DC fault by providing artificial zero-crossing using the auxiliary circuit under fault conditions and thus, being an essential part of the DC protection design [64, 67]. The DCCBs can be formed from magnetic arc blowers and arc chutes [68]. DC fault current is interrupted by three parallel paths which are normal operation, commutation, and energy absorption path [69]. Because of the cost and technological issues involved with these CBs, fault interrupting devices continue to be a major problem in the expanding use of DC networks and the best approach to this issue is the use of power electronic devices like the Insulated Gate Bipolar Transistor (IGBT) and Integrated Gate-Commutated Thyristor (IGCT) [43, 70].

1.3.8 Inverters Low Fault Current Capacity

DC microgrids rely on Power Electronic converters for the integration of DGs and batteries. The converters have a limited ability to withstand a fault current, usually within a range of 2–3 times nominal load current for a few tens of microseconds [62]. In microgrids, inverter fault current capability is usually below half of the fault current for which it is built. As a result, at fault condition, if the permeation of distributed energy supplies (DERs) in the microgrid is low, the DC microgrid operating mode can be adjusted to islanding mode to reduce the level of fault current [53]. Subsequently, current relays' coordination and sensitivity would be compromised, potentially resulting in a delay or inconsistency between relays. The inverter connected to the AC grid and the DC microgrid is the main inverter that can play a restricting function and has the most fault current; hence, it has the most fault tolerance. Furthermore, the fault would be limited to a wide degree of tolerance by these inverters. One or two AC/DC converters power the DC distribution bus, with the AC side directly connected to the AC generator. A current source converter (CSC) or a voltage source converter (VSC) can be used as an AC/DC converter [71]. The thermal ability of the interfaced converters is much smaller than the components of the AC power grid, in such a way that current input from the AC side during faults can damage the interfaced converter within a very short time [71].

1.3.9 Constant Power Load (CPL) Impact

DC microgrids have two types of loads: active (constant) loads and passive loads. CPLs such as DC motor, energy storage charged by constant power, absorb a constant level of power independent of the input voltage [76]. The CPL comes from tight control when applied to power electronic devices. Constant control has negative incremental resistance characteristics [77]. Figure 1.4 shows the negative impedance characteristic of CPLs [77–80]. The bus voltage in the DC microgrid decreases during a fault condition, and the current increases. The negative impedance of the CPL persists and accelerates the current of the fault. Besides, during islanding mode, CPLs induce oscillation in voltage and current, leading to a breakdown of the protection devices, and the instability of the DC system [80–82].

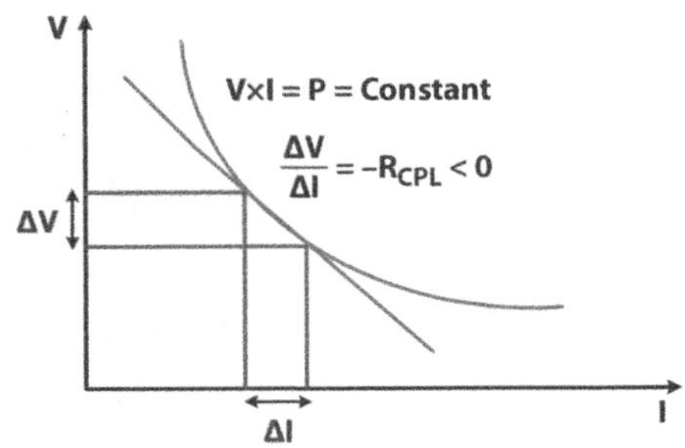

Figure 1.4 V-I curve of CPLs (Negative impedance characteristic of CPL).

1.3.10 Grounding

The grounding configuration in DC power network is a very critical problem which affects both the transient short-circuit current, the overvoltage of the system, protection capabilities and fault detecting [72]. Furthermore, it is important for personal and equipment safety to decrease the Common Mode Voltage (CMV) and the sustainable operation of the power distribution network [73]. The CMV can be described as the offset voltage that is specific to both polarity conductors that is independent of the L-L voltage and appears to be neutral to the network devices causing loop currents between converters. Grounding affects the stability of the system by allowing the system to filter noise so that the equipment is not disturbed, and the protection devices are not unduly tripped. Fault detection and localization systems must be selected to accommodate the grounding system and the degree of protection required [62, 74]. The principal purposes of the DC microgrid grounding system were clarified in [75] as follows:

1. Maximize the safety of personnel/equipment by decreasing both CMV and LG current fault value,
2. Simplicity of fault identification,
3. Reduce DC stray current and corrosion under normal operating conditions, and
4. Minimize voltage stress on system apparatuses.

In IEC 60364 standard, the two-letter codes TN, TT, and IT are used to describe three grounding layout groups where the first letter shows the ground connection to the power supply system and the second letter reveals the relation between the ground or network and the equipment which is energized. Figures, definitions, applications, merits, and demerits of DC microgrid grounding configurations are given in Table 1.1.

Table 1.1 Grounding configurations comparison [11, 30, 71–75].

Configuration	**TN**	**TT**	**IT**
Figure	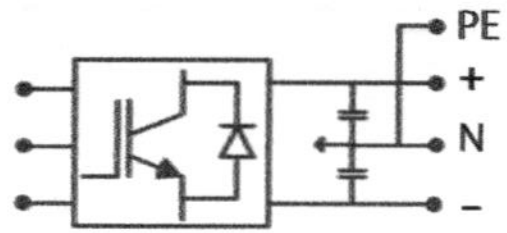	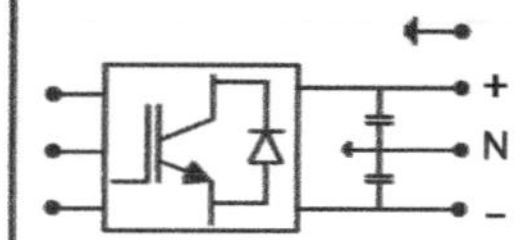	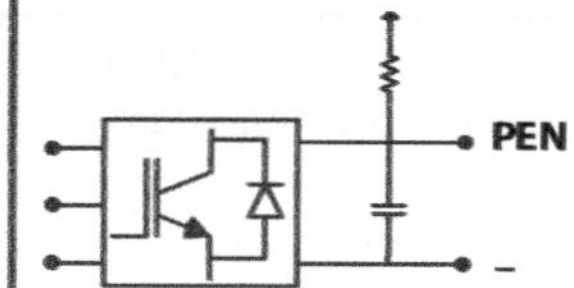
Description	- A most acceptable DC microgrid grounding configuration. - The central point of the converter is attached to the ground and the enclosure of the equipment is connected to the neutral and protective earth. - Low-resistance grounding is needed, and high-resistance LG faults can be detected by relays using over/undervoltage devices or a combination of leakage current detection and ground potential raise detecting.	Both converter middle point and the equipment body are grounded at many points	- The neutral point is either ungrounded or grounded by a significant resistance. The equipment's frame is grounded separately. -The configuration restricts the value of the LG fault currents, where a suitable grounding resistor will be picked.
Applications	LVDC loads for residential, commercial, and industrial applications.	LVDC microgrids	- Telecommunication power systems. - Isolated and non-isolated unipolar DC distribution systems.

(Continued)

Table 1.1 Grounding configurations comparison [11, 30, 71–75]. (*Continued*)

Configuration	TN	TT	IT
Merits	Minimized possible contact potential and improved safety because of the low-resistance return path. LG fault current is large enough to enable fault detection. Implementation of well-defined P-G voltage and leakage currents paths. No stress overvoltage isolation of equipment.	Simple and easy installing. Effective and efficient protection for LG faults	Under LG faults, current transients are lower. Stable load operation with LG faults.
Demerits	Because of the low-resistance return direction, high current transients are subject to LG faults.	Circulating currents passing through devices. High voltage stress.	In second LG, a massive fault current is flow. Requirement of isolation monitoring system. Hard to spot LG fault because of the unpretentious fault current. Over-voltages caused by large LG faults.

The grounding scheme allows for a network ground fault to be passed. Corrosion related to stray current is a big concern related to DC grids. In an ungrounded or a high resistance grounded network, stray current is reduced; however, the network is defenseless against transient overvoltage [62]. DC network grounding techniques and their critical consequences for the architecture, operation and protection of DC microgrids have been implemented in [22]. The effects of grounding strategies for residential DC microgrids on system behavior against ground faults and electrical shocks were also highlighted in [26]. So, design of a suitable grounding network is essential for the effective protection of DC microgrids. Table 1.2 illustrates the DC microgrid protection challenges, problems and proper methodology of solutions that have to be taken into consideration for designing and selecting the protection schemes and grounding configuration for each DC microgrid topology.

1.4 DC Microgrid Protection Schemes

The major challenges of a DC microgrid's effective protection system are its ability to identify, locate and isolate the faulty area with reasonable speed. The causes of these challenges are inverter-interfaced island microgrids bi-directional current flow and also the low fault current, which might not be able to initiate the conventional overcurrent relay or improve its operating time[83]. Moreover, the characteristics of DC power networks, including massive DC capacitors, small DC cable impedance, extreme transient current, and voltage. Numerous issues are related only to DC protection for these DC systems, which need a quick and efficient protection system because of the avoidance of any damages to voltage-source inverters (VSIs). The main design guidelines for any protection system are listed in [27, 84, 85] as follows:

1. Reliability: the ability to predict the protective system response to faults and dependability to not trip uncertainty on transients or noise.
2. Speed: eliminating the fault from the system quickly and rapidly restoring normal operating voltages.
3. Performance: continuity of energizing loads; where a lower performing system loses a substantial section of its loads when a fault happens.

Table 1.2 DC microgrids protection challenges, problems and solution methodology.

DC microgrid protection challenges	**Problems**	**Solution methodology**
Low Inductance of DC System.	High peak in the fault currents. The rate of DC fault current change is high.	Using suitable protection relay based on fault location and fast protective devices.
Fast Rise Rate of DC Fault Current.	Electronic devices damaging due to lowering of its withstanding rate.	Using fast protective devices (SSCBs and Z source CBs), differential protection strategy can be used
Difficulties of Overcurrent (O/C) Relays Coordination.	Protection miscoordination and low selectivity due to measure a high ratio of fault current and setting value for all series (O/C) relays simultaneously.	The communication connection between series O/C relays is used to provide selectivity and isolate only the faulty parts.
Detect and locate the fault.	Fault current requires a few milliseconds for the transient behavior of the DC system capacitors and diodes.	An appropriate protection scheme should be used based on fault resistance which can be computed from current and voltage characteristics.
Arcing Fault Detection and Clearing.	Serious hazard for personnel safety. Erosion of circuit breakers contacts.	SSCBs or hybrid CB approaches with a lower or no arcing and low fault clearance time should be used.

(*Continued*)

Table 1.2 DC microgrids protection challenges, problems and solution methodology. (*Continued*)

DC microgrid protection challenges	Problems	Solution methodology
Short Circuit (SC) Analysis and Change of Its Level.	Dropping the voltage of DC bus, overheating and damaged power electronics equipment.	Using proper over current protection Using very fast protective device to save power electronic devices from damage such as SSCBs. Suppression strategy should be used to illuminate the negative impacts of short-circuit current.
Non-Suitability of AC Circuit Breakers.	ACCBs cannot be completely implemented in DC systems due to the slow moving of its contacts and lack of cross zero point leading to the interfaced converter being unable to withstand the fault currents.	Proper circuit breakers must interrupt short-circuit currents very quickly to dissipate the vast amount of energy contained in the system inductances A reasonably fast and efficient protection scheme able to work in fractions of a second to protect the VSC free-wheeling diodes from damage.
Inverters Low Fault Current Capacity.	Converters have a limited fault current withstand ability so it can be damaged. The coordination and sensitivity of current relays will be impacted and may lead to a delay or discoordination between relays.	Proper control method must be used to decrease the penetration of DERs and convert DC microgrid to islanded mode. Fast DCCB have to use with suppuration methods to save power electronic devices.

(*Continued*)

Table 1.2 DC microgrids protection challenges, problems and solution methodology. (*Continued*)

DC microgrid protection challenges	**Problems**	**Solution methodology**
Grounding.	The L-G fault is directly influenced by the grounding scheme. The L-G fault decreases the faulted conductor voltage to near zero and, based on the grounding technique, can increase the healthy line voltage with respect to the ground. The L-G fault-related arc endangers the safety of personnel/equipment.	Developing an effective grounding scheme for dependable DC microgrid protection configurations and also, personnel/equipment safety.
Consist Power Load (CPL).	Under fault condition and due to the incremental negative resistance, the system voltage will decrease to maintain the constant power, and the current will increase and continue to accelerate the fault current. In islanding mode, the protection derives cannot differentiate between the current oscillation due to the CPL and the fault condition.	Using energy storage system to mitigate temporary faults. Compensation Techniques and control have to use to solve Instability Effects of CPL.

4. Economics: installation and occasionally costs in contrast to performance, which means a well-performing system commonly costs extra.
5. Simplicity: amount of sections, protection areas, level control decentralization required to ensure its reliability.

Protection strategies for a DC microgrid are proposed in many researches and can be classified as given in the following subsections.

1.4.1 The Differential Protection-Based Strategies

In AC networks, both the phasor and the current magnitude are determined by relays that induce a delay in time. However, in DC networks, only the current magnitude is needed, and the measuring unit output is in the form of a digital signal that does not require analog to digital exchange. So, the time of the DC protection relay choice process will remarkably decrease [86]. The differential protection strategy is a protection common approach that measures differential quantities such as energy, impedance and current. This strategy protects the network by utilizing and evaluating current magnitudes and directions at the designated unit's limits within the model to detect whether the fault occurred or not [87]. Differential protection is primarily used in islanding mode since the fault current is small and is hard to detect using conventional techniques [88].

In [89], a protection scheme using digital relays with a communication network was suggested for the microgrid protection system, which depends mainly on differential protection based on a sampling of the current waveform at 16 samples per-cycle or even more. It was mentioned that the proposed protection scheme detected high impedance faults (HIFs) in two methods. The first method was based on the high sensitivity of the current transformers. Whereas the HIF current value was at least 10% of the nominal current, the HIF was observed by using differential protection scheme. The other technique was based on relay programming to identify some of the HIF features that were detected and then tripping when those features were present in the differential current. In [90], in a microgrid system, a differential energy-based fault protection was proposed as a protection scheme through using time-frequency transformation like S-transform. The scheme is used as a pattern recognition method for microgrid protection. Fault currents at both ends of the specific feeder in the microgrid have been recovered and managed via the modified S-transform to determine the spectral energy content. The variation in the spectrum energy content of the time-frequency forms of both signals were the main predictors for capturing fault patterns in the microgrid. It was mentioned that the suggested technique efficiently protected the microgrid against shunt and HIF faults.

In [91], a new communication-based protection system for isolated microgrids was implemented, with a data mining technique used to determine relay settings and parameters. Current symmetrical components were used as differential quantities, which demonstrated extra efficient electrical quantities form that was used in relays for both remote and local functions. Through checking the data, current values, current total harmonic distortion, and phase angles placed second as the best features of fault detecting.

In [92], in smart DC distribution systems a high-speed current differential was represented, introducing a technique capable of detecting sub-millisecond faults. It was proposed to use a central processing device to compare current measurements. Firstly, the direct input of analog measurements sent to the central device, then the sum of the currents was transformed to digital data, as compared to the trip setting values and the decision sent to CBs. The proposed strategy has benefited from the natural properties of DC differential implementation. Eventually, in case of typical AC current differential schemes, a central microcontroller was used to transform all-analog measurements to digital data that were previously summed up and compared to the trip threshold. For minimizing the time of fault detection this method carried out the differential calculation on the microcontroller itself. It was stated that the results illustrated that differential protection identified DC faults very quickly and efficiently.

In [93], for multiple photovoltaic DC microgrids, a new differential current-based fast fault detection and position configuration was introduced. A new discrete time frame, non-iterative, rapid fault categorization, and fault distance calculation scheme based on the Moore-Penrose pseudo inverse methodology was represented. The proposed DC microgrid device was exposed to numerous faults: pole - pole, and pole - ground for the protection functionality validation. The worst scenario was DC arc fault because the PV system was not sufficiently grounded. The results revealed that the proposed method, based on differential protection, detected the location of the fault within a reasonable time.

Corresponding to the above research studies, the protection system based on differential protection that compares between input and output current of a zone showed resistance to the current flow direction and magnitude differences. So, it was suitable for bus protection and independent of microgrid architectures and can be used against subsequent faults. On the other hand, there are some issues related to unbalance and transients. Depending on the communication channel, protection relays are often costly.

1.4.2 The Voltage-Based Protection Strategies

The protection approaches in DC microgrids must be much faster than AC systems because high transient faults on them drop the voltage within a short time [94]. The voltage measurements are used to initiate protection for various faults in DC microgrids [88]. Also, the fault condition was detected using the voltage drop but this method may cause misoperation because voltage drop may be caused by load switching [95].

In [96], a protection scheme based on the conversion of measured signals from the abc operating frame to the dq frame was proposed to define the occurrence of S/C faults and allow for fault discrimination either inside or outside of the specified microgrid zone. The detection algorithm in the proposed protection scheme was implemented in two steps. Firstly, the set of input voltages was converted from abc to dq stationary frame. Afterward, the dq stationary frame was converted to a dq synchronous rotating frame (only DC values for the dq variables). It was mentioned that the proposed scheme prevented problems due to a change in potential fault currents related to the connected utility and isolating operation of the microgrid.

In [97], it was suggested that a system based on clearing DC fault current by opening all AC Circuit Breakers (ACCBs) that VSCs already have on the AC-sides be used. "Handshaking scheme" is the name of the process used to identify and isolate different L-L and L-G faults in the Multi-Terminal DC (MTDC) power grid. For clearing the DC fault, the fault current must be consumed long enough to extinguish the plasma, and there is no risk of re-striking when the voltage recovers. From the viewpoint of cost, DC circuit breakers are more expensive than AC circuit breakers, thus a combination between AC circuit breakers and fast DC switches was recommended. In multi-terminal configuration with a positive pole to ground fault on one line of the DC bus, the current direction of the fault line usually flows from the bus to the fault point, while the current flows in the negative direction on the other healthy lines of the DC bus.

In [98], for the multi-terminal AC/DC hybrid distribution network, a fast DC fault detection method based on the voltage change rate of the current-limiting inductor (CLI) was proposed. Also, a backup detection method using the directions and amplitudes of the voltage change rates of the double-ended CLIs was presented for high-resistance DC faults. During low-resistance faults, the voltage change rates of the single-ended CLIs were used to diagnose faulty line, fault category, and faulty poles. In the case of a high-resistance fault, the voltage change rate of the CLIs at two ends of the DC lines was then used to classify the faulty line. The voltage change rate of the CLI at one end of the fault line was also used to evaluate the fault type and the faulted poles.

In previous articles, a voltage-based protection system is appropriate for DC microgrid networks owing to the lack of phasor parameters and relies only on the measurement quantity. It is also optimal for both line and bus protection at an affordable cost. On the other hand, the voltage-based protection system relies on the configuration of the microgrid, and the HIF cannot be identified. In addition, voltage drops can cause errors and low accuracy in the connected grid.

1.4.3 The Adaptive Overcurrent Protection Schemes

An adaptive overcurrent protection depends on changing or adapting relay setting based on system different topologies. In [99], an adaptive overcurrent protection system that automatically compensates for all overcurrent relays protection settings in response to the effects of distributed generation (DG), active network management (ANM), and isolated operating was suggested. The proposal has been advanced using a three-layer-based architecture. The adaptive protection scheme was applied using a centralized methodology, where calculations and adjustment commands were handled by a single processing unit instead of an agent-based approach, which also usually requires the distribution of processing burdens and decision-making.

In [100], a novel dynamic Adaptive Overcurrent Relaying (AOCR) technique independent of external controllers was applied to determine the relay pickup value, which guaranteed less overhead communication. The scheme was designed for Active Distributed Networks (ADNs) with high distributed energy resource (DER) penetration capable of operating in grid-connected and microgrid configurations. The relay AOCR was modified, thus ensuring stable performance for changing line load and current fault capability by using the following approaches:

1. Use DER mode (ON/OFF) rather than multiple DER output levels. Dynamic AOCR only stands for DER status when calculating relay pickup value. When the DER was online, the maximum fault of the current contribution shall be regarded. No contribution from DER was noticed while offline.
2. Communication-based relay coordination. Communication was used to support relay coordination for low fault current in microgrid modes. The tests based on transient simulations in MATLAB/Simulink had coved all fault types and operating modes of a 15-bus distribution system with 55% DER penetration.

In [101], an adaptive relay coordination technique was implemented and protection guidelines relevant to the study of the different fault scenarios for DG-based distribution systems were provided. The methodology offered relay coordination in those various operating modes and rapid fault clearing relative to the conventional protection system, taking into consideration different online/offline DG unit cases and fault conditions. An algorithm was used to ensure adaptive relay coordination in the IEEE 4-node network was provided. It was stated that this adaptive approach made a reduction ratio of 10 to 31.25% for trip times during faults at three fault locations.

In [102], a dynamic technique for adaptive security and relay coordination was introduced. Protection commandments related to studying different fault cases during grid-connected and islanded mode for DG-based distribution systems were presented. The recommended methodology was applied to the IEEE 13-node system. The case studies considering DG units connected to many grid locations were assessed for different fault locations and the algorithm was described for various scenarios and schemes. It was reported that this adaptive approach outclassed other systems in both islanding and grid-connected operating modes, as well as active changes in DG status; moreover, early or late tripping of relays is a problem for the efficient operation of distribution systems, and, by the proposed approach, improving system performance and for this, a more efficient and secure system with reduced outages can be achieved. On the other hand, the proposal was complicated and didn't take into consideration microgrid configuration and detecting high impedance faults.

An adaptive protection is suitable for both line and bus protection and its cost is reasonable. But it can't be used for protection against subsequent faults. Also, it is dependent on microgrid configuration [103].

1.4.4 Impedance-Based Protection Strategy (Distance Protection)

Measured impedance from the fault point to specific measuring point is the base for distance protection in DC microgrid. If the impedance is below the defined distance value, a tripping signal will be sent to the appropriate CB after a time delay to obtain the expected protection selectivity. There is no need to use a time-consuming approach to closely locate the faulty point in the case of a fast distance protection scheme, and an estimate of the impedance is convenient for a relay assessment.

In [94], in a DC microgrid, a fault detection technique was developed based on the behavior of electronic power converters with bus segmenting contractors' action, depending solely on independent local measurements.

The equivalent circuit of given DC microgrid with short-circuit fault near load was simplified, in which voltage and current is computed at the converter terminals. The equivalent resistance between converter and any fault point was different from equivalent resistance between converter and another one. During low impedance short-circuit faults, the fault resistance was far less than cable resistance, thus required the protection algorithm to discriminate between faults near or far from the device and to take corrective action. An extra detection relays or implementations were applied to the presented detection relay to protect against high impedance faults and overloads. The fault currents were instantaneously limited, and faults can be isolated within 10ms to 20ms. The system can be reenergized within 40ms to 60ms. This approach was cost-effective sensors and filters. On the other hand, this method has a poor fault location performance during high impedance faults in short cables.

In [95], a developed non-unit protection scheme dependent on DC current natural characteristics and its first and second derivatives in fault transients were developed for DC microgrids protection. The suggested scheme based on locally measured system factors did not require any communication channel or system shutdown to clear the fault. The DC cables were protected by Protective Devices (PDs) connected on both the ends. The proposed PD contained analog to digital converters for converting measurements into digital form, a current transducer, digital relays configured with a microcontroller for a threshold setting, and a solid-state circuit breaker. A high bandwidth current transducer was used to sample the current and transformed it into digital form using analog to digital converters implanted into the digital relay. The current rate of change was calculated and compared to the threshold set to produce a trip signal. The first- and second-order current derivatives were used to monitor and identify low and high impedance faults in the system. The process based on the high sampling rate of the line current such that the sensitivity of the rapid noise calculation makes this method not feasible in practice.

In [104], in DC loop-type microgrids, a fault distance estimation-dependent protection system based on two-terminal electrical quantities (currents and voltages) to detect and locate faults was proposed. A small inductor was applied at each head of the line, so the current derivative can be simply obtained and calculated using the voltage drop on the small inductor during the fault. The steps for proposed protection were, Measure the voltage for bus, small inductance and the current of the line, then calculate and compare the voltage difference between bus voltage and small impedance voltage to set protection setting value of distance. If the calculated value was higher than the setting value and the current flow in opposite direction the pole to ground fault occurred, otherwise there was no fault. If the final calculated fault distance was higher than protection setting value, then no fault was found too, otherwise the faulted line circuit breaker would be opened. It was mentioned that the method was capable of quick detect fault location accurately. This proposal was applied only for pole-to-pole fault and the added small inductor which was implemented at each head of the line could be influenced by transient state.

In [105], for microgrids in distribution system, a Phasor Measurement Unit (PMU) based on the distance protection approach, dynamic monitoring and control system was proposed. The PMU was defined as a system used to measure electrical variables by embracing a common time source for these measured variables synchronization. Generic Substation Event (GSE) notifications were obtained with high-speed peer-to-peer communication. In the Control Center (CC) for fault detection and analysis, configurations set for IEDs were developed. The relays Setting groups for various modes of operation were determined and then recovered on their own. As soon as a new configuration was recognized in CC, the corresponding pre-settings generated for different modes of operation were activated. The impedance-based distance protection was determined from the measurement of voltage and current at the relay point. Based on calculated impedance, the fault was detected if it was inside the protection zone or not.

In [106], an adaptive distance protection scheme based on an improved wavelet algorithm (Chaari complex wavelet) for digital filtering was suggested as a backup for inverse time overcurrent protection to clear faults in both islanding and grid-connected operation modes. The microgrid under study was composed of (inverter-based distributed generation), PQ control approach was issued for controlling the micro-sources for grid-connected and islanding operation mode. Voltage-Frequency (u-f) control deception was used also to control micro-source. For both modes, different faults were simulated and it was clear that operation and control modes of micro-sources impact the short-circuit magnitudes which are spotted by the protection device. For an adaptive distance protection process, different mother wavelets were configured using an enhanced periodic wavelet construction model and discretion of a complex continuous wavelet sampled signal. By implementing distance protection as backup protection for current protection and simulating single-phase earth fault at 50% distance away from the inverter occurred, it was observed that the time of last action and the duration of the fault were shorter. Also, the discrimination between the distance protection and current protection was accomplished through the ant colony optimization algorithm.

Distance protection in a DC microgrid has simple algorithm and high speed for detecting faults and is suitable for the meshed distribution system. It is moderate for selectivity, reliability and price. On the other hand, it usually needs a back-up unit, high bandwidth communication link and has low performance in short lines.

1.4.5 Non-Conventional Protection Schemes (Data-Based Protection Scheme)

In [107], a new protection approach was introduced to identify, classify and locate faults in microgrids using an interval type-two fuzzy logic technique. The phase angle between the superimposed modal voltage and the modal current was used as an input to the fuzzy process to define the fault direction. The proposed approach used four primary functional modules which were data acquisition, pre-processor, processing, and output processor stages. It was mentioned that the proposed strategy protected the microgrid against all types of faults and offered backup protection in the event of failure of the main protection and could protect the microgrids against fault after single-phase tripping. Nonetheless, the proposed methodology is complex, costly and there is a requirement for communication channel bandwidth.

In [108], an adaptive Neuro-Fuzzy Interface System (ANFIS) based on a decentralized online adaptive protection strategy for microgrid protection during, prior and after fault occurrence periods was proposed. The ANFIS was used to develop each protection relay modeling in the proposed protection strategy. The IEC 60255 inverse definite minimum time (IDMT) relays were used to employ proposed strategy. The relay function was developed to detect the potential subsequent faults after the first permanent fault happening, and before fault occurrence period each relay received data of the network through communication channels to feed the embedded ANFIS. Also, the relays always observed measured current and voltage signals at their locations to detect the states whether there was a fault on the line section.

In [109], during the connected operating mode for microgrid, a wavelet energy fuzzy neural network-based strategy was applied to produce fault protection. The proposed technique was divided into three steps that were feature Extraction (FE), Feature Condensation (FC), and Disturbance Detection (DI) and was implemented on a microprocessor which was used to perform the identification of power disturbance. The microprocessor triggered the driver switch off the protective relay and turned the microgrid to the islanding operation mode when detecting a fault situation. Table 1.3 illustrates advantages and disadvantages of DC microgrid protection schemes and strategies.

Table 1.3 DC microgrid protection strategies merits and demerits.

Protection strategy	Merits	Demerits
Differential Protection	- High sensitivity, more reliable and its operation independent of fault resistance. - Independent on microgrid configuration. - Not affected by fast rise rate of dc fault current. - Simple and fast. - Not affected by the direction of fault current.	- Can be affected by unbalance and transients. - Its operation requires communication network. - Expensive due to using communication channel.
Voltage-Based Protection	- Depends only on measurement quantities. - Suitable for both Line and bus protection. - Reasonable cost.	- Dependent on microgrid configuration. - Can't detect high impedance faults. - Voltage drops can create errors and poor accuracy - Need backup protection
Adaptive Overcurrent Protection	- Suitable for both line and bus protection. - Reasonable cost. - Independent of current directions - Simple, fast and reliable	- Can't be used for protection against subsequent faults. - Dependent on microgrid configuration. - Cannot detect high-impedance faults - Need backup protection.

(Continued)

Table 1.3 DC microgrid protection strategies merits and demerits. (*Continued*)

Protection strategy	Merits	Demerits
Impedance-Based Protection	- Simple algorithm - High speed for detecting faults. - Suitable for the meshed distribution system. - Suitable for fault location detection	- Need backup protection unit. - Need high bandwidth communication link - High cost - Has low performance in short lines - Dependent on microgrid configuration
Non-Conventional Protection	- Detecting faults based on data processing - More accuracy - Continuously of monitor measured current and voltage signals	- Complex methodology - High cost - Need communication link

1.5 DC Microgrid Protective Devices (PDs)

During faults in DC microgrid, the current interruption leads to create an arc which cannot be easily extinguished without crossing through zero. Commercially viable technologies for the protection of DC microgrid schemes use fusing and mechanical circuit breakers (MCBs) that begin the process of opening the current flow path at an overcurrent arc state [84, 111]. The devices are constructed in such a way that the length of the arc increases as the overcurrent condition is maintained until the metal melts or the contacts are fully released, causing the current to drop to zero and the arc to extinguish [110]. The tripping device could be thermal-magnetic or electronic. Magnetic tripping devices identify instantaneous values while the thermal tripping devices measure the Root Mean Square (RMS) values of current. Overcurrent protection devices including magnetic electromechanical circuit breakers respond differently to overload and short-circuit currents because of their current-time characteristics or delay characteristics curves [112, 113]. Additionally, the DC microgrid load side has input capacitance that discharges from loads adjacent to faults contributes and as inrush current from loads is added to the grid, it has the ability to cause undesirable circuit breaker trips [110].

There are also some issues associated with CBs and fuses in LVDC networks, including significant time constants, long breaker operating time, and high-speed DCCB, that can start interrupting fault current within 10 milliseconds. In the DC distribution system, fuses are used to heat up and melt the conductor, depending on the thermal characteristics, but they are not quick enough to protect against short-circuit faults because the response time is within tens or hundreds of milliseconds, which caused damage in the DC link capacitors [113, 114]. Besides, the fuses must be replaced manually, increasing the time for system recovery [115]. Power electronic protection devices are proposed and developed to break down the drawback of fuses and MCBs usage. Using power-electronic switches such as Thyristors and Insulated Gate Bipolar Transistors (IGBT) improves operating speed, and the inductive current interruption capability can be increased [116]. On the other hand, the associated losses of the power-electronic approach are still much greater than the mechanical switch [117]. Often, SSCBs can open the fault path very rapidly, preventing large system stresses within the microsecond range, but this fast operation can contribute to overvoltage across semiconductor devices if not adequately protected. Many approaches are used to provide overvoltage protection cross SSCBs such as [118]:

1. Metal Oxide Varistor (MOV) interconnected SSCB terminals or between one terminal and grounding.
2. Resistor-Capacitor-Diode (RCD) snubber attached in parallel to the SSCBs.
3. Freewheeling diodes are used to provide continuity with currents flowing through inductive components.

To investigate the various protection coordination strategies that could be used in DC microgrid systems, it is crucial to analyze which protection devices are efficient for protecting main system components such as power converters, battery storage devices, DC-link capacitors, load feeders, and PV systems[12].

1.5.1 Z-Source DC Circuit Breakers (ZSB)

In DC systems, the arc is constant and more resistant to interruption due to continuity of voltage such that the DCCBs must provide additional arc-extinguishing methods. Usually, they provide a mechanism to lengthen and dissipate the electrical arc to make interruption smooth. DCCBs operate with the same concept of thermal protection and magnetic protection used in ACCBs as follows:

1. Thermal protection is carried out by the DCCB when the electrical current is present above the rating value. This protection mechanism is based on a bi-metallic contact that heats, extends, and trips the circuit breaker. It also operates better as the current increases, so more heat is produced to outspread and open the electrical contact. Thermal protection in DCCB protects against overloading current.
2. Magnetic protection sends a trip command signal to the DCCB when higher fault currents are detected, and the response is very often instantaneous. DCCB has a rated breaking capacity that reflects the highest fault current that can be disrupted. Magnetic protection in DCCB protects against short circuits and faults that are considerably greater than overload.

Figure 1.5 illustrates Z-source breaker designs, including classical and series-connected designs. The classical architecture provides isolation from the fault to the source through the SCR but does not provide a common path to the ground through the breaker. Series one has a common ground

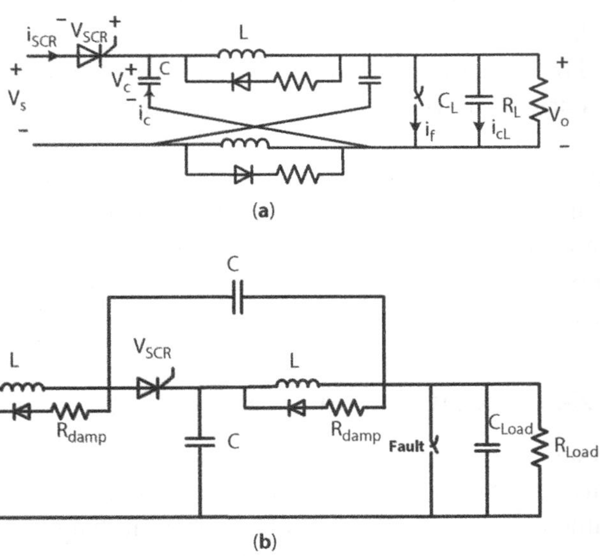

Figure 1.5 Z-source breaker designs (a) classic design, (b) series connected design [124].

path, but it has the downside of not having protection between input and output when the SCR was turned off [119]. The Z-Source network was classified as an impedance network composed of inductors and capacitors. In many publications, a Z-Source circuit breaker was implemented to force a DC current zero crossing with an impedance network to induce current fluctuation and voltage source isolation with power thyristor [120]. When a transient fault occurred, the Z-source breaker used Silicon-Controlled Rectifier (SCR) in series with separate LC circuit topologies to produce a zero crossover. When a zero crossing occurred, the SCR was certainly commutated off before the fault current go over the turn-off capability and the fault was removed from the source [121, 122].

In [120], a modification to the Z-Source DCCB was proposed to reduce power consumption by withdrawing the power thyristor from the main current path and replacing it with a speedy mechanical switch. The parallel- connected Z-Source CB was focused in the research which forced a zero-current crossing through resonance between the impedance network connected in the system. In this design, the thyristor was replaced by a fast-mechanical switch in the conventional Z-source breaker.

Besides, an extra inductor was put in series to achieve the required time interval between the current zero crossings in the case of a fault to reproduce practical circuit characteristics. The source inductor is smaller than the impedance network inductor. As a result, the circuit was assumed as a series of LC circuits and during the fault, the current did not change instantly because going through the large impedance network inductors, the fault current flowing through the series LC path. The result revealed that the fast mechanical switch increased the system's energy efficiency by reducing the state power consumption of the circuit breaker but was not limited to the unidirectional power flow.

In [123], for DC microgrids, a DCCB was implemented using a short conduction path between the breaker and a mutual coupling to automatically switch off in reply to a fault. The current flowed from the source to the load via the SCR and coupled inductors. The current passed from the source to the load via the SCR and coupled inductors for normal operation and during load side fault condition, an impulse current in the short path comprising the capacitor and the secondary winding of the coupled inductors. This current was reflected in the primary and effectively forced the SCR current to zero. The breaker was developed with certain parameters, such as the choice of wire size, the primary winding of the transformer, the overall SCR turn-off time as the resonance type, and eventually the resistance was adjusted to a low value so as not to interfere with the performance of the breaker.

In [124], a two Z-Source DCCB structures with bi-directional power flow abilities were introduced. The first technology was a Z-source breaker with two different inductors that used SCR in the primary conduction path that was commutated when the fault happened. It was also arranged in such a manner that the power flow could be controlled in any direction. The second technology was a bi-directional ZSB with a coupled inductor, which was designed as a coupled inductor rather than having two different inductors since the current through both inductors was similar in both modes of operation. In a coupled inductor, however, owing to the presence of mutual inductance, the current of one inductor influenced the current and voltage of the second inductor.

In [125], in DC microgrids, a bi-directional z-source DCCB topology was implemented to support short-circuit fault protection techniques by using two unidirectional z-source breakers connected in a parallel and backward direction with a common shunt capacitor. When zero crossing occurred, the fault was cleared due to the natural commutation of SCR. The bidirectional breaker was modified by using only two inductors instead of four; also, the number of an SCR was increased to six instead of two. It was mentioned that modification was for simplicity and was cheap. The proposal could detect and clear large transient fault currents. On the other hand, it did not detect small transient short circuits or provide protection for steady-state overloads. Also, the proposal didn't solve large reflected current to the source.

1.5.2 Hybrid DC Circuit Breakers (HCB)

In DC microgrid, the fault must be cleared as quickly as possible to mitigate DC bus disturbance and the operation of voltage source converters. A consistent and efficient DC breaker is therefore essential to isolate faulted sections and thus prevent the DC grid from collapsing [126]. Figure 1.6 illustrates a hybrid circuit breaker with a forced commutation to improve the performance of the mechanical switch such that the break time gets shorter and the peak value of the fault current is reduced.

In [126], a hybrid breaker with induced commutation was implemented to switch without arc during DC grid faults. The proposed DC breaker consisted of three parallel-connected branches as follows:

- Ultrafast mechanical switch connected in series with diode and forced commutation circuit (inductor, capacitor and thyristor),
- Full-controlled type semiconductors, and Varistor.

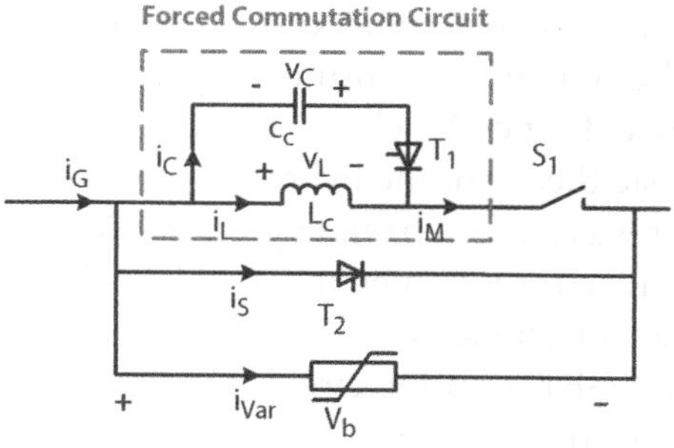

Figure 1.6 Hybrid circuit breaker with forced commutation circuit [126].

In normal operation, the DC current passed through the mechanic switch branch and the semiconductor branch was turned off. During DC fault, the current quickly increased to trigger value, which activated both the full-controlled semiconductor branch and the thyristor as well as the demagnetizing and discharging procedure for the inductor and capacitor. Then mechanical switch voltage became negative and the current in the semiconductor branch also was zero. To reverse the commutation capacitor voltage and surpass the on-state voltage of the semiconductor branch, the system current commutated from the mechanical switch branch to the semiconductor branch, the mechanical switch was completely opened and the magnetic energy in the system was fully dissipated by the varistor and the fault current was interrupted.

In [127], for a DC microgrid, an arc-less hybrid DCCB based on a natural zero current switch was presented. The solution consisted of four sections as follows:

1. Semiconductor branch having an IGBT of antiparallel diode,
2. The absorption branch has parallel MOV connections to capture the magnetic energy contained in the DC system,
3. L-C branch to produce resonance current during fault state to interrupt fault current at zero cross-point,
4. The MCB.

During fault, the fault current raised very steeply, but the inductor currents didn't change immediately. Thus, the capacitor was discharged into the impedance of the fault. As the current approached the threshold value of the fault detected by the controller, the trip signal was sent to the MCB and the arc was shut off when the current was first zero after opening the

MCB. When the arc was extinguished, the IGBT switched off and the stored energy of the system was consumed by the MOV.

In [128], the mixed type DC circuit breaker prototype was developed to illustrate the basic theory of the hybrid DC circuit breaker. The offered hybrid DC breaker was formed from High Speed Vacuum Circuit Breaker (HSVCB), Power Electronics Interrupting Unit (PEIU), control system and Resistor Capacitor Diode (RCD) buffer circuit. The PEIU contained parallel-connected IGBT transfer circuit worked as breaking units and ZnO arrester to absorb the overvoltage during IGBT switched off. The RCD buffer circuit built to eliminate line inductance energy and minimize IGBT voltage increase during shutdown status. The control system formed by breaker status identification, current and voltage detection, IGBT control circuit, synchronous control circuit, and repulsion control circuit. The proposed breaker was tested and closed for 400 V system and 200 A load. It was recorded that the IGBT closed at 1.9ms, and the current was fully adopted after 0.02ms. The fast switch closes at 2ms and the current flows from the IGBT to a fast repulsion switch, then the current is fully transferred to the fast vacuum switch after 0.1ms, and the entire circuit was closed at 1.92ms.

In [129], a DC microgrid HCB based on the current commutation between the vacuum arc and the IGBTs was addressed. The proposed HCB consisted of two parts HSVCB and PEIU. The HSVCB was represented by the vacuum circuit breaker, the insulated rod, the over-travel unit and the fast electro-dynamic drive. Whereas the parallel IGBTs, the snubber circuit and the voltage limit voltage described the PEIU. The current commutation between the VCB and the IGBTs was analyzed by controlling it for a range of 5ms, a voltage variation of less than 5%, and the current distribution between the HSVCB and the IGBTs relied on the resistance of the vacuum arc and on-state resistance of the IGBTs. The suggestion was studied experimentally, and the results revealed that the duration of the vacuum arc commutation could not be effective if the resistance of the IGBTs was greater than 10mΩ. It was stated that several studies suggested different technologies for HCB, but the research centered on the DC HCB principle and the simplicity of current commutation. Oppositely, the large size, high cost and complexity were disadvantages of this proposal.

In [130], A hybrid circuit breaker was implemented to increase the system's power quality by adopting various limiting behaviors for permanent and temporary faults and by passing the load current in both directions. A parallel combination of high current-carrying capacity MCB and high-speed arc-less interrupting capacity SSCB was used to model the proposed method. The proposal consisted of fast MCB with 1ms interruption speed with series inductors, semiconductor branch with two common-cathode SCRs with antiparallel diodes and current-limiting branch consisted of a limiting resistance and a low-speed mechanical switch. During temporary fault condition the rate of current flowed through series inductors increased until reaching threshold value; at this moment the MCB was tripped and its contacts started to open, the voltage cross SCRs increased, and the fault current transferred from MCB to the semiconductor branch. In permanent fault condition, the faulted current transferred from MCB to the limiting branch and capacitor discharged completely then the current was diverted to the semiconductor branch in a soft transition and low-speed breaker was opened. The result showed that the proposal protected DC power systems under fault condition and improved the power quality with reasonable cost. The disadvantages are the large size medium speed which may cause protection issues.

In [131], for hybrid DCCBs, a solid-state switching strategy was developed to dynamic control of the voltage potential through the mechanical switch and reduce the fault isolation time and decrease the peak fault current during the breaker action. The proposal consisted of Commutating Switch (CS) connected in series with ultrafast mechanical switch (UFMS) and multi-stage solid-state branch (SSCBs and MOVs) connected in parallel to gradually step up the voltage across the circuit breaker to balance the UFMS dielectric strength and restrict the fault current. DC current passed through UFMS and CS in normal operation. In the case of a fault, the breaker controller issued a gate signal to open the UFMS with active damping and, based on the dielectric medium, the UFMS obtained a specific voltage withstand capacity proportional to the contact displacement. At this time, if the voltage rating raised to the first stage MOV blocking value, the fault current flowed in it and the rest of the ON-switches. As the UFMS contact displacement increases, the controller turns OFF-each semiconductor switch stage matches the dielectric strength of the UFMS. When all the stages were turned off, the residual energy due to the system inductance was dissipated by the MOVs, the fault current was dissipated to zero, and the system was separated. The solution was tested, and the results revealed that the four-stage HCB voltage and current waveforms were gradually damped. Large size and control complex for several solid-state stages are drawbacks of this proposal.

In [132], unidirectional and bidirectional two coupled inductor-based hybrid DCCB schemes have been implemented to minimize significant conduction power losses due to the ON-state SSCB drop. A hybrid circuit breaker was developed using a mechanical breaker connected with a parallel branch having semiconductor devices. The unidirectional topology consisted of the primary winding of the coupled inductor connected in series with the mechanical switch and the secondary winding connected to the uncharged capacitor. Besides, the diode was connected in parallel with the contacts of the mechanical switch to form a negative current path through the breaker after a zero crossing. In the fault condition, the coupled inductor created a resonant circuit with a commutating capacitor to turn OFF - the mechanical switch and breaker contact opened at zero current, which reduced arcing. The bidirectional design topology consisted of a commutating capacitor that served as the input voltage source for the H-bridge circuit, two mechanical switches were used to provide bidirectional braking capability and antiparallel diode pairs that conducted while the capacitor was charging and provided a negative current path.

1.5.3 Solid State Circuit Breakers (SSCBs)

The Electromechanical DC breakers need long opening times, which contributes to the overestimation of system dimensions. The SSCBs can open the fault path very rapidly, within a microsecond, preventing oversizing and high system loading. Furthermore, this rapid operation can lead to over-voltages across semiconductor devices if they are not adequately protected. Figure 1.7 [118] shows a basic SSCB assembly consisting of an IGBT with an antiparallel diode that allows the current to be controlled in both directions. Figure 1.8 also illustrates the solid-state current switch that is usually required for each power line of the feeder [84].

In [133], a solid-state DCCB with low reflection current was presented to address the problems of the conventional mechanical circuit breaker and the Z-source circuit breaker by isolating the fault current automatically

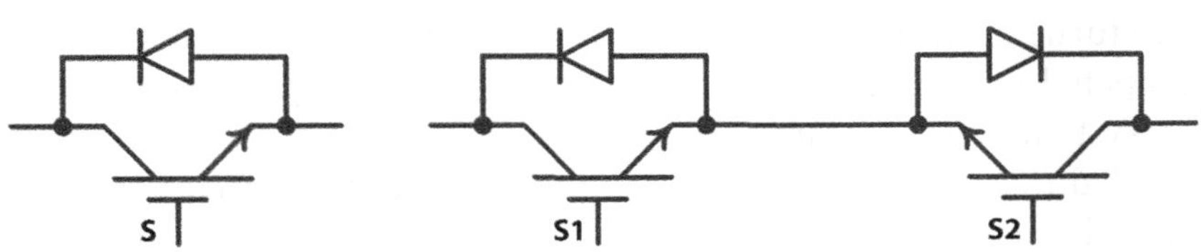

(a) Unidirectional SSCB (b) Bidirectional SSCB

Figure 1.7 Simple SSCB assembly is composed of an IGBT [118].

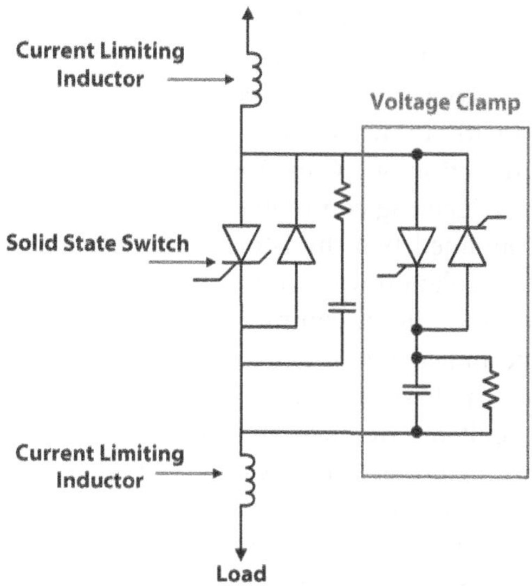

Figure 1.8 Solid state current interrupter [84].

without using control circuits. It consisted of SCR, a transformer whose windings were used to replace the LC legs of the Z-source circuit breaker, a capacitor, and two branches consisting of a diode and a resistor in parallel with the two windings of the transformer. In steady-state operation, both currents passing through the primary and secondary coils, and through the thyristor was equivalent and under fault condition, the instantaneous enormous current streamed through the secondary coils, which were connected to the load. The performance was tested experimentally under fault conditions, and the results showed that the thyristor quickly became OFF and switched to a forward-blocking state with an increase in the inductive current on the primary side of the transformer.

In [134], the RCD snubber was proposed to protect the SSCB from overvoltage caused by its tripping when a DC-bus fault occurs in the DC microgrid. The snubber had capacitor in series a resistor, and diode, connected in series with another resistor. The capacitor, resistor series branch was attached to either side of SSCB, while the diode resistor branch was connected between the DC bus side of the SSCB and the neutral line. The snubber diode and an additional snubber resistor were used to provide a path for the fault energy instead of the snubber capacitor. Therefore, the overvoltage at the SSCB was suppressed without increasing the value of the snubber capacitor. The L-L fault was presented for the bipolar DC bus, which has a low fault impedance. Also, intensive electrical stress was formed when SSCB trips. The RCD snubber was constructed in MATLAB/Simulink and the results showed that the overvoltage across SSCB of the grid-connected converter side was attenuated and the SSCB voltage reached 200 Volts without overshoot.

In [135], for protection of DC microgrids a new thyristor-based DCCB with a single-shot triggering commutating circuit was proposed. The suggested breaker contained two thyristors. The first thyristor provided the main path for fault isolation with inductance to limit the rise of fault current, and the second was in the commutation circuit as a trigger to generate the reverse voltage and current required to block the first thyristor. During normal operation, the first thyristor was ON, while the second was OFF. When a short-circuit or overcurrent fault occurred, the output current increased the current protection threshold, whereupon the Analog real-time Protection Unit (APU) generated a second thyristor trigger signal that was amplified by the thyristor driver to make the second thyristor ON, so that the first thyristor was reverse-biased and switched OFF to isolate the fault. Two fully charged capacitors were introduced into the proposal to maintain the same voltage level as the source during normal operation, but during the fault condition both capacitors were connected in series and their total voltage would be almost twice the source voltage. All the energy stored in the inductors and capacitors during fault was consumed by circuit resistors.

In [136], a method for selecting suitable DCCB in an LVDC microgrid was proposed using the schwarz black box arc model, which is one of the analysis methods used to determine the threshold values of DCCB and to describe their behavior in the electrical circuit during a power interruption. A commercial DCCB manufactured by Eaton was modeled to test the method. The LVDC microgrid connected to the utility grid through a bidirectional converter was simulated, and four L-L fault scenarios (worst case instead of L-G) were assumed at different points in the DC microgrid. For each fault scenario, the voltage and current stresses subjected to which the Eaton CBs are subjected were analyzed. The simulation results showed that the probable fault currents were fully interrupted and the maximum voltage and current stresses to which the Eaton CBs were exposed were well within the safe operating range as shown in the Eaton CB datasheet and trip curve.

In [137], short-circuit protection issues of LVDC microgrids were analyzed and methodology based on SSCBs to provide Fault Current Limiting (FCL) was predestined. A 1-kV DC microgrid was modeled and simulated to supply 3.75 MW two DC load centers protected by SSCBs and free-wheeling diodes to ensure current continuity due to load-side inductances when the load SSCB was opened. The proposed SSCB based on 2.5-kV Reverse Blocking-Integrated Gate Commutated Thyristor (RB-IGCT) was simulated and modeled in MATLAB using the following elements:

1. GTO block with a diode in series.
2. An impedance connected in parallel with the GTO block.
3. A series inductor to provide the critical rate of current rise during turn-on of the RB-IGCT.

The MOV was fixed in parallel with the diode and the RB-IGCT to protect against potential overvoltage that occurs during the SSCB turn-off. The FCL control circuit for overcurrent protection was introduced based on the comparison between measured current and threshold value, and then sends the output which is inverted and transmitted to the gate of the SSCB. Overcurrent threshold, undervoltage threshold, apparent resistance, current derivative threshold, and overcurrent combined with current derivative threshold were discussed as fault detection techniques for clearing a fault and activating FCL function. In addition, the permitting energy for previous fault detection techniques (calculated as the integral of the square of the current from the start time of the fault to the isolation of the fault by the SSCB) was compared.

In [138], an ultrafast silicon carbide junction gate field-effect transistor (SiC JFET)-based self-powered SSCB in radial distribution systems (340-V residential DC community microgrid) was proposed to detect short-circuit faults by noticing the abrupt voltage increase between its two power terminals and drawing current from the fault condition itself to switch off the SiC JFETs. The behavior of the unidirectional SiC JFETs during different L-L fault scenarios was analyzed using switch current, inductor currents, sensing and driving voltages. The proposed hardware was tested, and the results showed that the SSCB-JFET detected the fault current and returned it to zero within a few microseconds. The proposal was only for radial systems, which may not be applicable for different configurations of DC microgrid. Moreover, it was only suitable for unidirectional SSCB.

1.5.4 Arc Fault Current Interrupter (AFCI)

In [139], a development of AFCI testing for DC circuits operating at 24-80 volts was introduced to identify an effective method to determine DC arcing occurrence. The arc is *"a continuous luminous discharge of electricity across an insulation medium, usually accompanied by the partial volatilization of the electrodes*" based on IEEE definition [139]. Also, National Electric Code (NEC) describes AFCI as "*advice intended to provide protection from the effects of arc faults by recognizing characteristics unique to arching and by functioning to de-energize the current when arc fault is detected*" [140]. An experimental test was introduced to detect and find the finest solution to extinguish an arc before damage. Four categories of arcs (series, parallel load, parallel conducting and grounding) arching were tested by using two different arc generators: UL/Eaton design and knife switch design.

In [141], for DC arc fault protection in PV systems a safety requirements evolution was investigated. The DC arc faults were categorized into two kinds: series and parallel arc faults, in relation to their current direction position in the electrical circuit. Series arc occurred in a PV system when the continuity of a connection or conductor was compromised while the DC current was being generated by the system. The most likely arc fault to occur is a series arc because of the large number of series connections in an array and the fact that thermal cycling and aging can lead to a poor electrical connection over time. A new PV-AFCI approach was developed to identify and interrupt arcing faults caused by faults in the expected continuity of conductor and terminal, module, or even other system apparatuses in DC PV source and output circuits.

In [142], the performance of arc fault circuit interrupter management was improved by adding network function into the existing arc fault circuit interrupters. The proposal structure of AFCI consisted of arc detection unit, overload detection unit, and short-circuit detection unit (which detected arc fault, overloads, and ground fault) connected with network unit. All units sent the detection signals to the microprocessor which received the signals and sent them to the trip function unit to turn OFF the power and sent the signals to the network unit at the same time. The proposal allowed the circuit to be interrupted by remote control signal to prevent both electrical accidents and fires. The proposal was dependent on communication network which can be expensive and needs backup protection strategy. Moreover, the proposal approach may not detect HIF.

In [143], a DCCB was proposed in which the two parts, current limiting and interrupting, were connected in series to interrupt a high short-circuit current in a DC system. The interrupting part had conventional gas CB and the current limiting part contained superconductor tapes and parallel resistance. The ZnO arrester was used to absorb the residual energy after fault interrupting. The proposal was tested experimentally under different conditions.

1.5.5 Fuses

The fuses are the most conventional element in protection schemes by melting the fuse link in the circuit. The fuse-blowing time is divided into the melting and arcing time where the melting time is described as the time between the start of the accident and the time when the element melts and the arcing time is defined as the time between the element starting to melt and the current reaching zero. In a shortening or overcurrent state, the current rises sharply and then falls. The shape of the rising curve is mainly determined by the inductance and resistance of the short circuit. Owing to the current going through the fuse element, the amount of heat is deposited in the fuse and as the heat exceeds a specific value, the fuse element melts and the current decreases dramatically. After element melting, an arc is generated as a result, the current reduces, and the voltage continues to rise. So, the arc ends, and the fuse is isolated from the circuit such that the current approaches zero and the voltage becomes the system level [144]. The key advantages of fuses are: simple, inexpensive, standardized, and with low steady-state losses [145]. Fuses are characterized by several factors, which can be found in their datasheet and allow a correct selection for a particular application. To properly select DC fuses, the challenge is to find the operating time of DC fuses based on the fault characteristic [146]. Semiconductor fuses are used in DC distribution systems instead of fast solid-state DCCBs, which are expensive for wide application [146].

In [145], a fuse-based SC protection strategy for radial LVDC microgrids was developed. The developed methodology was presented based on the system conditions of voltage balancing converters, load converters, fuse and allowed voltage dips. The following four characteristics of the fuse were focused on:

- ($I^2 t$) value, which represents the amount of thermal energy in joules transferred to the system when the SC happened before the fuse was opened,
- The peak current in DC grids, in other words, the maximum current that a fuse can safely interrupt without being damaged,
- The selectivity, and the voltage dips.

In [146], the operation time of DC fuse based on the fault characteristic is a major challenge for choosing DC fuse, so the functioning of DC fuses and reclosers during the fault in DC Microgrids was studied based on DC microgrid structure either ring or radial systems. Also, a local fuse saving method by proper coordination between fuse and recloser switch in DC microgrids was presented. A set of constraints was used to provide a stable coordination scheme for multiple operating modes and conditions. The performance of DC fuses and reclosers was illustrated during a fault in DC microgrids, and also the behavior of the DC network during a fault was summarized and classified as three-phase capacitor-discharge, free-wheeling diode, and converter current. The proposed approach used the magnitude of fault current and time to achieve the maximum for recloser switch setting using local measurements and independent communication channels.

In [147], a thermal model of the fuse was provided to explain the melting and cooling impact of the fuse. The model parameters were extracted from the fuse datasheet Time Current Curve (TCC). It is mentioned that the proposal can be used in computer simulation as protective elements against short-circuit faults and simulation data can be used to design selectivity of the fuses under short-circuit fault and overall protection scheme of DC distribution system. The proposal did not take into consideration the DC network configuration, which may cause the fuse to work in an improper way.

In [148] for PV systems below 750 VDC, fuse type Surge protective devices (SPDs), which are defined as devices that protect electronic devices and equipment from lightning surges, were introduced. A fusible Disconnect Switch (FDS) was developed to satisfy the dimensions of fuse links with a cylindrical contact cap conforming to IEC60269-4. Unpredictable effects of fuse opening on microgrid voltage and distributing cable length constraints make the fuses not a suitable solution for DC microgrid protection [84]. In addition, it must be replaced after its correcting operation.

Table 1.4 illustrates the difference between previous PDs based on cost, efficiency, reliability, and fast operation based on [12, 27, 75].

1.6 Conclusions

In this paper, the DC microgrid protection challenges are introduced. Also, the protection strategies and schemes are discussed for fault detection and location methods based on the differential, voltage measurements, an

Table 1.4 PDs assessment based on Cost, Efficiency, Reliability, and Fast operation.

	Cost	Efficiency	Reliable	Fast of operation
Z-source DC circuit breakers	- Moderate expensive and lower cost than SSCBs	- large transient fault is needed to be triggered.	- could not support long-term protection	- Fast operation
Hybrid DC circuit breakers	- Extremely costly	- Low power loss - No arcing on mechanical contacts	- High reliability	- Quick response interval time
Solid state circuit breakers	- Expensive, because it has high power loss so, heatsink must be implemented lead to big size of breaker.	- High voltage blocking and current carrying capability - having arc during operation	- Fully controllable - IGBTs can withstand in high SC currents	- Fastest response - High switching speed
Mechanical circuit breaker	- Relatively low cost	- Very low power loss - Limited interruption current capability [27]	- Moderate reliability	- Long operating time
Fuses	- Low cost due to Simple structure. - extremely low power loss	- Low efficacy due to limited ability to classify between a transient and a permanent fault [27]	- Challenge in selecting an appropriate fuse and coordinating with other equipment - difficulty on predicting the voltage transient state of the fuse opening and a rise-time constant of current transient	- Slow operation time

adaptive overcurrent, distance, and non-conventional protection schemes. An efficient protection scheme contributes to more accurate and fast detection of fault location and reduces the duration of outages, but conventional relays might present modest performance in DC microgrid protection. So, fault location algorithms are presented based independently of sensor error and communication systems by used intelligent electronic devices. Furthermore, the protection devices in DC microgrids are explained. Due to the DC fault current nature and absence of zero-crossing points, DC protective devices must be designed to solve these obstacles. DC fault current interrupting devices including MCB, ZSB, HCB, SSCB, AFCI, and fuse are presented. SSCB has shown a fast time response compared with other types of protective devices. It is observed that DC microgrids need future work on the following challenges:

1. The effects of output filter components on power electronic conversion systems during low impedance faults.
2. Selection of an appropriate grounding layout to mitigate the impact of short-circuit current on DC microgrids.
3. Proper protection approach for detection and location of high impedance fault.
4. Developing protection and control strategy for a microgrid containing unbalanced and nonlinear loads.

References

1. N.W.A. Lidula and A.D. Rajapakse, "Microgrids research: A review of experimental microgrids and test systems", *Renewable and Sustainable Energy Reviews*, Vol. 15, No. 1, January 2011, pp. 186–202.
2. Jackson JohnJusto, FrancisMwasilu, JuLee and Jin-WooJung, "AC-microgrids versus DC-microgrids with distributed energy resources: A review", *Renewable and Sustainable Energy Reviews*, Vol. 24, 2013, pp. 387–405.
3. Z. W. Khan, H. Minxiao, C. Kai, L. Yang and A. u. Rehman, "State of the Art DC-DC Converter Topologies for the Multi-Terminal DC Grid Applications: A Review", *IEEE International Conference on Power Electronics, Smart Grid and Renewable Energy*, Cochin, India, 2020.
4. Tomislav Dragicevic, Pat Wheeler and Frede Blaabjerg, *DC Distribution Systems and Microgrids*, The Institution of Engineering and Technology, 2018.
5. Li, Xiangkun, Wu, Weimin, Wang, Houqing, Gao, Ning Chung, Henry and Blaabjerg F, "A New Buck-Boost AC/DC Converter with Two-Terminal Output Voltage for DC Nano-Grid", *Energies*. Vol. 12, Oct 2019.

6. S. M. Aaqib, M. J. ur Rehman, H. M. Bilal, M. Usman, M. Z. Shabbir and H. E. Gelani, "Protection Schemes for Low Voltage DC Networks", *International Conference on Electrical, Communication, and Computer Engineering (ICECCE), Swat, Pakistan, 2019*, pp. 1–5.
7. AL-Nussairi, M.K., Bayindir, R., Padmanaban, S. *et al.* "Constant power loads (CPL) with microgrids: problem definition, stability analysis and compensation techniques", *Energies*, Vol. 10, 2017.
8. U. Ghisla, I. Kondratiev and R. Dougal, "Protection of medium voltage DC power systems against ground faults and negative incremental impedances", *Proceedings of the IEEE SoutheastCon, Concord, NC, 2010*, pp. 259–263.
9. X. Lai, F. Liu, K. Deng, Q. Gao and X. Zha, "A short-circuit current calculation method for low-voltage DC microgrid", *International Power Electronics and Application Conference and Exposition, Shanghai, 2014*, pp. 365–371.
10. J. C. Das and R. H. Osman, "Grounding of AC and DC low-voltage and medium-voltage drive systems", *IEEE Transactions on Industry Applications*, Vol. 34, No. 1, Jan.-Feb. 1998, pp. 205–216.
11. J. Mohammadi, F. Badrkhani Ajaei and G. Stevens, "Grounding the DC Microgrid", *IEEE Transactions on Industry Applications*, Vol. 55, No. 5, Sept.-Oct. 2019, pp. 4490–4499.
12. D. M. Bui, S. Chen, C. Wu, K. Lien, C. Huang and K. Jen, "Review on protection coordination strategies and development of an effective protection coordination system for DC microgrid", *IEEE Asia-Pacific Power and Energy Engineering Conference, Hong Kong, 2014*, pp. 1–10.
13. Jianwei Li, Rui Xiong, Qingqing Yang, FeiLiang, MinZhang and WeijiaYuan, "Design/test of a hybrid energy storage system for primary frequency control using a dynamic droop method in an isolated microgrid power system", *Applied Energy*, Vol. 201, Sept. 2017, pp. 257–269.
14. J. Park, J. Candelaria, L. Ma and K. Dunn, "DC Ring-Bus Microgrid Fault Protection and Identification of Fault Location", *IEEE Transactions on Power Delivery*, Vol. 28, No. 4, Oct. 2013, pp. 2574–2584.
15. Wooin Choi and Bo-Hyung Cho, "Topology and control design of converters for short-circuit fault protection in DC microgrids", *IECON 39th Annual Conference of the IEEE Industrial Electronics Society, Vienna, 2013, pp. 1386–1391.*
16. T. R. de Oliveira, A. S. Bolzon and P. F. Donoso-Garcia, "Grounding and safety considerations for residential DC microgrids", *IECON 2014 - 40th Annual Conference of the IEEE Industrial Electronics Society, Dallas, TX, 2014*, pp. 5526–5532.
17. A. Meghwani, S. Chakrabarti, S. C. Srivastava and S. Anand, "Analysis of fault characteristics in DC microgrids for various converter topologies", *IEEE Innovative Smart Grid Technologies - Asia (ISGT-Asia), Auckland, 2017*, pp. 1–6.

18. M. K. Bucher, R. Wiget, G. Andersson and C. M. Franck, "Multiterminal HVDC Networks—What is the Preferred Topology?" in *IEEE Transactions on Power Delivery*, Vol. 29, No. 1, Feb. 2014, pp. 406–413.
19. A. Berizzi, A. Silvestri, D. Zaninelli, and S. Massucco, "Short-circuit current calculations for DC systems," *IEEE Transactions on Industry Applications*, Vol. 32, No. 5, 1996, pp. 990–997.
20. Short-Circuit Currents in D.C. Auxiliary Installations in Power Plants and Substations, Part 1: Calculation of Short-Circuit Currents, IEC61660-1, 1997.
21. E. Dehghanpour and R. Kheirollahi, "Developing a new fault location topology for DC microgrid systems", *7th Power Electronics and Drive Systems Technologies Conference (PEDSTC), Tehran, 2016*, pp. 297–301.
22. D. Salomonsson and A. Sannino, "Low-voltage dc distribution system for commercial power systems with sensitive electronic loads", *IEEE Trans. Power Delivery*, Vol. 22, No. 3, July 2007, pp. 1620–1627.
23. D. Kumar, F. Zare and A. Ghosh, "DC Microgrid Technology: System Architectures, AC Grid Interfaces, Grounding Schemes, Power Quality, Communication Networks, Applications, and Standardizations Aspects", *IEEE Access*, Vol. 5, June 2017. pp. 12230–12256.
24. Zhang, L., Tai, N., Huang, W. *et al.* "A review on protection of DC microgrids," *J. Mod. Power Syst. Clean Energy*, 6, March 2018, pp. 1113–1127.
25. A. Meghwani, S. C. Srivastava and S. Chakrabarti, "A new protection scheme for DC microgrid using line current derivative", *IEEE Power and Energy Society General Meeting, Denver, CO, 2015*, pp. 1–5.
26. N. Bayati, A. Hajizadeh and M. Soltani, "Localized Fault Protection in the DC Microgrids with Ring Configuration", *IEEE 28th International Symposium on Industrial Electronics (ISIE), Vancouver, BC, Canada, 2019*, pp. 136–140.
27. S. Beheshtaein, R. M. Cuzner, M. Forouzesh, M. Savaghebi and J. M. Guerrero, "DC Microgrid Protection: A Comprehensive Review," in *IEEE Journal of Emerging and Selected Topics in Power Electronics, 2019.*
28. Ali Abdali, Kazem Mazlumi and Reza Noroozian, "High-speed fault detection and location in DC microgrids systems using Multi-Criterion System and neural network", *Applied Soft Computing*, Vol. 79, June 2019, pp. 341–353.
29. H. Nian and L. Kong, "Transient Modeling and Analysis of VSC Based DC Microgrid During Short Circuit Fault", *IEEE Access*, Vol. 7, 2019, pp. 170604–170614.
30. J. Park and J. Candelaria, "Fault Detection and Isolation in Low-Voltage DC-Bus Microgrid System", *IEEE Transactions on Power Delivery*, Vol. 28, No. 2, April 2013, pp. 779–787.
31. R. Mohanty and A. K. Pradhan, "Protection of Smart DC Microgrid with Ring Configuration Using Parameter Estimation Approach", *IEEE Transactions on Smart Grid*, Vol. 9, No. 6, Nov. 2018, pp. 6328–6337.
32. A. A. Sheikh, S. A. Wakode, R. R. Deshmukh and M. S. Ballal, "A Protection Scheme for Fault Detection, Location and Isolation in DC Ring Microgrid",

45th Annual Conference of the IEEE Industrial Electronics Society, Lisbon, Portugal, 2019, pp. 2109–2114.

33. G. Madingou, M. Zarghami and M. Vaziri, "Fault detection and isolation in a DC microgrid using a central processing unit", *IEEE Power and Energy Society Innovative Smart Grid Technologies Conference (ISGT), Washington, DC, 2015*, pp. 1–5.
34. R. Bhargav, B. R. Bhalja and C. P. Gupta, "Algorithm for fault detection and localization in a mesh-type bipolar DC microgrid network", *IET Generation, Transmission and Distribution*, Vol. 13, No. 15, 2019, pp. 3311–3322.
35. F. Paz and M. Ordonez, "Embedded fault location in DC microgrid systems based on a Lock-In Amplifier", *IEEE 8th International Symposium on Power Electronics for Distributed Generation Systems (PEDG), Florianopolis, 2017*, pp. 1–6.
36. K. A. Saleh, A. Hooshyar and E. F. El-Saadany, "Ultra-High-Speed Traveling-Wave-Based Protection Scheme for Medium-Voltage DC Microgrids", *IEEE Transactions on Smart Grid*, Vol. 10, No. 2, March 2019, pp. 1440-1451.
37. C. Diendorfer *et al.*, "Graph traversal-based automation of fault detection, location, and recovery on MVDC shipboard power systems", *IEEE Second International Conference on DC Microgrids (ICDCM), Nuremburg, 2017*, pp. 119–124.
38. S. Mirsaeidi, X. Dong, S. Shi and D. Tzelepis, "Challenges, advances and future directions in protection of hybrid AC/DC microgrids", *IET Renewable Power Generation*, Vol. 11, No. 12, 2017, pp. 1495–1502.
39. T. Sakagami, Y. Asai and H. Kitano, "Simulation to optimize a DC microgrid in Okinawa", *IEEE International Conference on Sustainable Energy Technologies (ICSET), Hanoi, 2016*, pp. 214–219.
40. C. Strobl, "Arc fault detection in DC microgrids", *IEEE First International Conference on DC Microgrids (ICDCM), Atlanta, GA, 2015*, pp. 181–186.
41. K. Gajula and L. Herrera, "Detection and Localization of Series Arc Faults in DC Microgrids using Kalman Filter", *IEEE Journal of Emerging and Selected Topics in Power Electronics*, 2020.
42. F. M. Uriarte *et al.*, "A DC Arc Model for Series Faults in Low Voltage Microgrids", *IEEE Transactions on Smart Grid*, Vol. 3, No. 4, Dec. 2012, pp. 2063–2070.
43. L. Yue, V. Le, Z. Yang and X. Yao, "A Novel Series Arc Fault Detection Method using Sparks in DC Microgrids with Buck Converter Interface", *IEEE Energy Conversion Congress and Exposition (ECCE), Portland, OR, 2018*, pp. 492–496.
44. G. Seo, J. Ha, B. Cho and K. Lee, "Series arc fault detection method based on statistical analysis for dc Microgrids", *IEEE Applied Power Electronics Conference and Exposition (APEC), Long Beach, CA, 2016*, pp. 487–492.
45. L. Mackay, A. Shekhar, B. Roodenburg, L. Ramírez-Elizondo and P. Bauer, "Series arc extinction in DC microgrids using load side voltage drop

detection", *IEEE First International Conference on DC Microgrids (ICDCM), Atlanta, GA, 2015*, pp. 239–244.
46. Z. Wang and R. S. Balog, "Arc Fault and Flash Signal Analysis in DC Distribution Systems Using Wavelet Transformation", *IEEE Transactions on Smart Grid*, Vol. 6, No. 4, July 2015, pp. 1955–1963.
47. X. Yao, L. Herrera and J. Wang, "Impact evaluation of series dc arc faults in dc microgrids", *IEEE Applied Power Electronics Conference and Exposition (APEC), Charlotte, NC, 2015*, pp. 2953–2958.
48. X. Lai, F. Liu, K. Deng, Q. Gao and X. Zha, "A short-circuit current calculation method for low-voltage DC microgrid", *International Power Electronics and Application Conference and Exposition, Shanghai, 2014*, pp. 365–371.
49. S. A. Arefifar, Y. A. I. Mohamed and T. H. M. EL-Fouly, "Optimum Microgrid Design for Enhancing Reliability and Supply-Security", *IEEE Transactions on Smart Grid*, Vol. 4, No. 3, Sept. 2013, pp. 1567–1575.
50. D. Velasco de la Fuente, C. L. Trujillo Rodríguez, G. Garcerá, E. Figueres and R. Ortega Gonzalez, "Photovoltaic Power System with Battery Backup with Grid-Connection and Islanded Operation Capabilities", *IEEE Transactions on Industrial Electronics*, Vol. 60, No. 4, April 2013, pp. 1571–1581.
51. S. Ahmad, M. J. Uddin, I. H. Nisu, M. M. Ahsan, I. Rahman and N. H. Samrat, "Modeling of grid connected battery storage wave energy and PV hybrid renewable power generation", *International Conference on Electrical, Computer and Communication Engineering (ECCE), Cox's Bazar, 2017*, pp. 375–380.
52. S. Hemavathi and M. Singh, "Microgrid Short Circuit Studies", *IEEE 8th Power India International Conference (PIICON), Kurukshetra, India, 2018*, pp. 1–6.
53. N. Bayati, A. Hajizadeh and M. Soltani, "Protection in DC microgrids: a comparative review", *IET Smart Grid*, vol. 1, No. 3, 2018, pp. 66–75.
54. X. Yang, Z. Shuai, D. He and W. Wang, "Short-Circuit Fault Analysis of Energy Storage System Converter with Different Control in DC Microgrid", *IEEE 3rd Conference on Energy Internet and Energy System Integration (EI2), Changsha, China, 2019*, pp. 2534–2539.
55. D. Patii, A. K. Rathore, D. Srinivasan and S. K. Panda, "High-frequency soft-switching LCC resonant current-fed DC/DC converter with high voltage gain for DC microgrid application", *IECON 2014 - 40th Annual Conference of the IEEE Industrial Electronics Society, Dallas, TX, 2014*, pp. 4293–4299.
56. S. Hemavathi and M. Singh, "Microgrid Short Circuit Studies," *IEEE 8th Power India International Conference (PIICON), Kurukshetra, India, 2018*, pp. 1–6.
57. A. Virdag, T. Hager and R. W. De Doncker, "Estimation of short-circuit currents in future LVDC microgrids", in *CIRED - Open Access Proceedings Journal*, Vol. 2017, No. 1, pp. 1098–1101.
58. Ming Yu, Y. Wang, L. Zhang and Ziguang Zhang, "DC short circuit fault analysis and protection of ring type DC microgrid", *IEEE 8th International Power*

Electronics and Motion Control Conference (IPEMC-ECCE Asia), Hefei, 2016, pp. 1694–1700.
59. X. Feng, L. Qi and Z. Wang, "Estimation of short circuit currents in mesh DC networks", *IEEE PES General Meeting Conference & Exposition, National Harbor, MD, 2014*, pp. 1–5.
60. X. Zha, H. Ning, X. Lai, Y. Huang and F. Liu, "Suppression strategy for short-circuit current in loop-type DC microgrid", *IEEE Energy Conversion Congress and Exposition (ECCE), Pittsburgh, PA, 2014*, pp. 758–764.
61. X. Lai, F. Liu, K. Deng, Q. Gao and X. Zha, "A short-circuit current calculation method for low-voltage DC microgrid", *International Power Electronics and Application Conference and Exposition, Shanghai, 2014*, pp. 365–371.
62. Z. J. Shen, Z. Miao and A. M. Roshandeh, "Solid state circuit breakers for DC micrgrids: Current status and future trends", *IEEE First International Conference on DC Microgrids (ICDCM), Atlanta, GA, 2015*, pp. 228–233.
63. A. A. Elserougi, A. S. Abdel-Khalik, A. M. Massoud and S. Ahmed, "A New Protection Scheme for HVDC Converters Against DC-Side Faults with Current Suppression Capability", *IEEE Transactions on Power Delivery, vol. 29, No. 4, Aug. 2014*, pp. 1569–1577.
64. C. M. Franck, "HVDC Circuit Breakers: A Review Identifying Future Research Needs", *IEEE Transactions on Power Delivery*, Vol. 26, No. 2, April 2011, pp. 998–1007.
65. L. Qi, A. Antoniazzi and L. Raciti, "DC Distribution Fault Analysis, Protection Solutions, and Example Implementations", *IEEE Transactions on Industry Applications*, Vol. 54, No. 4, July-Aug. 2018, pp. 3179–3186.
66. J. Yang, J. E. Fletcher and J. O'Reilly, "Short-Circuit and Ground Fault Analyses and Location in VSC-Based DC Network Cables", *IEEE Transactions on Industrial Electronics*, Vol. 59, No. 10, Oct. 2012, pp. 3827–3837.
67. Saurabh Kulkarni and Surya Santoso, "Interrupting Short-Circuit Direct Current Using an AC Circuit Breaker in Series with a Reactor", Hindawi Publishing Corporation, *Advances in Power Electronics*. Vol. 2012.
68. S. R. Samantaray, G. Joos and I. Kamwa, "Differential energy based microgrid protection against fault conditions", *2012 IEEE PES Innovative Smart Grid Technologies (ISGT), Washington, DC, 2012*, pp. 1–7.
69. M. Abedrabbo, W. Leterme and D. Van Hertem, "Systematic Approach to HVDC Circuit Breaker Sizing", *IEEE Transactions on Power Delivery*, Vol. 35, No. 1, Feb. 2020, pp. 288–300.
70. B. Kakarla, T. Ziemann, R. Stark, P. Natzke and U. Grossner, "Short Circuit Ruggedness of New Generation 1.2 kV SiC MOSFETs", *IEEE 6th Workshop on Wide Bandgap Power Devices and Applications (WiPDA), Atlanta, GA, 2018*, pp. 118–124.
71. K. Satpathi, A. Ukil and J. Pou, "Short-Circuit Fault Management in DC Electric Ship Propulsion System: Protection Requirements, Review of

Existing Technologies and Future Research Trends", *IEEE Transactions on Transportation Electrification*, Vol. 4, No. 1, March 2018, pp. 272–291.
72. Mustafa Farhadi and Osama A. Mohammed, "Protection of multi-terminal and distributed DC systems: Design challenges and techniques", *Electric Power Systems Research* 143, 2017, pp. 715–727.
73. M. Mobarrez, D. Fregosi, S. Bhattacharya and M. A. Bahmani, "Grounding architectures for enabling ground fault ride-through capability in DC microgrids", *IEEE Second International Conference on DC Microgrids (ICDCM), Nuremburg, 2017*, pp. 81–87.
74. D. K. J. S. Jayamaha, N. W. A. Lidula and A. D. Rajapakse, "Ground Fault Analysis and Grounding Design Considerations in DC Microgrids", *IEEE 4th Southern Power Electronics Conference (SPEC), Singapore, Singapore, 2018*, pp. 1–8.
75. S. Beheshtaein, R. Cuzner, M. Savaghebi and J. M. Guerrero, "Review on microgrids protection", in *IET Generation, Transmission & Distribution*, Vol. 13, No. 6, 2019, pp. 743–759.
76. X. Lu, K. Sun, J. M. Guerrero, J. C. Vasquez, L. Huang and J. Wang, "Stability Enhancement Based on Virtual Impedance for DC Microgrids with Constant Power Loads", *IEEE Transactions on Smart Grid*, Vol. 6, No. 6, Nov. 2015, pp. 2770–2783.
77. M. K. Al-Nussairi and R. Bayindir, "DC-DC Boost Converter Stability with Constant Power Load", *IEEE 18th International Power Electronics and Motion Control Conference (PEMC), Budapest, 2018*, pp. 1061–1066.
78. A. B. Jusoh, "The instability effect of constant power loads", *Proceedings National Power and Energy Conference, Kuala Lumpur, Malaysia, 2004*, pp. 175–179.
79. E. Hossain, R. Perez, A. Nasiri and S. Padmanaban, "A Comprehensive Review on Constant Power Loads Compensation Techniques", *IEEE Access*, Vol. 6, 2018, pp. 33285–33305.
80. AL-Nussairi, M.K.; Bayindir, R.; Padmanaban, S.; Mihet-Popa, L., Siano P., "Constant Power Loads (CPL) with Microgrids: Problem Definition, Stability Analysis and Compensation Techniques", *Energies*, 2017.
81. L. Herrera, W. Zhang and J. Wang, "Stability Analysis and Controller Design of DC Microgrids with Constant Power Loads", *IEEE Transactions on Smart Grid*, Vol. 8, No. 2, March 2017, pp. 881–888.
82. P. Magne, B. Nahid-Mobarakeh and S. Pierfederici, "Dynamic Consideration of DC Microgrids With Constant Power Loads and Active Damping System—A Design Method for Fault-Tolerant Stabilizing System", *IEEE Journal of Emerging and Selected Topics in Power Electronics*, Vol. 2, No. 3, Sept. 2014, pp. 562–570.
83. Hamed Karimi, Ghazanfar Shahgholian, Bahador Fani, Iman Sadeghkhani and Majid Moazzami, "A protection strategy for inverter-interfaced islanded microgrids with looped configuration", *Electrical Engineering* Vol. 101, 2019, pp. 1059–1073.

84. R. M. Cuzner and G. Venkataramanan, "The Status of DC Micro-Grid Protection", *IEEE Industry Applications Society Annual Meeting, Edmonton, AB, 2008*, pp. 1–8.
85. Jincheng Li and J. L. Kohler, "New insight into the detection of high-impedance arcing faults on DC trolley systems", *IEEE Transactions on Industry Applications*, Vol. 35, No. 5, Sept.-Oct. 1999, pp. 1169–1173.
86. D. Liu, D. Tzelepis and A. Dyśko, "Protection of Microgrids with High Amounts of Renewables: Challenges and Solutions", *54th International Universities Power Engineering Conference (UPEC), Bucharest, Romania, 2019*, pp. 1–6.
87. M. M. A. M. Aly and M. A. H. El-Sayed, "Enhanced fault location algorithm for smart grid containing wind farm using wireless communication facilities", *IET Generation, Transmission and Distribution*, Vol. 10, No. 9, 2016, pp. 2231–2239.
88. Senarathna, Sisitha and Hemapala, K T M U, "Review of adaptive protection methods for microgrids", *AIMS Energy*, Vol. 11, September 2019.
89. E. Sortomme, M. Venkata and J. Mitra, "Microgrid protection using communication-assisted digital relays," IEEE PES General Meeting, Providence, RI, 2010.
90. S. R. Samantaray, G. Joos and I. Kamwa, "Differential energy based microgrid protection against fault conditions", *IEEE PES Innovative Smart Grid Technologies (ISGT), Washington, DC, 2012*, pp. 1–7.
91. E. Casagrande, W. L. Woon, H. H. Zeineldin and D. Svetinovic, "A Differential Sequence Component Protection Scheme for Microgrids with Inverter-Based Distributed Generators", *IEEE Transactions on Smart Grid*, Vol. 5, No. 1, Jan. 2014, pp. 29–37.
92. S. D. A. Fletcher, P. J. Norman, K. Fong, S. J. Galloway and G. M. Burt, "High-Speed Differential Protection for Smart DC Distribution Systems", *IEEE Transactions on Smart Grid*, Vol. 5, No. 5, Sept. 2014, pp. 2610–2617.
93. S. Dhar, R. K. Patnaik and P. K. Dash, "Fault Detection and Location of Photovoltaic Based DC Microgrid Using Differential Protection Strategy", *IEEE Transactions on Smart Grid*, Vol. 9, No. 5, Sept. 2018, pp. 4303–4312.
94. A. Meghwani, S. C. Srivastava and S. Chakrabarti, "A Non-Unit Protection Scheme for DC Microgrid Based on Local Measurements", *IEEE Transactions on Power Delivery*, Vol. 32, No. 1, Feb. 2017, pp. 172–181.
95. A. Shekhar, L. Ramírez-Elizondo, S. Bandyopadhyay, L. Mackay and P. Bauera, "Detection of Series Arcs Using Load Side Voltage Drop for Protection of Low Voltage DC Systems", *IEEE Transactions on Smart Grid*, Vol. 9, No. 6, Nov. 2018, pp. 6288–6297.
96. H. Al-Nasseri, M. A. Redfern and F. Li, "A voltage based protection for microgrids containing power electronic converters", IEEE Power Engineering Society General Meeting, Montreal, Que., 2006.

97. L. Tang and B. Ooi, "Locating and Isolating DC Faults in Multi-Terminal DC Systems", *IEEE Transactions on Power Delivery*, Vol. 22, No. 3, July 2007, pp. 1877–1884.
98. Xiaomin Qi , Wei Pei , Luyang Li and Li Kong, "A Fast DC Fault Detection Method for Multi-Terminal AC/DC Hybrid Distribution Network Based on Voltage Change Rate of DC Current-Limiting Inductor", *Energies, MDPI, Open Access Journal*, Vol. 11, July 2018, pp. 1–22.
99. F. Coffele, C. Booth and A. Dyśko, "An Adaptive Overcurrent Protection Scheme for Distribution Networks", *IEEE Transactions on Power Delivery*, Vol. 30, No. 2, April 2015, pp. 561–568.
100. R. Jain, D. L. Lubkeman and S. M. Lukic, "Dynamic Adaptive Protection for Distribution Systems in Grid-Connected and Islanded Modes", *IEEE Transactions on Power Delivery*, Vol. 34, No. 1, Feb. 2019, pp. 281–289.
101. Yavuz Ates , Mehmet Uzunoglu , Arif Karakas , Ali Rifat Boynuegri, Abdullah Nadar and Bulent Dag, "Implementation of adaptive relay coordination in distribution systems including distributed generation", *Journal of Cleaner Production*. Vol. 112, Part 4, 20 Jan. 2016, pp. 2697–2705.
102. Ates Y., Boynuegr A.R., Uzunoglu M., Nadar A., Yumurtacı R., Erdinc O., Paterakis N.G. and Catalão J.P.S., "Adaptive Protection Scheme for a Distribution System Considering Grid-Connected and Islanded Modes of Operation", *Energies* 2016.
103. Wen-Jun Tang and Hong-Tzer Yang, "Data Mining and Neural Networks Based Self-Adaptive Protection Strategies for Distribution Systems with DGs and FCLs", *Energies, MDPI, Open Access Journal*, Vol. 11(2), Feb. 2018, pp. 1–13.
104. L. Zhang, N. Tai, W. Huang and Y. Wang, "Fault distance estimation-based protection scheme for DC microgrids", *Journal of Engineering*, Vol. 2019, No.16, 2019, pp. 1199–1203.
105. H. Lin, J. M. Guerrero, J. C. Vásquez and C. Liu, "Adaptive distance protection for microgrids", 41st *Annual Conference of the IEEE Industrial Electronics Society, Yokohama, 2015*, pp. 725–730.
106. L. Jin, M. Jiang and G. Yang, "Fault analysis of microgrid and adaptive distance protection based on complex wavelet transform", *International Power Electronics and Application Conference and Exposition, Shanghai, 2014*, pp. 360–364.
107. Syed Basit Ali Bukhari, Raza Haider, Muhammad Saeed Uz Zaman, Yun-Sik Oh, Gyu-Jung Cho and Chul-Hwan Kim, "An interval type-2 fuzzy logic based strategy for microgrid protection", *Electrical Power and Energy Systems* No. 98, 2018, pp. 209–218.
108. M. J. Daryani, A. E. Karkevandi and O. Usta, "A Hybrid Approach for Microgrid Protection System Based on Neural Network and Fuzzy Logic", *IEEE PES Innovative Smart Grid Technologies Europe (ISGT-Europe), Bucharest, Romania, 2019*, pp. 1–5.

109. Chen Cheng-I, Lan Chien-Kai, Chen Yeong-Chin, Chen Chung-Hsie and Chang Yung-Ruei, "Wavelet Energy Fuzzy Neural Network-Based Fault Protection System for Microgrid", *Energies*, Vol. 13, 2020.
110. M. Baran and N. R. Mahajan, "PEBB Based DC System Protection: Opportunities and Challenges", *IEEE/PES Transmission and Distribution Conference and Exhibition, Dallas, TX, 2006*, pp. 705–707.
111. Aswani J and P. Kanakasabapathy, "Protection of a low-voltage DC ring microgrid system", *International Conference on Energy Efficient Technologies for Sustainability (ICEETS), Nagercoil, 2016*, pp. 17–22.
112. M. Komatsu, "Approach and basic evaluation for the DC circuit breaker with fault current limiting feature", *IEEE International Telecommunications Energy Conference (INTELEC), Austin, TX, 2016*, pp. 1–5.
113. D. Salomonsson, L. Soder and A. Sannino, "Protection of Low-Voltage DC Microgrids", *IEEE Transactions on Power Delivery*, Vol. 24, No. 3, July 2009, pp. 1045–1053.
114. I. Almutairy and J. Asumadu, "Examination of breaker-based protection systems for implementation in LVDC SSCB applications", *IEEE 8th Annual Computing and Communication Workshop and Conference (CCWC), Las Vegas, NV, 2018*, pp. 509–514.
115. A. Kwasinski, "Advanced power electronics enabled distribution architectures: Design, operation, and control", *8th International Conference on Power Electronics - ECCE Asia, Jeju*, 2011, pp. 1484–1491.
116. M. Baran and N. Mahajan, "Overcurrent protection on voltage-source converter- based multiterminal dc distribution systems," *IEEE Trans. Power Del.*, Vol. 22, No. 1, Jan. 2007, pp. 406–412.
117. J.-M. Meyer and A. Rufer, "A DC hybrid circuit breaker with ultra-fast contact opening and integrated gate-commutated thyristors (igcts)", *IEEE Trans. Power Del.*, Vol. 21, No. 2, April 2006, pp. 646–651.
118. W. A. Martin, C. Deng, D. Fiddiansyah and J. C. Balda, "Investigation of low-voltage solid-state DC breaker configurations for DC microgrid applications", *IEEE International Telecommunications Energy Conference (INTELEC), Austin, TX, 2016*, pp. 1–6.
119. A. Maqsood, A. Overstreet and K. A. Corzine, "Modified Z-Source DC Circuit Breaker Topologies", *IEEE Transactions on Power Electronics*, Vol. 31, No. 10, Oct. 2016, pp. 7394–7403.
120. L. Mackey, M. Rifat Kaisar Rachi, C. Peng and I. Husain, "Z-Source circuit breaker utilizing Ultra-Fast Mechanical Switch for high efficiency DC circuit protection", *IEEE Second International Conference on DC Microgrids (ICDCM), Nuremburg, 2017*, pp. 452–458.
121. A. Overstreet, A. Maqsood and K. Corzine, "Modified z-source DC circuit breaker topologies", *Clemson University Power Systems Conference, Clemson, SC, 2014*, pp. 1–6.

122. K. Corzine, "Dc micro grid protection with the z-source breaker," *39th Annual Conference of the IEEE Industrial Electronics Society, Vienna, 2013*, pp. 2197–2204.
123. K. A. Corzine, "Circuit breaker for DC micro grids", *IEEE First International Conference on DC Microgrids (ICDCM), Atlanta, GA*, 2015, pp. 221–221.
124. S. Savaliya, S. Singh and B. G. Fernandes, "Protection of DC system using bi-directional Z-Source Circuit breaker,", *42nd Annual Conference of the IEEE Industrial Electronics Society, Florence, 2016*, pp. 4217–4222.
125. M. W. Aslam and N. A. Zaffar, "A novel bidirectional Z-source DC circuit breaker for DC microgrids," *IEEE 7th Power India International Conference (PIICON), Bikaner, 2016*, pp. 1–5.
126. Y. Bingjian, G. Yang, W. Xiaoguang, H. Zhiyuan, C. Longlong and S. Yunhai, "A hybrid circuit breaker for DC-application," *IEEE First International Conference on DC Microgrids (ICDCM), Atlanta, GA, 2015*, pp. 187–192.
127. D. Keshavarzi, E. Farjah and T. Ghanbarih, "A hybrid DC circuit breaker for DC microgrid based on zero current switching", *Iranian Conference on Renewable Energy and Distributed Generation (ICREDG), Mashhad, 2016*, pp. 45–49.
128. M. Liao, S. Wang, G. Ge, J. Huang and X. Duan, "Investigation on breaking characteristic of hybrid DC switch for DC micro-grid", *27th International Symposium on Discharges and Electrical Insulation in Vacuum (ISDEIV), Suzhou, 2016*, pp. 1–4.
129. J. Huang, G. Ge, M. Liao, X. Duan, L. Pu and J. Zou, "The current commutation characteristics of DC micro-grid hybrid circuit breakers", *27th International Symposium on Discharges and Electrical Insulation in Vacuum (ISDEIV), Suzhou, 2016*, pp. 1–4.
130. D. Keshavarzi, E. Farjah and T. Ghanbari, "Hybrid DC Circuit Breaker and Fault Current Limiter with Optional Interruption Capability", *IEEE Transactions on Power Electronics*, Vol. 33, No. 3, March 2018, pp. 2330–2338.
131. L. Mackey, C. Peng and I. Husain, "Progressive Switching of Hybrid DC Circuit Breakers for Faster Fault Isolation", *IEEE Energy Conversion Congress and Exposition (ECCE), Portland, OR*, 2018, pp. 7150–7157.
132. A. Ray, S. N. Banavath and K. Rajashekara, "Coupled Inductor Based Hybrid DC Circuit Breaker Topologies for DC Grid Application", *44th Annual Conference of the IEEE Industrial Electronics Society, Washington, DC, 2018*, pp. 1116–1121.
133. W. Li, Y. Wang, X. Wu and X. Zhang, "A Novel Solid-State Circuit Breaker for On-Board DC Microgrid System", *IEEE Transactions on Industrial Electronics*, Vol. 66, No. 7, July 2019, pp. 5715–5723.
134. Wenjun Liu, Hua Yang, Fei Liu, Jianjun Sun and Xiaoming Zha, "An improved RCD snubber for solid-state circuit breaker protection against bus fault in low-voltage DC microgrid", *IEEE 2nd International Future Energy Electronics Conference (IFEEC), Taipei, 2015*, pp. 1–5.

135. Z. Zhou, M. Chen, J. Jiang, D. Zhang, S. Ye and C. Liu, "Analysis and Design of a Novel Thyristor-Based Circuit Breaker for DC Microgrids", *IEEE Transactions on Power Electronics*, Vol. 35, No. 3, March 2020, pp. 2959–2968.
136. I. Pramudya, U. A. Khan, Suwarno, J. Y. Koo and B. W. Lee, "Modelling and evaluation of low voltage DC circuit breaker for the protection of DC microgrid", *International Conference on Condition Monitoring and Diagnosis (CMD), Xi'an, 2016*, pp. 295–298.
137. S. Munasib and J. C. Balda, "Short-circuit protection for low-voltage DC microgrids based on solid-state circuit breakers", *IEEE 7th International Symposium on Power Electronics for Distributed Generation Systems (PEDG), Vancouver, BC, 2016*, pp. 1–7.
138. K. Palaniappan, W. Sedano, N. Hoeft, R. Cuzner and Z. J. Shen, "Fault discrimination using SiC JFET based self-powered solid state circuit breakers in a residential DC community microgrid", *IEEE Energy Conversion Congress and Exposition (ECCE), Cincinnati, OH, 2017*, pp. 3747–3753.
139. C. Aarstad, T. Taufik, A. Kean and M. Muscarella, "Development of arc fault interrupter laboratory testing for low voltage DC electricity", *International Seminar on Intelligent Technology and Its Applications (ISITIA), Lombok, 2016*, pp. 583–588.
140. National Electrical Code 2017 Edition, Chapter 1, Article 100, pp. 70–33.
141. D. A. Dini, P. W. Brazis and K. Yen, "Development of Arc-Fault Circuit-Interrupter requirements for Photovoltaic systems", *37th IEEE Photovoltaic Specialists Conference, Seattle, WA, 2011*, pp. 1790–1794.
142. H.-S. Konga and W.-J. Rab, "Usability of arc fault circuit interrupters with network function", *Scientia Iranica. Transaction A, Civil Engineering*, Vol. 27, No.2, April 2020, pp. 607–613.
143. B. Xiang, Z. Liu, Y. Geng and S. Yanabu, "DC Circuit Breaker Using Superconductor for Current Limiting", *IEEE Transactions on Applied Superconductivity*, Vol. 25, No. 2, April 2015, pp. 1–7.
144. Toru Tanaka, Hiroshi Kawaguchi, Takashi Terao, Tadatoshi Babasaki and Mikio Yamasaki, "Modeling of fuses for DC power supply systems including arcing time analysis", *INTELEC 07 - 29th International Telecommunications Energy Conference, Rome, 2007*, pp. 135–141.
145. S. Ravyts, G. V. d. Broeck, L. Hallemans, M. D. Vecchia and J. Driesen, "Fuse-Based Short-Circuit Protection of Converter Controlled Low-Voltage DC Grids", *IEEE Transactions on Power Electronics*, Vol. 35, No. 11, Nov. 2020, pp. 11694–11706.
146. N. Bayati, H. R. Baghaee, A. Hajizadeh and M. Soltani, "A Fuse Saving Scheme for DC Microgrids with High Penetration of Renewable Energy Resources", *IEEE Access*, Vol. 8, 2020, pp. 137407–137417.
147. Seong-Yong Lee ,Young-Kwang Son , Hyung-June Cho, Seung-Ki Sul ,Sang-Hyun Kim , Nam-Sook Kang and Woo-Jae Park, "Simplified Thermal Model of Semiconductor Fuse for DC Distribution System", *10th International*

Conference on Power Electronics and ECCE Asia , Busan, Korea (South), 2019, pp. 2641–2646.

148. A. Sato and Y. Oda, "Application of FDS fuses to DC 1000V photovoltaic systems", *34th International Conference on Lightning Protection (ICLP), Rzeszow, 2018*, pp. 1–6.

2

Control Strategies for DC Microgrids

Bhabani Kumari Choudhury* and Premalata Jena

Department of Electrical Engineering, Indian Institute of Technology Roorkee, Roorkee, India

Abstract

The excellent compatibility of direct current (DC) microgrids with photovoltaics, electric cars, and dc loads make it widely admired for low-voltage distribution networks. A power-electronic converter with a DC link capacitor is used to connect distributed generation (DG) units to the microgrid. As a result, new control mechanisms for these converters must be devised in order to fully use the DG units in the event of an islanded microgrid. This chapter provides a comprehensive overview of different existing control schemes for DC Microgrids, along with the motivation and challenge behind it. It covers the basic and multi-level control schemes in detail, along with the converter control scheme for solar, wind, battery, and fuel cell-based systems. Finally, the concluding remarks and the future research scope are highlighted.

Keywords: DC microgrids, maximum power point tracking, solar system, wind energy system, battery-based energy storage system, fuel cell

2.1 Introduction: The Concept of Microgrids

Nowadays, the deployment of microgrids (MGs) has resulted in significant advantages to electrical consumers and power grid operators. Some of them are improvement of power quality, reduction in transmission and conversion losses, economical and energy-efficient [1]. MGs have proven to be the most promising power system structure for the electricity demand of remote locations. Also, the MGs have a unique islanding proficiency in response to any fault or disturbance on the grid side to strengthen the

**Corresponding author*: bkumarichoudhury@ee.iitr.ac.in

Nikita Gupta, Mahajan Sagar Bhaskar, Sanjeevikumar Padmanaban and Dhafer Almakhles (eds.) DC Microgrids: Advances, Challenges, and Applications, (63–82) © 2022 Scrivener Publishing LLC

reliability and resilience of the system [2–5]. MGs can be characterized according to the location, such as the campus, military, residential, commercial, and industrial. Secondly, MGs can be described based on sizes, such as small, medium, and large scales. Thirdly, they can be characterized by connectivity, such as remote and grid-connected. However, they can be identified as AC MGs, DC MGs, and hybrid MGs based on the output type to the customers. In the AC MGs, the power generated by the distributed energy resource (DERs) is converted to AC with the help of DC-AC converters to supply the AC loads. In DC MGs, rectifiers are used to connect DERs, and inverters are used to provide power to AC-type loads. Moreover, both AC and DC buses are deployed in hybrid MGs connected to each bus, depending on the loads and DERs joined [6].

This chapter aims to give readers some insights into different control topologies and challenges of DC MGs and some motivation in this field. For the sake of understanding, some existing control schemes are addressed along with their significant limitations. Then the pros and cons of the methods are discussed with their concluding remarks. Finally, future research trends on the requirements of the new control schemes to increase the accuracy of the protection of DC MGs are highlighted.

2.1.1 DC Microgrids

MGs are not a new concept in the power system. They already exist from past times for supplying power to small islands. Despite this explanation, MGs are still difficult to define. In 2005, the Department of Energy (DoE) and the Commission of the European Communities (CEC) jointly commissioned a Navigant consulting report, which defined MGs based on only two points of universal agreement. (1) MGs have interconnected distributed energy resources to supply sufficient power to the loads continuously, which are connected in that area. (2) MGs perform independent control and intentional islanding activities with minimum interruption of service.

The revolution in the field of power electronic devices made voltage regulation easy. However, a boost in the penetration of DERs and DC loads also encouraged the research interests in the field of DC MGs to get a highly efficient system. A DC MG supplies power to a residential building, mining area, ships, or college campuses. The various voltage levels of the DC distribution are 380V, 48V, and 24V DC. These voltage levels can be obtained by either 120/230V 3-phase four wire AC distributions or 277/415V used primarily on the DC distribution systems [7].

2.2 Introduction: The Concept of Control Strategies

The connectional diagram of distributed generation (DGs) units with loads is shown in Figure 2.5. Here, two DGs are taken into account. The internal resistance of two DGs is represented as r1, and r2, respectively. Vdc1 and Vdc2 are the voltage drop across DGs. io1 and io2 are the output current from the DGs. RL1 and RL2 are the line resistance from DG to load. RLoad and VL are the load resistance and load voltage, respectively. It is the desired voltage for interconnection with the DC bus. It is also the operating voltage. Circulating current may exist between two converters if the voltage drop is different. Figure 2.1 can express that if Vdc1 ≠ Vdc2, then circulating current will appear in between them. To solve this problem, a high-speed digital communication-based control is required [8].

DC MGs have the following advantages as compared to AC MGs.

- The control strategy is not required for reactive power and frequency regulation.
- The problem related to grid synchronization is not present.
- Inrush current is absent in this case because the transformer is not present.
- The number of converters is reduced, so the conversion losses are reduced.
- It has its own fault-ride-through capability.

The following are the disadvantages of DC MGs.

- The private DC distribution line construction is required.
- As zero crossing is absent in DC systems, its protection is more challenging.

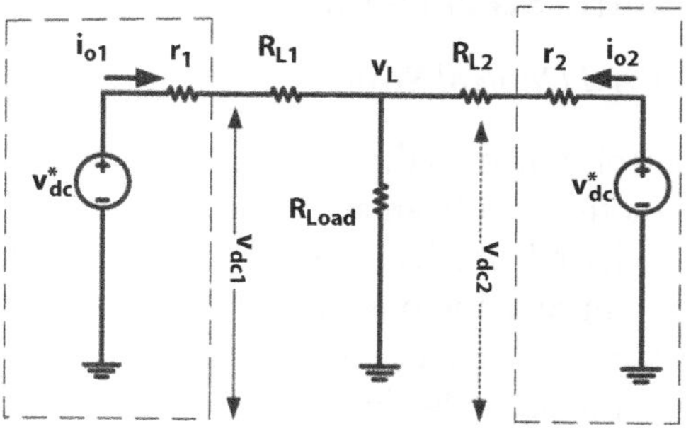

Figure 2.1 Simple representation of DC MGs.

- The voltage stability of DC systems is only affected by active power flow. In AC cases, voltage and stability are regulated using reactive power flow without disturbing the active power flow.

The objectives of control techniques for DC MG are as follows.

- To maintain proportional load sharing between different sources
- To maintain proper regulation of voltage in the system
- Both voltage and current control can be done efficiently in different modes.
- Stable operation with different loads
- The proper coordination can be maintained between DGs and ESS.
- For proper synchronization of DC MGs with conventional grid
- Proper power flow control can be maintained between DC MGs and utility.
- A smooth transition can be possible between the two operation modes.
- The economic load dispatch and optimization of generation cost can be possible.
- The DER potentials can be utilized properly.
- The transmission losses can be minimized.

The DC MGs control schemes have been categorized into two types: basic control schemes and multilevel control schemes. The detailed classification diagram is given in Figure 2.2.

2.2.1 Basic Control Schemes for DC MGs

2.2.1.1 Centralized Control Strategy

A centralized control transmits the information between distributed generators and collects the data from all the components of the MGs. Based on the statistics collected, efficient control action is initiated. The central control is used to improve the observability and controllability, which is shown in Figure 2.3. However, it causes the whole system failure whenever it faces a failure at any point, leading to maloperate the central controller [9]. The reduced flexibility, reduced expandability, and a lack of sufficient

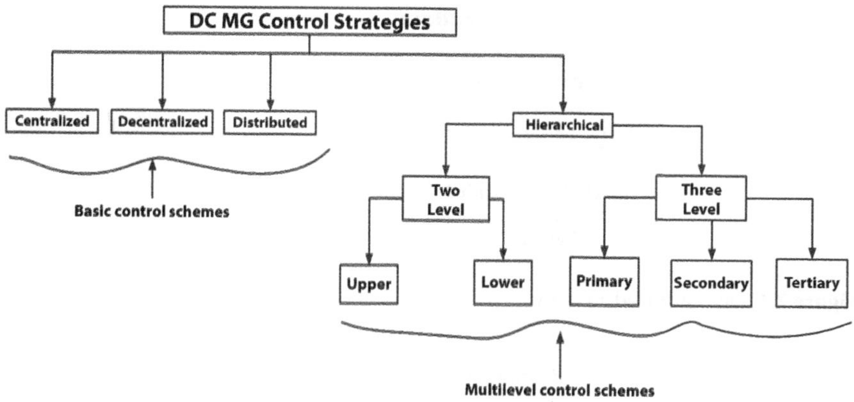

Figure 2.2 Classification of DC MG control strategies.

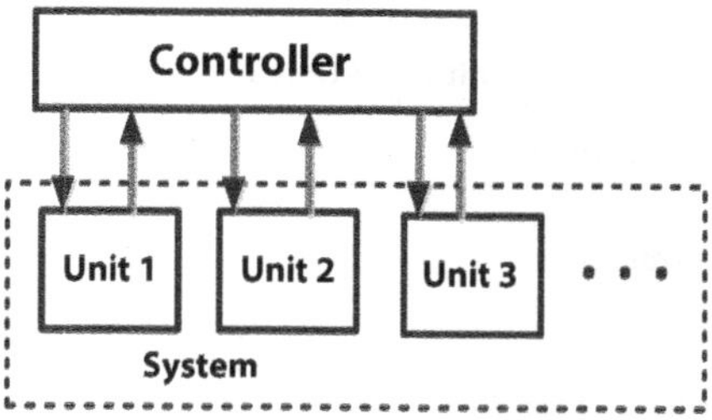

Figure 2.3 Centralized control strategy.

computational resources are other disadvantages of centralized control. Usually, centralized control is possible with fewer communication channels and less computational cost, where the information gathered is less with centralized optimization. An example of a centralized control strategy is the master and slave control strategy.

2.2.1.2 Decentralized Controller

As shown in Figure 2.4, the decentralized controller requires only local information, which does not require any real-time data and communication from other parts. However, global coordination is missing due to the lack of coordination between local regulators. The drop controller is an example of a decentralized controller whose accuracy depends on the system configuration, controller, and electrical parameters [10]. An example of this type of control strategy is droop control-based schemes or DC bus signaling techniques.

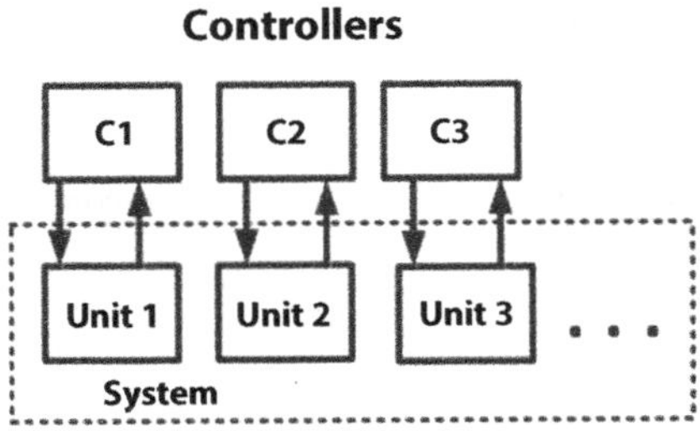

Figure 2.4 Decentralized strategy.

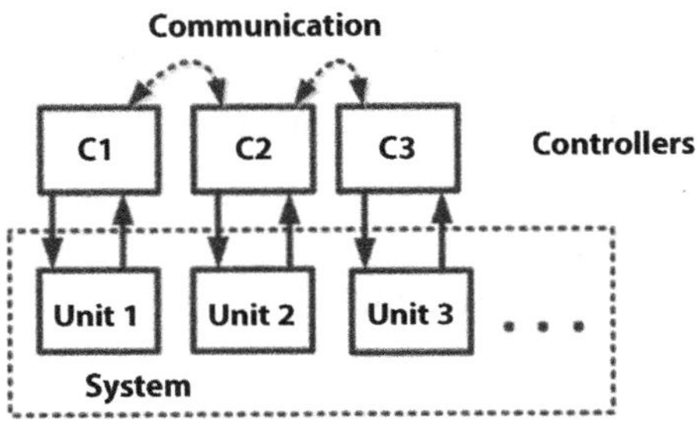

Figure 2.5 Distributed strategy.

2.2.1.3 Distributed Control

In recent years, the evolution of communication technology has enabled the information exchange to the distributed controller and management systems, which is reliable for practical applications. A distributed controller structure contains a centralized controller, as given in Figure 2.5. The controllers communicate with each other for sharing information between them using a communication network to facilitate the coordinated behavior of all the units. Modern power systems, integrating renewables, storage, and DGs, are becoming more sophisticated and require brilliant control schemes. However, distributed and decentralized control schemes may not fulfill all the requirements during a complicated decision-making process [11]. The consensus-based and agent-based methods are used in this case.

2.2.2 Multilevel Control

A multilevel control scheme is shown in Figure 2.6. It functions simply and is widely used as its local control regulates the basic operation of the MGs. Also, the central controller contains advanced controller and management functions. The stages of the control scheme, namely primary controller,

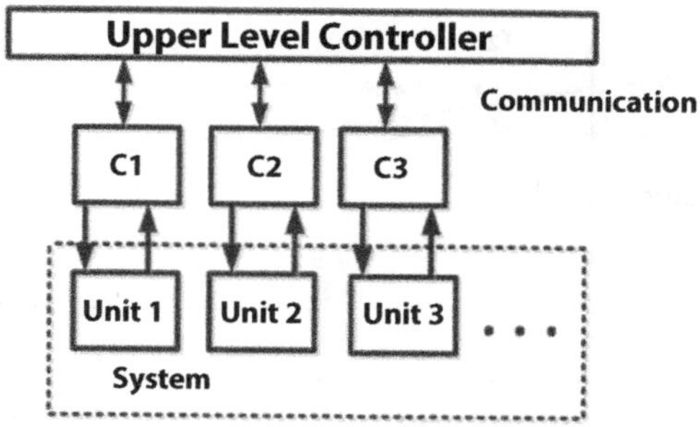

Figure 2.6 Higher-level control strategy.

secondary controller, and tertiary controller, are explained in brief in the upcoming subsections. The conventional droop controllers are not efficient in achieving the objectives because of the error in nominal voltages and load distribution. A widely accepted and standardized solution, i.e., a multilevel control scheme with high-speed communication, has been proposed to manage the DC MGs [12].

2.2.2.1 Primary Control

Local current and voltage can be controlled to achieve the operation and stability requirements. However, a decentralized controller is also present in it for attaining the proper load and generator sharing. The various methods used in the primary control are discussed in detail in the subsequent subsections.

2.2.2.1.1 Conventional Droop Controller and Virtual Impedance

The above-mentioned current-sharing method for the DC-to-DC converter was based on the high bandwidth communication channel. It is used in small centralized DC systems, such as DC electrical aircraft and DC severe systems. Usually, in the MGs, loads and DERs are connected at the common coupling point, where high bandwidth communication is unsuitable and costly. Hence, researchers are giving attention to the droop controller as a decentralized method for control.

The droop controller works on the principle of linearly reduced DC voltage reference for increasing the output current [13]. Hence, the droop controller is also known as the adaptive voltage positioning method. The current sharing of the adjacent converter is achieved by adjusting the voltage deviation within specified limits, given in Figure 2.7. The level of current sharing depends upon the droop coefficient. If the droop coefficient is more significant, then the current sharing is high, but voltage deviation

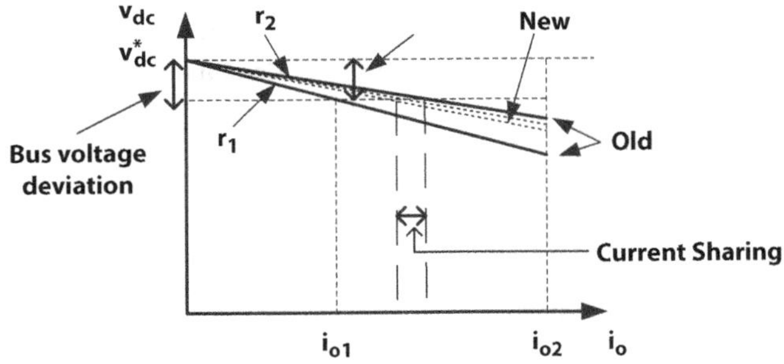

Figure 2.7 Droop control characteristics for DC MGs.

increases accordingly. It is essential to select the largest droop gain while limiting the variation in voltage during the maximum load condition.

$$V_{dci} = V_{dc}^{*} - r_i \times i_{oi} \tag{2.1}$$

$$\Delta V_{dc} = \left|V_{dc}^{*} - V_{dc}\right| \leq \Delta V_{dc\max} \tag{2.2}$$

where V_{dc}^{*} is the desired DC voltage, i_{oi} is the converter output current, and as r_i is the droop coefficient of ith converter, ΔV_{dc} is the deviation of DC voltage and ΔV_{dcmax} is the maximum. From the Equation (2.2), the droop coefficient is the resistor r_i showing a relationship between the output voltage, and current, r_i is also known as the virtual resistance of DC to DC converter and $\Delta V_{dcmax} \leq 5\%$. From Figure 2.5, the converter output current can be given as follows.

$$i_{oi} = \frac{V_{dc}^{*} - V_L}{r_i + R_{Li}} \tag{2.3}$$

The relation between load sharing and virtual resistances is given as follows.

$$\frac{i_{o1}}{i_{o2}} = \frac{r_2 + R_{L2}}{r_1 + R_{L1}} \tag{2.4}$$

Load-sharing error ($\Delta\xi$) in case of two converters is i_{o1}-i_{o2}, which is calculated as follows.

$$\Delta\xi = \frac{(V_{dc}^{*} - V_L)[(r_2 + R_{L2}) - (r_1 + R_{L1})]}{(r_1 + R_{L1})(r_2 + R_{L2})} \tag{2.5}$$

Hence, the higher the virtual resistance, the better is the sharing of power that is achieved, but it may lead to a rise in the output voltage deviation of the converter. However, bus voltage deviation and current sharing accuracy deterioration are the significant drawbacks of the conventional droop control technique.

2.2.2.1.2 Improved Droop Controller

Conventional droop controller has constant droop gain, while this droop controller has adaptive droop gain, which is variable in nature. In this case, power-sharing and voltage stability have been improved by adding **±Δr** to the conventional droop Equation, which is given as follows.

$$\Delta V_{dci} = V_{dc}^{*} - (r_i + \Delta r) \times i_{oi} \tag{2.6}$$

The nominal voltage values of converters should be identical to achieve accuracy in load sharing [14]. Here, one converter with a low current sharing value has low nominal voltage and higher voltage deviation. Based on the DC bus voltage variation and current sharing ability, the DC bus nominal voltage is increased with the help of the controller. The improved droop control characteristics are as shown in Figure 2.8. Another converter has a high DC bus nominal voltage and lower voltage deviation. To reach up to the voltage level of the second converter, the first converter increases the nominal voltage. This process reduces the error due to current sharing.

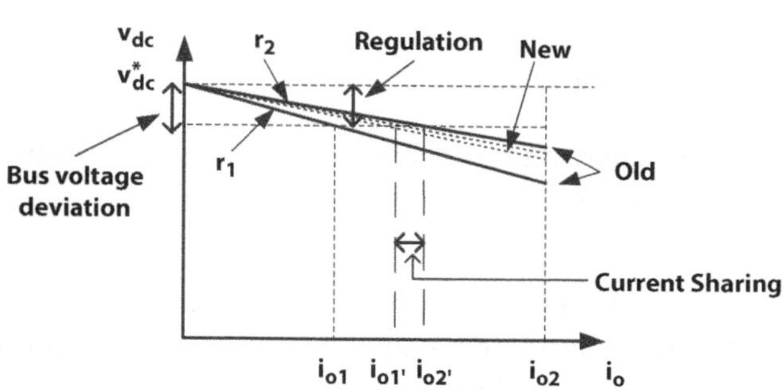

Figure 2.8 Improved droop control characteristics for DC MGs.

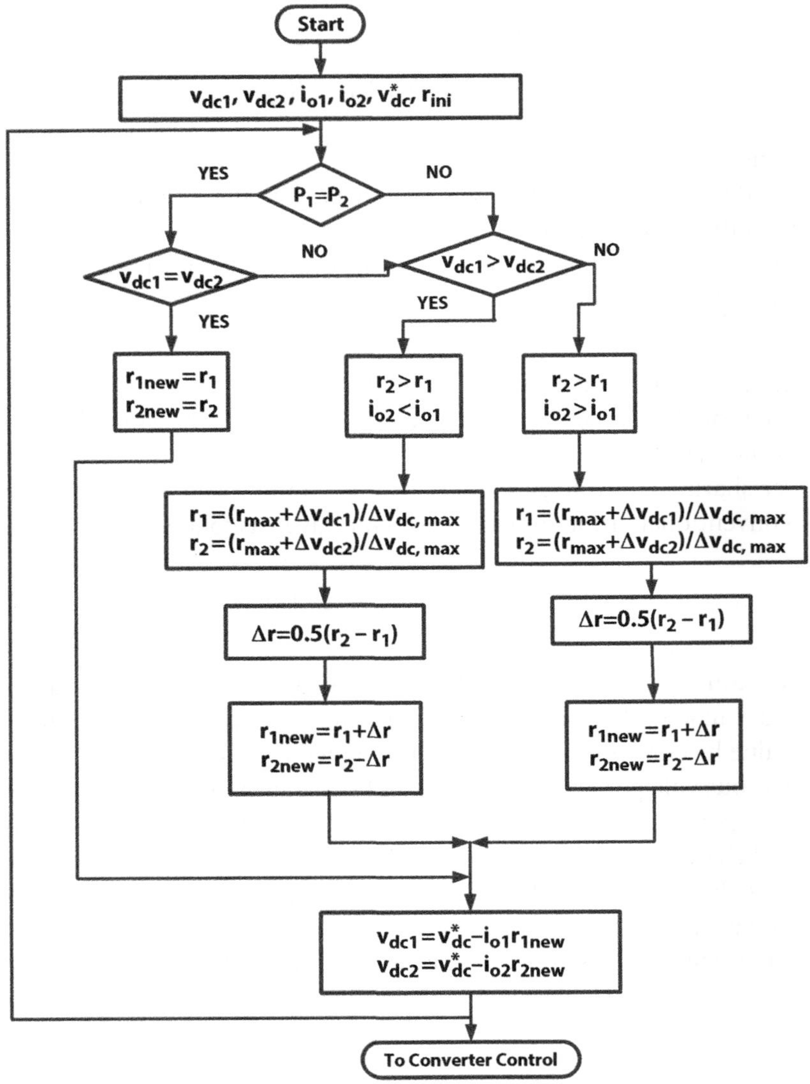

Figure 2.9 Droop control algorithm for DC MGs.

The complete flow chart representation of the improved droop controller is provided in Figure 2.9.

2.2.2.1.3 Active Current Controller

In MGs, load power-sharing depends on the current and power ratings. This technique was proposed years ago for DC-DC converters. Here, the converters are connected in parallel, and a master-slave control scheme is used to achieve accurate current sharing. The first one is selected as the master controller and the other one as a slave controller. While the master controller stabilizes the voltage, the slave controller operates in the DC bus feeding mode. The disadvantages of this technique are as follows.

- There is a possibility of single-point failure.
- Supervisory control is required.
- Shorter battery life is due to poor scalability.

2.2.2.1.4 DC Bus Signaling

Another distributed controller is DC bus signaling. This controller can be used for accurate power management between the sources and loads by measuring the local voltage at the point of common coupling. The voltage at the coupling point decides the operating mode of the system. When the voltage of the bus system is decreased, a source is considered to establish the DC bus voltage. Commonly, three different modes of operation are employed; namely, utility-dominated mode of operation, dominating storage mode and generating dominating mode of operation. In these modes of operation, utility grids, ESS, and DGs dominate the DC MGs, which are responsible for establishing the voltage at the DC bus [15].

2.2.2.2 Secondary Control

The secondary controller control the voltage, power-sharing, and power quality in DC MGs. The various fundamental methods used in the secondary controller are voltage boundary controller, current sharing controller, and energy storage management in DC MGs.

Secondary control loop is used to reduce the deviation in voltage during off grid mode. Both desired and actual voltage of MGs is compared and the error is reduced using a PI controller, which is given by

$$\Delta V_{MG} = K_{P\_sec}(V_{MG}^{*} - V_{MG}) + K_{I\_sec}\int (V_{MG}^{*} - V_{MG})dt \tag{2.7}$$

where K_{P_sec} and K_{I_sec} are the PI controller parameters of the secondary loop. In order to bring the bus voltage to the desired value, the droop characteristics of each converter is shifted by ΔV_{MG}. The updated droop characteristic is given by

$$\Delta V_{dci} = V_{dc}^{*} - (r_i \times i_{oi}) + \Delta V_{MG} \tag{2.8}$$

2.2.2.3 Tertiary Control

Energy scheduling and load flow management are important aspects of power system operation and management, and these aspects are usually treated separately. However, tertiary control also provides an optimized solution for managing power and energy issues, like balanced supply, energy storage, reduced power flow losses, minimum operating costs, etc. During grid connected mode, the utility grid takes the responsibility of proper voltage control. Power flow between the DC MGs and the utility grid is controlled by the tertiary control loop. Here, the comparison between the actual and desired current of MGs is performed, and the error is reduced using a PI controller. The expression of deviation in voltage is given as follows.

$$V_{MG} = K_{P\_ter}(I_{MG}^{*} - I_{MG}) + K_{I\_ter}\int (I_{MG}^{*} - I_{MG})dt \tag{2.9}$$

where K_{P_ter} and K_{I_ter} are the PI controller parameters of the tertiary loop. In order to supply the DC grid with required power and voltage, the droop characteristics of each converter is shifted by ΔV_{MG}. Figure 2.10 and Figure 2.11 shows the timing diagram and control stages for three level control schemes, respectively. The above-mentioned controller levels are implemented in a hierarchical fashion as centralized, decentralized, and distributed, as shown in Figure 2.11.

2.2.2.4 Current Sharing Loop

The problems related to current sharing accuracy in DC MGs cannot be solved by tertiary and secondary control loops. In order to improve the accuracy in current sharing, the power supplied by each converter should depend on the rated value. The local and desired current of each converter is compared using the current sharing loop in both off grid and grid connected modes. The error is then minimized using a PI control loop to improve the accuracy.

2.2.2.5 Microgrid Central Controller (MGCC)

The main function of MGCC is to control and manage the MGs. Here, coordinated operation and control of all the sources is achieved to maintain the stability of the system. It monitors the whole MGs by collecting the

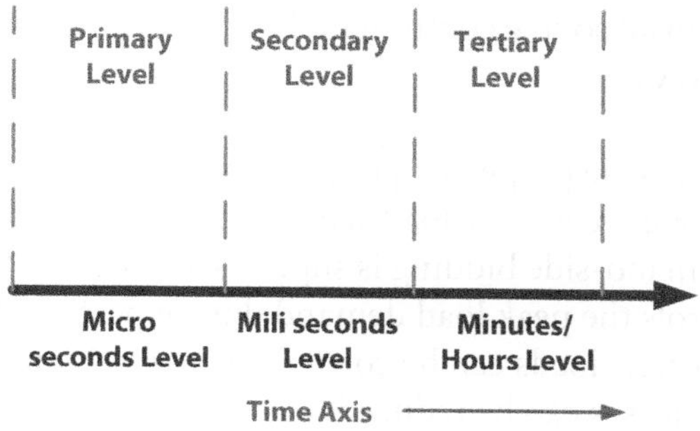

Figure 2.10 Timing diagram for three levels.

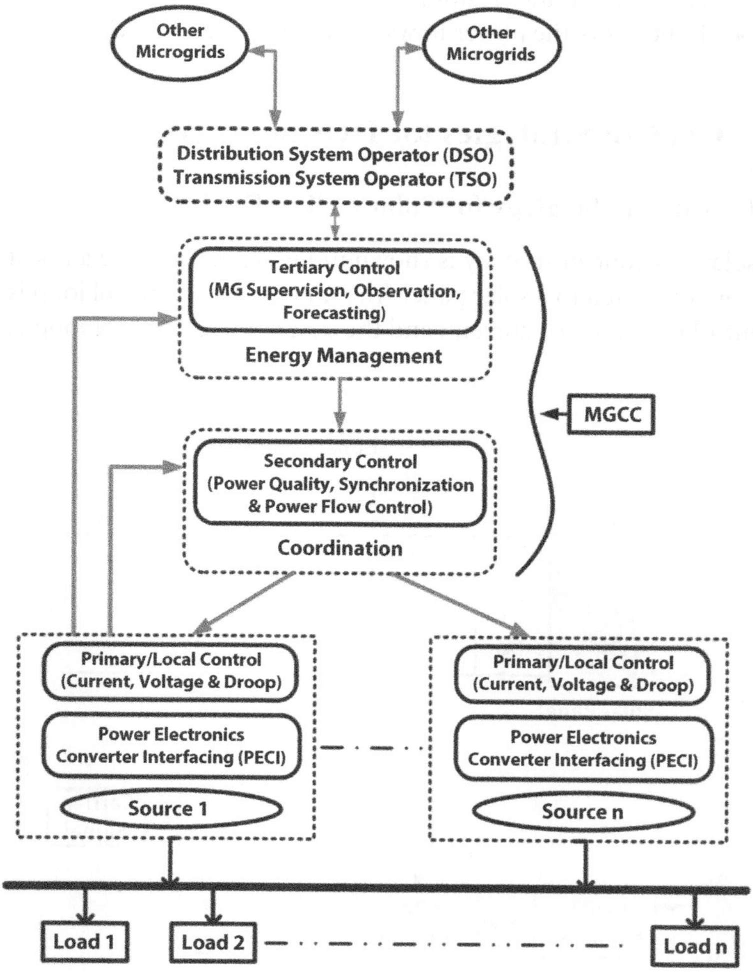

Figure 2.11 Three-level control strategy.

information from all source controllers. The major functions of the MGCC are listed as follows.

- It can provide proper set points for all DGs.
- The proper economic load scheduling can be maintained.
- The demand-side bidding is supervised properly.
- It controls the peak load demand during peak load periods.
- Non-critical loads can be controlled during islanding,
- System losses can be reduced.
- It can detect islanding by monitoring the PCC voltage.
- Resynchronization of the grid with local DGs can be initiated quickly after restoration.
- It monitors the power flows between DGs and PCC.

2.3 Control Strategies for DGs in DC MGs

2.3.1 Control Strategy for Solar Cell in DC MGs

The solar cell control strategy is shown in Figure 2.12, where a boost converter is connected to a solar panel. The inner current control loop is used to control the inductor current, and the outer voltage control loop is used

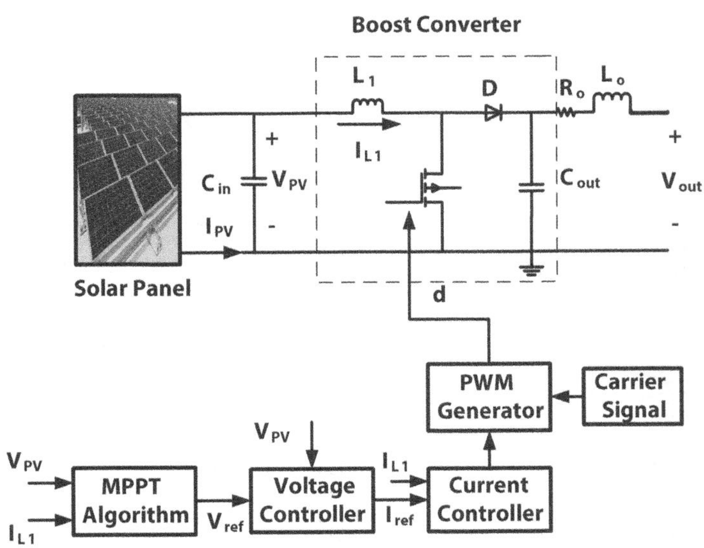

Figure 2.12 Control loop for a solar system in DC MGs.

to manage the converter input voltage. Decoupling of the solar panel operating conditions during load fluctuations can be obtained using this type of configuration. The crossover frequency is maintained higher for the inner loop with respect to the outer loop for error reduction. This condition also helps to avoid propagation of disturbances towards the solar panel side. Here V_{out} is the output voltage of the converter. MPPT algorithm is used to maximize power extraction for all the prevailing conditions [16].

2.3.1.1 Control Strategy for Wind Energy in DC MGs

As wind energy is intermittent in nature, the MPPT is used to track the maximum power points of the wind energy conversion system, as shown in Figure 2.13 [17]. The MPPT algorithm in the case of wind energy is of three kinds, which is given as follows [18].

Tip speed ratio control - It regulates the rotational speed of the generator to extract maximum power.

Power signal feedback control - It uses MPPT to track and regulate the maximum power.

Hill climb search control - It uses a search algorithm to track the peak power of the wind turbine. This method overcomes the disadvantages of the above two methods.

2.3.1.2 Control Strategy for Fuel Cell in DC MGs

The fuel cell control strategy is depicted in Figure 2.14. A boost converter is connected to the fuel cell. The control strategy consists of two closed

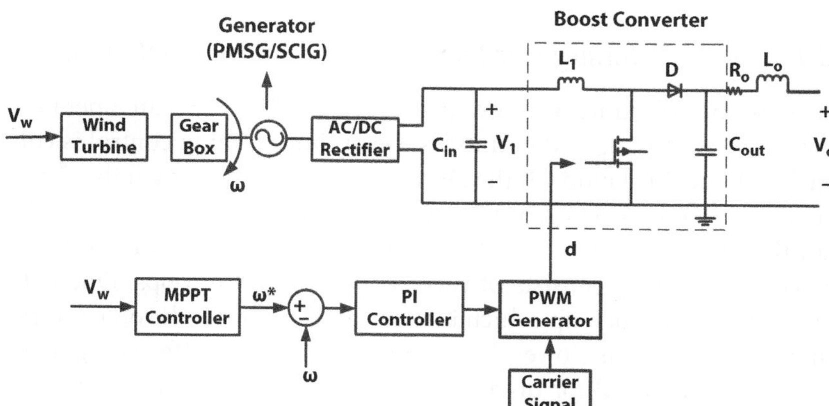

Figure 2.13 Control loop for wind energy system in DC MGs.

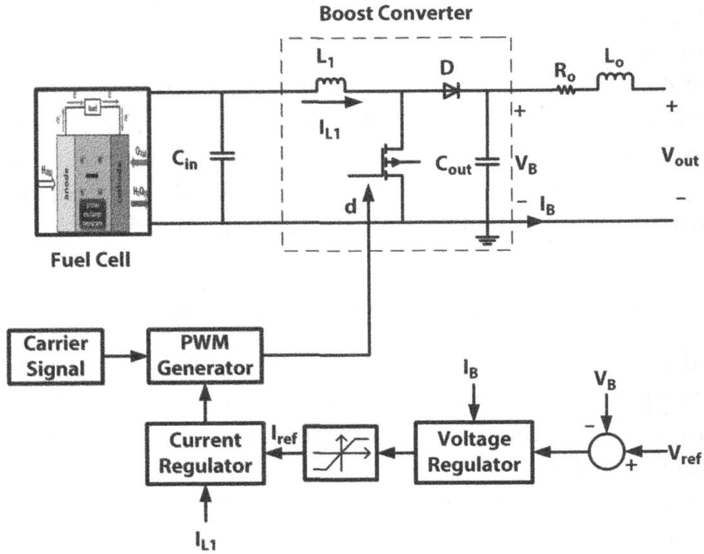

Figure 2.14 Control loop for fuel cells in DC MGs.

loops. One is the inner current control loop, which is faster, and another is the outer voltage control loop. The output of the current control loop provides duty cycles to the boost converter. The lower limit of I_{ref} is zero. The fuel can operate in discharging, and off modes. If the DC bus voltage is higher than the threshold, the fuel cell is in off state. When the other DERs cannot meet the load demands and the DC bus voltage reaches below the discharge reference, the fuel comes into operation and starts discharging to fulfill the load demand [19].

2.3.1.3 *Control Strategy for Energy Storage System in DC MGs*

The ESS control strategy is given in Figure 2.15. ESS can operate in charging, discharging, and off modes. Dead time running control is combined with the third mode. If the DC bus voltage is higher than the threshold, the charging process generally takes place. If the DC bus voltage is less than the threshold, then ESS stops working and leads towards off state. The control strategy consists of two closed, controlled loops. One is the inner current control loop, which is faster, and another is the outer voltage control loop, which is slower. V_{refj} is the threshold value (for charging the subscript $j=1$ and $j=2$ for discharging. k is a double-valued function. Its value can be 1 and -1. The inner loop generates the duty cycle through the

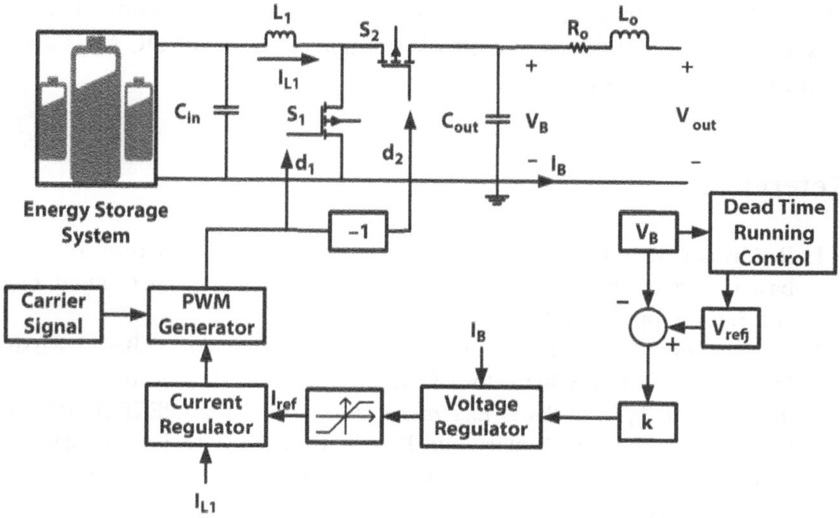

Figure 2.15 Control loop for battery-based storage systems in DC MGs.

PWM generator. The duty cycles for MOSFET S_1 and S_2 are d_1 and d_2 (negative transformation of d_1), respectively. Zero is the lower limit of I_{refj}. Both discharging and off modes are controlled through the discharge control scheme, and charging are controlled through the charge control scheme. V_{ref2} is the threshold value of voltage for discharge with $k = 1$ in this case. During discharging, the interfacing converter can operate in three modes: constant power, constant voltage, and off modes. When the output energy of solar cells is not sufficient to meet the load demand, then the bus voltage falls below the threshold value of voltage assigned for discharge control. As a result, the ESS circuit switches on and plays its role like a constant voltage source. During the discharging period, the converter limits the power of the power output of the source crosses the maximum power output of ESS. The working principle of charging is the same as the discharging [20].

2.4 Conclusions and Future Scopes

DC MGs have attained significant attention due to the easy integration of DERs, energy storage devices, and DC loads. Several advantages of DC MGs include high efficiency, less cost, and reduced conversion. However, the intermittent nature of renewable energy requires an efficient control system, which is a challenging area of research due to the lack of standards and guidelines. This chapter highlights the requirement of control strategies in DC MGs and different control strategies present for DC MGs along with their advantages and disadvantages.

References

1. R. Ahshan, M. T. Iqbal, G. K. I. Mann, and J. E. Quaicoe, "Micro-grid system based on renewable power generation units," *Can. Conf. Electr. Comput. Eng.*, vol. 3, no. 8, pp. 1–5, 2010.
2. B. K. Choudhury and P. Jena, "Superimposed Impedance Based Passive Islanding Detection Scheme for DC Microgrids," in *IEEE Journal of Emerging and Selected Topics in Power Electronics*, doi: 10.1109/JESTPE.2021.3076459.
3. Choudhury, Bhabani Kumari, and Premalata Jena. "A review on islanding detection schemes for dc microgrids." *Recent Advances in Power Systems*. Springer, Singapore, 2021. 15–26.
4. B. K. Choudhury and P. Jena, "Islanding Detection in DC Microgrids using System Impedance Monitoring Techniques," *2020 IEEE 17th India Council International Conference (INDICON)*, 2020, pp. 1-6, doi: 10.1109/INDICON49873.2020.9342549.
5. B. K. Choudhury and P. Jena, "Analysis and Comparison of Various Islanding Detection Schemes for DC Microgrids," *2020 3rd International Conference on Energy, Power and Environment: Towards Clean Energy Technologies*, 2021, pp. 1-6, doi: 10.1109/ICEPE50861.2021.9404504.
6. A. T. Elsayed, A. A. Mohamed, and O. A. Mohammed, "DC microgrids and distribution systems: An overview," *Electr. Power Syst. Res.*, vol. 119, pp. 407–417, 2015.
7. P. Fairley, "DC Versus AC: The Second War of Currents Has Already Begun In My View]," *IEEE Power Energy Mag.*, vol. 10, no. 6, pp. 104–103, 2012.
8. R. Scattolini, "Architectures for distributed and hierarchical Model Predictive Control - A review," *J. Process Control*, vol. 19, no. 5, pp. 723–731, 2009.
9. I. V. Prasanna, D. Srinivasan, and S. K. Panda, "Design, Analysis and Implementation of a Four-Tier Centralized Control Architecture for Intelligent Operation of Grid-connected Microgrids," *IEEE Int. Conf. Power Electron. Drives Energy Syst. PEDES 2016*, vol. 2016, January, pp. 1–6, 2017.
10. J. M. Guerrero, M. Chandorkar, T. Lee and P. C. Loh, "Advanced Control Architectures for Intelligent Microgrids—Part I: Decentralized and Hierarchical Control," in *IEEE Transactions on Industrial Electronics*, vol. 60, no. 4, pp. 1254-1262, April 2013, doi: 10.1109/TIE.2012.2194969.
11. K. Sun, L. Zhang, Y. Xing, and J. M. Guerrero, "A distributed control strategy based on DC bus signaling for modular photovoltaic generation systems with battery energy storage," *IEEE Trans. Power Electron.*, vol. 26, no. 10, pp. 3032–3045, 2011.

12. Xiaoliang Yang, Fen Tang, Xuezhi Wu and Xinmin Jin, "Hierarchical control strategy of grid-connected DC microgrids," *2016 IEEE 8th International Power Electronics and Motion Control Conference (IPEMC-ECCE Asia), Hefei,* 2016, pp. 3723–3727, doi: 10.1109/IPEMC.2016.7512891.
13. S. E. Mhankale and A. R. Thorat, "Droop Control Strategies of DC Microgrid: A Review," 2018 *International Conference on Current Trends towards Converging Technologies (ICCTCT), Coimbatore, 2018,* pp. 372–376, doi: 10.1109/ICCTCT.2018.8550854.
14. E. G. Shehata, J. Thomas, R. M. Mostafa and M. A. Ghalib, "An Improved Droop Control for a Low VoltageDC Microgrid Operation," *2018 Twentieth International Middle East Power Systems Conference (MEPCON), Cairo, Egypt,* 2018, pp. 850-855, doi: 10.1109/MEPCON.2018.8635258.
15. F. Li, Z. Lin, Z. Qian and J. Wu, "Active DC bus signaling control method for coordinating multiple energy storage devices in DC microgrid," *2017 IEEE Second International Conference on DC Microgrids (ICDCM), Nuremburg,* 2017, pp. 221-226, doi: 10.1109/ICDCM.2017.8001048.
16. F. Musavi, K. Al-Haddad, and H. Y. Kanaan, "A large signal averaged modelling and control of paralleled DC/DC converters with automatic load sharing," *Conf. Proc. - IEEE Appl. Power Electron. Conf. Expo. - APEC,* vol. 2, pp. 1353–1358, 2005.
17. B. Singh, B. K. Panigrahi and G. Pathak, "Control of wind-solar microgrid for rural electrification," *2016 IEEE 7th Power India International Conference (PIICON), Bikaner,* 2016, pp. 1–5, doi: 10.1109/POWERI.2016.8077263.
18. K. Gowtham, C. V. Sivaramadurai, P. Hariprasath and B. Indurani, "A Management of power flow for DC Microgrid with Solar and Wind Energy Sources," *2018 International Conference on Computer Communication and Informatics (ICCCI), Coimbatore,* 2018, pp. 1–5, doi: 10.1109/ICCCI.2018.8441324.
19. R. K. Sharma and S. Mishra, "Dynamic Power Management and Control of a PV PEM Fuel-Cell-Based Standalone ac/dc Microgrid Using Hybrid Energy Storage," in *IEEE Transactions on Industry Applications,* vol. 54, no. 1, pp. 526–538, Jan.-Feb. 2018, doi: 10.1109/TIA.2017.2756032.
20. A. Jahangir and S. Mishra, "Autonomous Battery Storage Energy System Control of PV-Wind Based DC Microgrid," *2018 2nd International Conference on Power, Energy and Environment: Towards Smart Technology (ICEPE), Shillong, India,* 2018, pp. 1–6, doi: 10.1109/EPETSG.2018.8658550

3

Protection Issues in DC Microgrids

Bhabani Kumari Choudhury* **and Premalata Jena**

Department of Electrical Engineering, Indian Institute of Technology Roorkee, Roorkee, India

Abstract
In recent years, the design, development, and distribution of DC systems have favored a direct current (DC) microgrid. One of the most typical issues with this sort of DC system is DC microgrid protection and operation. Several research and scientific publications have been published to investigate and fix this problem. The quest for an appropriate solution to this problem is still ongoing. This chapter describes different basic fault detection, location, and islanding detection methods for DC Microgrids, along with the advantages and disadvantages of the schemes. Finally, the concluding remarks and the future research scope are highlighted.

Keywords: DC microgrids, constant power loads, fault detection, power probe units, fault location, islanding detection

3.1 Introduction

DC MGs have gained considerable attention in the electricity industry due to the clean and green DC-based Renewable energy source (RES) and the presence of DC loads intrinsically [1]. DC MGs are a better alternative than the AC ones because of several advantages, like the absence of a synchronization process, zero reactive power abortion, improved overall efficiency, more straightforward control strategies, low cost, etc. [2]. However, DC MG protection remains one of the challenging issues as there is a lagging concept related to the control strategies, standards, protection schemes, DC MGs grounding strategies, topologies, and operating

**Corresponding author*: bkumarichoudhury@ee.iitr.ac.in

Nikita Gupta, Mahajan Sagar Bhaskar, Sanjeevikumar Padmanaban and Dhafer Almakhles (eds.)
DC Microgrids: Advances, Challenges, and Applications, (83–100) © 2022 Scrivener Publishing LLC

voltage. Another unavoidable issue is the high fault current, which should be cleared quickly. But the AC protection devices are unsuitable in DC due to the absence of zero-crossing nature [3]. Some of the protection issues of DC MGs are fault detection, isolation, islanding detection, and coordination [4–6].

The low inertia of DC MGs and converter behavior make the MGs sensitive to disturbance and faults [7]. Disturbance in DC MGs is mainly due to the input power variation, fluctuation in load, and change in load-sharing proportions. The temporary faults and failures, delay in communications, and the dynamics degrade the system performance. The voltage regulation in the DC bus is achieved with the help of a proper control strategy and communication protocol [8]. Hence, the time taken to clear the fault depends on the line parameter affecting the stability of the system. Another problem in the protection of DC MGs is the fault discrimination strategy to discriminate the faults from other non-fault disturbances. For better protection system performance, it should categorize the disturbance and faults as temporary and permanent [9]. Therefore, the design of a DC MG protection is a challenging task due to these issues. For reliable and safe operation of electrical power, a robust protection system is needed.

The objective of the proposed chapter is to give the readers some insights into different protection issues and challenges of DC MGs from different points, like fault detection, location schemes, islanding, etc. For the sake of understanding, some existing protection schemes are addressed along with their major challenges. Then the pros and cons of the techniques are discussed with the concluding remarks. Finally, future research trends on the requirements of the new protection schemes to increase the accuracy of the protection of DC MGs are highlighted.

3.1.1 Protection Challenge

Interruption of fault using existing circuit breakers (CBs) is a challenging task due to the absence of zero crossing in DC. The following subsections explain some of the protection challenges faced in DC MGs.

3.1.1.1 Arcing and Fault Clearing Time

The size of the DC MGs, components, and configurations are the significant factors for selecting the protection devices in DC MGs. The fast increase in the fault current and arcing are the major drawbacks in the traditional CBs in DC MGs protection [10, 11]. To overcome these drawbacks solid-state CBs or hybrid CBs are used. They interrupt the fault with less clearing time and with less or no arc. The economic feasibility of the CB is essential in the protection system for DC MGs.

3.1.1.2 Stability

The power variation in distributed generators, disturbance in the utility side, and load changes in the system cause the disturbance in DC MGs, which also causes a temporary fault. This will result in instability issues of DC MGs during the restoration of supply and fault. However, the main reasons for instability in DC MGs are the controllability of power converters, the resistive nature of the network, and the low physical inertia of DC MGs, which needs better control strategies (like virtual inertia and virtual impedance). Therefore, good protection schemes are needed to operate the DC MGs in stable mode [12].

3.1.1.3 Multiterminal Protections

The experience from the traction power systems is helpful when designing low voltage DC MGs. To protect the faults, the existing DC power system uses the current limiting rectifier, which allows the current flow only in one direction. But, in the DC MGs, power exchange between both AC and DC grid results in the bidirectional flow of current. That's why new protection schemes are required to detect the fault in multiple terminals along with multidirectional power flow [13].

3.1.1.4 Ground Fault Challenges

The power converters, both AC and DC, usually contain a capacitive output filter to decrease the output voltage ripple. If a fault occurs in the capacitive circuit, it will discharge very quickly through the fault, resulting in significant surge current flows. The fault current level is dependent on the fault location, fault resistance, and size of installed capacity. Usually, surge currents can have an amplitude of 10,000A to 50,000A. The main challenge of the circuit breaker, while discharging the high capacitance, is the coordination as it can cause the tripping of upstream and downstream breakers. The load impact will increase due to the tripping of the upstream circuit breaker. Also, in DC MGs, loads have a specific input capacity. The capacitor discharges the loads due to adjacent faults, which causes the unwanted tripping of the circuit breaker [14].

3.1.1.5 Communication Challenges

In DC MGs protection, a dedicated communication channel must transfer the voltage and current signal from one end to another in a data-based protection scheme. The protection scheme is faster than the controller of DC MGs. Therefore, the communication system in DC MGs should be quick and accurate [15].

3.1.2 Effect of Constant Power Loads (CPLs)

The loads in DC MGs are like constant current, constant voltage, constant power loads, etc. As most of the DC MGs load consists of several converters, the constant power load behavior comes into the picture. The undesirable properties of constant power loads generate incremental negative resistance and voltage drop. The rapid increase of DC loads causes the CPL study to become more predominant [16]. The CPL expression is as follows.

$$P_o = K \tag{3.1}$$

where P_o is the output power, which is constant (K) throughout the operation.

$$R_{CPL} = Slope = \frac{-v}{i} \tag{3.2}$$

where R_{CPL} is the resistance of CPL. It is calculated from the slope of the characteristics curve, as shown in Figure 3.1.

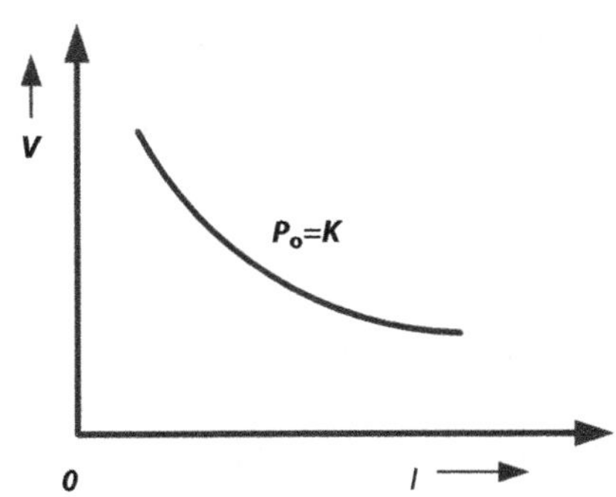

Figure 3.1 Constant power load characteristics.

Because of the incremental negative resistance, a high current is required to maintain power constant during abnormal situations in the DC system. Hence, the fault current accelerates. Another disadvantage of constant power load is that it introduces oscillations in current and voltage signals in islanding situations. It is a challenging task to discriminate between fault and non-fault cases using a protection device. In order to avoid it, one easy method is to alter the negative resistance during a fault to maintain the security of the system.

3.2 Fault Detection in DC MGs

DC MG is increasing its popularity in the current scenario from the generation level to the consumption level. The several faults, like short circuit fault, which includes L-G fault and L-L fault, and arc faults, like series and parallel arc faults, in the system network should be isolated. In DC MGs, the values of line impedance are very less. So, the current during fault is very high. The current is increased more than or equal to 100A within a few ms. This steep increase in the current may cause severe damage to the equipment of DC MGs. To avoid this, proper fault detection is required for DC MGs, which must be sensitive, reliable, and accurate to detect a fault within a specific time [17, 18]. This section describes different fault detection methods, including current derivative, over current, differential, directional, and distance, along with their advantages and disadvantages.

3.2.1 Principles and Methods of Fault Detection

There are several articles where the discussion on fault detection and protection schemes [19–21, 29] is carried out based on different measurements and calculations. This section summarizes the different research work done on DC protection, including DC MGs and high voltage DC (HVDC) line protection systems.

3.2.1.1 Voltage Magnitude-Based Detection

Analyzing the variations in the system voltage is one of the easiest and simplest ways of fault detection. A threshold needs to be set for this scheme. The system voltage drops below the threshold during the fault, and a trip signal is generated. This method is one of the fast detection methods as it depends only on the system voltage magnitude. But the demerit of this method is that it is unable to distinguish between temporary and permanent faults. As a result, the application of this scheme is limited by type and size of connected loads and rating of DC MGs.

3.2.1.2 Current Magnitude-Based Detection

This method is similar to the voltage magnitude method. Here the parameter which is taken into consideration is the magnitude of current deviations. Here the trip signal is generated when the current exceeds the threshold value during fault. A method is proposed for load fault protection, which utilizes a load-dependent current limiter based on solid-state switches. A solid-state switchboard could use multiple current limiters for a cluster of higher current loads.

3.2.1.3 Impedance Estimation Method

Impedance is estimated directly using the current and voltage samples. This method generates trip signals during fault scenarios when the variation in line impedance exceeds a specific threshold setting. The active impedance estimation is another scheme under this category, which is explained in [21, 22]. Here, in order to estimate line impedance, a small disturbance is injected through the DC-DC converters. Fast Fourier transform (FFT) is used to estimate line impedance using voltage and current samples. During any abnormality in the system, this method injects current spikes. However, the major disadvantage of this scheme is more DC-DC converters are required, which increases the fault clearing time.

3.2.1.4 Power Probe Unit (PPU) Method

For low voltage DV (LVDC) MGs systems, an external PPU circuit-based fault location scheme is proposed in [23] to analyze the fault. The PPU consists of a circuit breaker (CB), power source, inductor, and internal capacitor circuit of the PPU, as shown in Figure 3.2.

Here, DC MGs are divided into different zones. During a fault, intelligent electronic devices (IEDs) are used to isolate and detect fault locations based on a threshold setting. After isolating the fault location, the PPU examines the fault status and location using the probe current. The mathematical modeling of the fault location method based on the PPU is given as follows.

When the fault is isolated, the switches S_1 and S_2 are opened. The remaining network forms the series RL circuit, Now, switch SPF is closed, and it forms a series RLC circuit as shown in Figure 3.3. Here, l is the

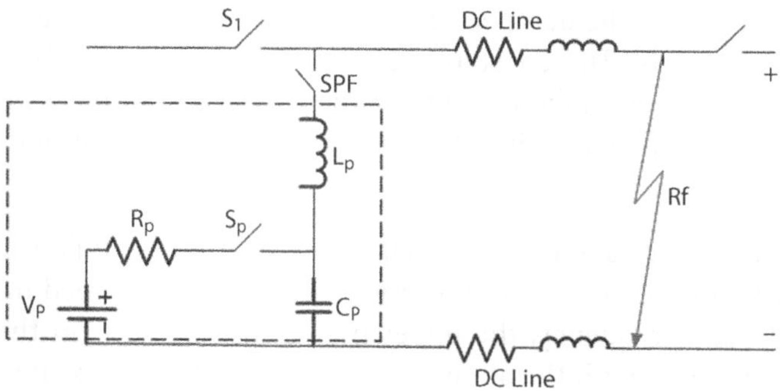

Figure 3.2 Internal circuit diagram of the PPU.

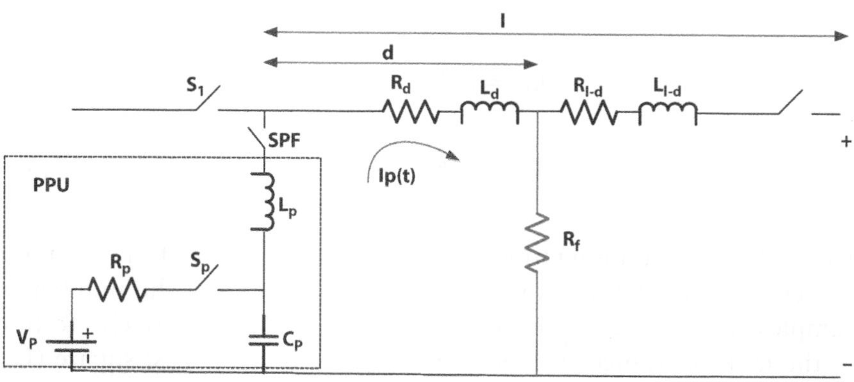

Figure 3.3 PPU equivalent circuit during fault.

length of the transmission line, d is the distance up to the fault point, R_d and L_d are the resistance and inductance of the faulted section and $R_{l\text{-}d}$ and $L_{l\text{-}d}$ are the remaining section inductance and resistance, C_p and L_p are the probe capacitance and inductance, and V_p is voltage source connected to the probe power unit. The current flow in the probe is given as follows.

$$\frac{d^2 i_p(t)}{dt^2} + \frac{R}{L}\frac{di_p}{dt} + \frac{1}{LC} i_p(t) = 0 \tag{3.3}$$

The solution of the above equation probe current can be written as follows.

$$i_p(t) = K_1 e^{-\alpha t} \cos(w_d t) + K_2 \sin(w_d t) \tag{3.4}$$

where w_d and α are the damped resonant frequency and attenuation of the probe, respectively. The constants K_1 and K_2 are computed by using the initial probe current different values. When a voltage source is not present in an RLC circuit, the zero input response attenuation factor of that circuit will be $\alpha = \frac{R}{2L}$.

The attenuation factor (α) calculation is discussed as follows. The current flowing in the probe is the response of the underdamped natural frequency. In this frequency, the upper part of the signal is in the form of exponential decay with the attenuation α. The positive peak of probe current can be written as follows.

$$kt_1 = e^{-\alpha n_1 \Delta t} = e^{-\alpha t_1} \tag{3.5}$$

$$kt_2 = e^{-\alpha n_2 \Delta t} = e^{-\alpha t_2} \tag{3.6}$$

$$kt_n = e^{-\alpha n \Delta t} = e^{-\alpha t_n} \tag{3.7}$$

where Δt is the sampling time interval, $t_1, t_2, \ldots, t_n$ are the time instances corresponding peak magnitudes of prob current, and the number of samples are $n_1, n_2, \ldots, n_n$ from 1st peak to n^{th} peak, respectively. To find α, the fault resistance and location are needed using least square (LS) method as follows. The response of probe current envelope is given as follows.

$$i_p env\,(t) = ke^{-\alpha t} \tag{3.8}$$

It is expanded mathematically as follows.

$$i_p env(t) = k\left[1 - \alpha t + \frac{(\alpha t)^2}{2!} - \frac{(\alpha t)^3}{3!} + \ldots\right] \tag{3.9}$$

Neglecting the higher-order terms can be modified as follows.

$$i_p env\,(t) = k\,[1 - \alpha t] \tag{3.10}$$

$$i_p env\,(t) = [A]\,[x] \tag{3.11}$$

where

$$A = [1 - t]$$

$$x = \begin{bmatrix} k \\ k\alpha \end{bmatrix}$$

Then x is estimated by using the LS method, which is given as follows.

$$x = A \times M \tag{3.12}$$

where

$$M = [kt]$$

The size of the M matrix depends on the length of the window. The value of α can be calculated from matrix M. The damped resonant probe current is stated as follows.

$$w_d = \sqrt{w_n^2 - \alpha^2} = w_n\sqrt{1 - \xi^2} \tag{3.13}$$

where

$$w_n = \frac{1}{\sqrt{(L_p + L_d)C}} \tag{3.14}$$

The damping coefficient becomes

$$\xi = \frac{\alpha}{w_n} = \frac{R}{2}\sqrt{\frac{C_p}{L}}$$

The fault distance calculation is stated as follows. Here, probe inductance is represented as L_d, probe capacitance as C_p and line inductance as L_l per km. The line inductance up to the location of fault is unknown. The natural frequency of the system is given by

$$w_n^2 = w_d^2 + \alpha^2 \tag{3.15}$$

where

$$w_n = \frac{1}{\sqrt{(L_p + L_u)C}}$$

Thus, the location of fault distance 'd' is given by

$$d = \frac{1 - L_p(w_d^2 + \alpha^2)C_p}{L_u(w_d^2 + \alpha^2)C_p} \tag{3.16}$$

The above-mentioned proposed method gives an accurate result if there is significant attenuation in the probe current. If the fault is close in, then there is less attenuation in the probe. To overcome this problem, when the fault is close in, an extra line is added to the line for estimating accurate fault location.

3.3 Fault Location

Fault in DC MGs should be located accurately, and quickly. So, the faulty sections can be repaired and replaced to restore the operation and to maintain the reliability of the system [24–26]. This section describes different fault location methods, namely passive and active approaches, along with their advantages and disadvantages.

3.3.1 Passive Approach

This approach uses the local signals to find the location of the fault. Examples of this technique are differential, traveling waves, and local measurement schemes. It is a straightforward approach as there is no need for an external device to extract the position of the fault.

3.3.1.1 Traveling Wave-Based Scheme

This scheme utilizes the traveling wave feature of current and voltage for the quick location of the fault. Once the fault occurs, this scheme is rapid.

But it requires data acquisition, which is costly. Once a fault occurs, it can locate the fault easily by the property of the traveling wave, like subsequent arrival time at one terminal, time intervals at different locations, measurement of first arrival times of waves at the converter stations, the time difference between the first arriving waves at both terminals, etc. But it is not suitable for sophisticated MGs. Some short distribution lines are used, which may suffer many reflections and degrade accuracy. Here, surge arrivals time is low. In order to satisfy this, a high sampling time is required.

3.3.1.2 *Differential Fault Location*

The scheme is widely used for the detection of fault and relay operation. But fault location also can be possible using this scheme. In order to perform this, current measurement is performed at both line terminals. Here, an Ethernet cable is used to transfer the measured signal. Here, differential current samples are used to identify fault using the cumulative sum method. Then currents and voltages samples of both ends are applied to Moore-Penrose pseudo-inverse scheme. It is a non-iterative scheme and calculates the impedance from a particular side to the fault location. Here, the drawback is the requirement of a fast communication system.

3.3.1.3 *Local Measurement-Based Fault Location*

This technique is easy and economical to implement. Here, local parameters are used to calculate the fault distance. The different local schemes are as follows.

a. Voltage-Based Scheme
   This scheme utilizes voltage samples for fault location. The major disadvantage of this scheme is the requirement of some additional voltage sensor, which increases the error during high impedance fault distance calculation.
b. Inductance Estimation Scheme
   Here, voltage, current, and rate of change in current signals are used for inductance estimation. The moving window least squares method is used here to extract the data samples. The rate of change in current-based estimations is speedy and accurate. It is also sensitive to noisy environments. But this method is only applicable to the DC MGs, where at one end of a line, capacitor is connected.

c. Transient Voltage Based Scheme
Here, the transient voltage ratio concept is used to locate the fault in the time domain. This is calculated by measuring the ratio of voltages at each terminal of the inductor. This scheme can be utilized in two ways. One is a single terminal, and the other is two terminals-based fault locations. The single-terminal location schemes are less accurate during high fault resistance. The two-terminal location schemes are highly reliable due to the use of a communication system along with additional voltage sensors.

3.3.2 Active Approach for Fault Location

This is an accurate approach to extract fault position irrespective of the topology of the MGs. Here, an external device is required. The impedance measurement-based scheme and power probe unit-based active approaches are commonly used in this scheme.

3.3.2.1 Injection-Based Fault Location

Here, two types of schemes are used: active impedance estimation and a power probe unit-based location scheme. In the first scheme, a triangular current signal is injected using a power converter. The noise is added after the fault is identified, and then the next step is to calculate the impedance at the point of coupling. In probe power unit-based schemes, the location of the fault is estimated using a second-order RCL network. Here, the fault distance is estimated by analyzing the current probe response and damped resonant frequency concept. The advantage of this fault location scheme is that it is accurate without using any communication systems. But the implementation cost is high due to the additional equipment.

3.4 Islanding Detection (ID)

ID is essential for both AC and DC MGs. With the failure in the detection, system breakdown may cause instability in the system. Both load-side devices and consumers get affected by the detection failure. Several research studies have been proposed to detect islanding in DC MGs [27–30]. It is of two types, passive and active detection. In the passive islanding detection technique, the system parameters of DC MGs, like the voltage, current, impedance, active power, etc., are used continuously to make the

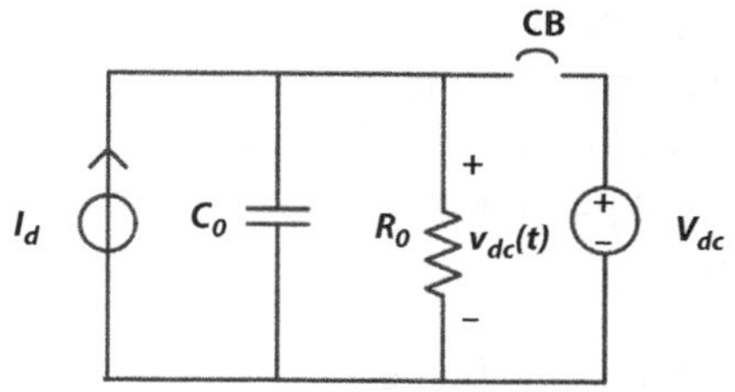

Figure 3.4 DC MGs equivalent circuit.

final decision. The types of IDSs in DC MGs are highlighted along with their advantages and challenges in this section.

Islanding is a situation in which an MG becomes isolated from the rest of the grid electrically. The connectional diagram of distributed generation units and the grid is as shown in Figure 3.4. Here, *Id* is the output current from the distributed source. *C*0 and *R*0 are the output circuit capacitors and resistance, respectively. The grid voltage is represented as *Vdc*. *CB* is the circuit breaker closed during the grid-connected case and opened during the islanding case. In this case, the conventional grid workers may not notice the disconnection, which causes dangerous shocks. Here, the balance should be maintained between generation and demand to achieve stability. To overcome the above problems, ID is essential.

3.4.1 Types of IDSs

A threshold value needs cross in all methods to get a condition of grid interruption, which brings a "non-detection zone" (NDZ) in the picture. NDZ is the range of conditions that are used to filter out real grid failures to maintain the stability and reliability of utility. The effectiveness of the IDS should be evaluated properly. Some new protection schemes are able to separate the system within 0.125 seconds of stability [31]. The IDSs for DC MGs are explained as follows.

Intentional/Planned Detection Schemes

In this IDS generation and loads are scheduled before a particular time. The main reason of this IDS is a preplanned switching event [32]. Here, the transients are less dangerous.

Unintentional/Unplanned Detection Schemes

These IDSs are unplanned. Hence, the transients generated in this case are uncertain and dangerous. Therefore, it is essential to detect islanding quickly to maintain reliability and stability. Depending on the pre-islanding operating condition, type of fault, location of the fault, detection time, switching actions, type of DGs, and the control strategies are different. These IDSs are of two types, remote detection techniques, and local detection techniques [33].

Remote/Resident Detection Schemes

Communication channel is the key factor in such kind of detection. The information should be transmitted between DGs and utility using these communication channels. Communication channels are costly because of the use of high-speed and advanced technologies. Some of the commonly used schemes are transfer trip, impedance insertion, and power line carrier communication methods. These schemes are reliable and accurate but the cost is very high in nature.

Local Detection Schemes

These schemes utilize measurement and monitoring of system parameters on the DGs side. The local detection schemes can further be classified as follows.

3.4.2 Passive Detection Schemes (PDSs) for DC MGs

Different system parameters monitoring is performed in the case of PDSs to detect islanding. Impedance, current, voltage, frequency, active and reactive powers are examples of system parameters in the AC MGs. In contrast, frequency and reactive power are absent in the case of the DC system. Threshold violation of system parameters is termed islanding. PDSs are simple, cheap, and accessible. However, non-detection zone (NDZ), threshold selection are the main challenges here [31–34].

3.4.3 Active Detection Schemes (ADS) for DC MGs

These schemes are performed by injecting an external signal. After the application of the signal, variations in the response are observed to detect islanding. This small disturbance creates significant variations during off-grid modes and can be compensated during the grid-connected case. By observing that variation, the objective of ADS is fulfilled. Some of the existing ADSs in DC MGs are listed in [28–30].

3.5 Protection Coordination Strategy

The failure of protection devices or delays of fault current interruption in DC MGs causes serious issues, which is the reason for system damage. The main reasons for the failure of protection devices are the non-suitability of AC protection devices in the case of DC, the absence of proper standardization, and lack of experience and guidelines [35, 36]. The primary protection schemes are not reliable due to the compact nature of DC MGs. So, it is hard to discriminate between external and internal faults. In order to overcome the above disadvantages, a proper coordination strategy is required to reduce the fault clearing time and also to reduce the delay between primary and backup schemes. In this section, an existing coordination strategy for DC MGs is described.

3.6 Conclusion and Future Research Scopes

The distinct fault current characteristics of DC MGs require an efficient protection strategy to maintain the reliability of the system. The proper implementation of the protection strategy can make the DC MGs suitable for various applications. The key components of the protection strategy include different factors, like proper grounding, topology, protection device, a proper interactive channel, etc., to communicate among different power electronics converters during fault inception events. In this chapter, a detailed analysis of different protection challenges, fault detection, and location methods along with the protection devices used for DC MGs are highlighted along with the merits and demerits of each case. Therefore, from the above study, the challenges and future scopes of DC MGs are visible clearly to do further research in this field.

References

1. J. M. Carrasco *et al.*, "Power-Electronic Systems for the Grid Integration of Renewable Energy Sources: A Survey," in *IEEE Transactions on Industrial Electronics*, vol. 53, no. 4, pp. 1002-1016, June 2006, doi: 10.1109/TIE.2006.878356.
2. T. Dragičević, X. Lu, J. C. Vasquez and J. M. Guerrero, "DC Microgrids—Part I: A Review of Control Strategies and Stabilization Techniques," in *IEEE Transactions on Power Electronics*, vol. 31, no. 7, pp. 4876-4891, July 2016, doi: 10.1109/TPEL.2015.2478859.

3. T. Dragičević, X. Lu, J. C. Vasquez and J. M. Guerrero, "DC Microgrids—Part II: A Review of Power Architectures, Applications, and Standardization Issues," in *IEEE Transactions on Power Electronics*, vol. 31, no. 5, pp. 3528–3549, May 2016, doi: 10.1109/TPEL.2015.2464277.
4. D. Salomonsson, L. Soder and A. Sannino, "Protection of Low-Voltage DC Microgrids," in IEEE Transactions on Power Delivery, vol. 24, no. 3, pp. 1045–1053, July 2009, doi: 10.1109/TPWRD.2009.2016622. [5] A. T. Elsayed, A. A. Mohamed, and O. A. Mohammed, "DC microgrids and distribution systems: An overview," *Electr. Power Syst. Res.*, vol. 119, pp. 407–417, 2015.
5. J. Park and J. Candelaria, "Fault Detection and Isolation in Low-Voltage DC-Bus Microgrid System," in *IEEE Transactions on Power Delivery*, vol. 28, no. 2, pp. 779-787, April 2013, doi: 10.1109/TPWRD.2013.2243478.
6. S. Dhar, R. K. Patnaik and P. K. Dash, "Fault Detection and Location of Photovoltaic Based DC Microgrid Using Differential Protection Strategy," in *IEEE Transactions on Smart Grid*, vol. 9, no. 5, pp. 4303–4312, Sept. 2018, doi: 10.1109/TSG.2017.2654267.
7. D. Wu, F. Tang, T. Dragicevic, J. M. Guerrero and J. C. Vasquez, "Coordinated Control Based on Bus-Signaling and Virtual Inertia for Islanded DC Microgrids," in *IEEE Transactions on Smart Grid*, vol. 6, no. 6, pp. 2627–2638, Nov. 2015, doi: 10.1109/TSG.2014.2387357.
8. H. Choi and J. Jung, "Enhanced Power Line Communication Strategy for DC Microgrids Using Switching Frequency Modulation of Power Converters," in *IEEE Transactions on Power Electronics*, vol. 32, no. 6, pp. 4140–4144, June 2017, doi: 10.1109/TPEL.2017.2648848.
9. D. Keshavarzi, E. Farjah and T. Ghanbari, "Hybrid DC Circuit Breaker and Fault Current Limiter With Optional Interruption Capability," in *IEEE Transactions on Power Electronics*, vol. 33, no. 3, pp. 2330–2338, March 2018, doi: 10.1109/TPEL.2017.2690960.
10. C. Strobl, "Arc fault detection in DC microgrids," *2015 IEEE First International Conference on DC Microgrids (ICDCM), Atlanta, GA, 2015*, pp. 181–186, doi: 10.1109/ICDCM.2015.7152035.
11. Z. Liu, A. Shekhar, L. Ramírez-Elizondo and P. Bauer, "Characterization of series arcs in LVdc microgrids," *2017 IEEE Second International Conference on DC Microgrids (ICDCM), Nuremburg,* 2017, pp. 297–301, doi: 10.1109/ICDCM.2017.8001060.
12. J. Liu, W. Zhang and G. Rizzoni, "Robust Stability Analysis of DC Microgrids With Constant Power Loads," in *IEEE Transactions on Power Systems*, vol. 33, no. 1, pp. 851–860, Jan. 2018, doi: 10.1109/TPWRS.2017.2697765.
13. Z. Lin, J. Du, J. Wu and X. He, "Novel communication method between power converters for DC micro-grid applications," *2015 IEEE First International Conference on DC Microgrids (ICDCM), Atlanta, GA,* 2015, pp. 92–96, doi: 10.1109/ICDCM.2015.7152016.

14. J. Mohammadi, F. Badrkhani Ajaei and G. Stevens, "Grounding the DC Microgrid," in *IEEE Transactions on Industry Applications*, vol. 55, no. 5, pp. 4490–4499, Sept.-Oct. 2019, doi: 10.1109/TIA.2019.2928278.
15. M. Farhadi and O. A. Mohammed, "Event-Based Protection Scheme for a Multiterminal Hybrid DC Power System," in *IEEE Transactions on Smart Grid*, vol. 6, no. 4, pp. 1658–1669, July 2015, doi: 10.1109/TSG.2015.2396995.
16. X. Lu, K. Sun, J. M. Guerrero, J. C. Vasquez, L. Huang and J. Wang, "Stability Enhancement Based on Virtual Impedance for DC Microgrids with Constant Power Loads," in *IEEE Transactions on Smart Grid*, vol. 6, no. 6, pp. 2770–2783, Nov. 2015, doi: 10.1109/TSG.2015.2455017.
17. S. Dhar, R. K. Patnaik and P. K. Dash, "Fault Detection and Location of Photovoltaic Based DC Microgrid Using Differential Protection Strategy," in *IEEE Transactions on Smart Grid*, vol. 9, no. 5, pp. 4303–4312, Sept. 2018, doi: 10.1109/TSG.2017.2654267.
18. J. Park and J. Candelaria, "Fault Detection and Isolation in Low-Voltage DC-Bus Microgrid System," in *IEEE Transactions on Power Delivery*, vol. 28, no. 2, pp. 779–787, April 2013, doi: 10.1109/TPWRD.2013.2243478.
19. P. Cairoli and R. A. Dougal, "Fault Detection and Isolation in Medium-Voltage DC Microgrids: Coordination Between Supply Power Converters and Bus Contactors," in *IEEE Transactions on Power Electronics*, vol. 33, no. 5, pp. 4535–4546, May 2018, doi: 10.1109/TPEL.2017.2724145.
20. R. Bhargav, B. R. Bhalja and C. P. Gupta, "Algorithm for fault detection and localisation in a mesh-type bipolar DC microgrid network," in *IET Generation, Transmission & Distribution*, vol. 13, no. 15, pp. 3311–3322, 6 8 2019, doi: 10.1049/iet-gtd.2018.5070.
21. E. Christopher, M. Sumner, D. W. P. Thomas, X. Wang, and F. De Wildt, "Fault location in a zonal DC marine power system using active impedance estimation," *IEEE Trans. Ind. Appl.*, vol. 49, no. 2, pp. 860–865, 2013.
22. D. Thomas, M. Sumner, D. Coggins, X. Wang, J. Wang, and R. Geertsma, "Fault location for DC marine power systems," *IEEE Electr. Sh. Technol. Symp. ESTS 2009*, pp. 456–460, 2009.
23. R. Mohanty, U. S. M. Balaji, and A. K. Pradhan, "An Accurate Noniterative Fault-Location Technique for Low-Voltage DC Microgrid," *IEEE Trans. Power Deliv.*, vol. 31, no. 2, pp. 475–481, 2016.
24. C. H. Sebastian and R. M. Shereef, "Location of Fault in a DC Microgrid using State Space Model Based Approach," *2020 IEEE International Conference on Power Electronics, Smart Grid and Renewable Energy (PESGRE2020), Cochin, India,* 2020, pp. 1–6, doi: 10.1109/PESGRE45664.2020.9070597.
25. R. Kheirollahi and E. Dehghanpour, "Developing a new fault location topology for DC microgrid systems," *2016 7th Power Electronics and Drive Systems Technologies Conference (PEDSTC), Tehran,* 2016, pp. 297–301, doi: 10.1109/PEDSTC.2016.7556877.
26. A. Meghwani, S. Chakrabarti and S. C. Srivastava, "An on-line fault location technique for DC microgrid using transient measurements," *2017 7th*

International Conference on Power Systems (ICPS), Pune, 2017, pp. 386–391, doi: 10.1109/ICPES.2017.8387325.

27. A. Makkieh *et al.*, "Assessment of passive islanding detection methods for DC microgrids," *15th IET International Conference on AC and DC Power Transmission (ACDC 2019), Coventry, UK,* 2019, pp. 1–6, doi: 10.1049/cp.2019.0016.
28. F. Paz and M. Ordonez, "An Impedance-Based Islanding Detection Method for DC Grids," 2018 *9th IEEE International Symposium on Power Electronics for Distributed Generation Systems (PEDG), Charlotte, NC,* 2018, pp. 1–7, doi: 10.1109/PEDG.2018.8447714.
29. A. M. I. Mohamad and Y. A. I. Mohamed, "Impedance-Based Analysis and Stabilization of Active DC Distribution Systems With Positive Feedback Islanding Detection Schemes," in *IEEE Transactions on Power Electronics,* vol. 33, no. 11, pp. 9902–9922, Nov. 2018, doi: 10.1109/TPEL.2018.2790945.
30. A. M. I. Mohamad and Y. A. I. Mohamed, "Assessment and Performance Comparison of Positive Feedback Islanding Detection Methods in DC Distribution Systems," in *IEEE Transactions on Power Electronics,* vol. 32, no. 8, pp. 6577–6594, Aug. 2017, doi: 10.1109/TPEL.2016.2618220.
31. Choudhury, Bhabani Kumari, and Premalata Jena. "Superimposed Impedance Based Passive Islanding Detection Scheme for DC Microgrids." *IEEE Journal of Emerging and Selected Topics in Power Electronics* (2021).
32. Choudhury, Bhabani Kumari, and Premalata Jena. "A review on islanding detection schemes for dc microgrids." *Recent Advances in Power Systems.* Springer, Singapore, 2021. 15–26.
33. Choudhury, Bhabani Kumari, and Premalata Jena. "Islanding detection in dc microgrids using system impedance monitoring techniques." *2020 IEEE 17th India Council International Conference (INDICON).* IEEE, 2020.
34. Choudhury, Bhabani Kumari, and Premalata Jena. "Analysis and Comparison of Various Islanding Detection Schemes for DC Microgrids." *2020 3rd International Conference on Energy, Power and Environment: Towards Clean Energy Technologies.* IEEE, 2021.
35. D. M. Bui, S. Chen, C. Wu, K. Lien, C. Huang and K. Jen, "Review on protection coordination strategies and development of an effective protection coordination system for DC microgrid," *2014 IEEE PES Asia-Pacific Power and Energy Engineering Conference (APPEEC), Hong Kong, 2014,* pp. 1–10, doi: 10.1109/APPEEC.2014.7066159.
36. V. Nougain, S. Mishra and A. K. Pradhan, "MVDC Microgrid Protection Using a Centralized Communication with a Localized Backup Scheme of Adaptive Parameters," in *IEEE Transactions on Power Delivery,* vol. 34, no. 3, pp. 869–878, June 2019, doi: 10.1109/TPWRD.2019.2899768.

4

Dynamic Energy Management System of Microgrid Using AI Techniques: A Comprehensive & Comparative Study

Priyadarshini Balasubramanyam and Vijay K. Sood*

Faculty of Engineering and Applied Sciences, Ontario Tech University, Oshawa, ON, Canada

Abstract

Economic scheduling in a microgrid aims to minimize the cost associated with energy production using Distributed Energy Resources (*DERs*). This work proposes an optimized Energy Management System (*EMS*) for a microgrid consisting of solar and wind generation units, an energy storage system and a diesel generator. The microgrid is capable of operating in either grid-connected or islanded modes. An optimization model for the *EMS* was modelled for minimizing the overall operating and maintenance costs for the real power flow in the microgrid. The *EMS* looks for the economically optimal way of meeting the load demand. The *DERs* in the microgrid were coordinated with the main grid. The optimization model was run on a typical microgrid considering real-time pricing data and ensured energy dispatch from/to the microgrid for a 24-hour period. The optimization was done using Genetic Algorithm (*GA*) and Pattern Search Algorithm (*PSA*) and a comparison is made based on the total operating and maintenance costs for real power flow.

***Keywords*:** Microgrid, distributed energy resources, optimization, genetic algorithm, pattern search algorithm, particle swarm optimization, economic dispatch, battery storage

**Corresponding author*: vijay.sood@ontariotechu.ca

Nikita Gupta, Mahajan Sagar Bhaskar, Sanjeevikumar Padmanaban and Dhafer Almakhles (eds.) DC Microgrids: Advances, Challenges, and Applications, (101–120) © 2022 Scrivener Publishing LLC

Nomenclature

α	Coefficient of cost function of diesel generator ($/hour)
β	Coefficient of cost function of diesel generator ($/hour)
γ	Coefficient of cost function of diesel generator ($/hour)
Γ_{ess}	Coefficient of cost function of energy storage system
Γ_{solar}	Coefficient of cost function of solar generation unit
Γ_{wind}	Coefficient of cost function of wind generation unit
κ_{buy}	Rate of power bought from utility
κ_{sell}	Rate of power sold to utility
C_{buy}	Cost function of power bought from utility
C_{dg}	Cost function of diesel generator
C_{ess}	Cost function of energy storage system
C_{sell}	Cost function of power sold to utility
C_{solar}	Cost function of solar generation unit
C_{total}	Cost function of total power of the microgrid
C_{wind}	Cost function of wind generation unit
DER	Distributed Energy Resources
DG	Diesel Generator
EMS	Energy Management System
ESS	Energy Storage System
GA	Genetic Algorithm
P_{demand}	Power demand
P_{dg}	Power output of diesel generator
P_{dg}^{max}	Maximum power output of diesel generator
P_{dg}^{min}	Minimum power output of diesel generator
P_{ess}	Power exchanged between microgrid and *ESS*
P_{ess}^{max}	Maximum power exchanged by *ESS*
P_{ess}^{min}	Minimum power exchanged by *ESS*
P_{solar}	Power output of solar generation unit
P_{solar}^{max}	Maximum power output of solar generation unit
P_{solar}^{min}	Minimum power output of solar generation unit
$P_{transaction}$	Power exchanged between microgrid and utility
$P_{transaction}^{max}$	Maximum power exchanged between microgrid and utility
$P_{transaction}^{min}$	Minimum power exchanged between microgrid and utility
P_{wind}	Power output of wind generation unit
P_{wind}^{max}	Maximum power output of wind generation unit
P_{wind}^{min}	Minimum power output of wind generation unit
PSA	Pattern Search Algorithm

4.1 Introduction

4.1.1 Background and Motivation

In recent years there has been a tremendous growth globally in renewable energy (such as solar and wind) due to adverse environmental effects caused by climate change [1]. These clean energy sources maintain a low carbon footprint but are characteristically intermittent. Solar and wind generation are weather dependent, and due to their intermittency, these renewable energy sources are coupled with energy storage equipment and secondary generation sources. These *DERs* are highly coordinated with the grid and this has led to the concept of a microgrid. A microgrid is a small-scale power system having *DERs* and loads which can operate in both grid-connected and islanded modes [2]. Advancements in power electronics and communication have aided in adopting DC microgrids over AC microgrids [3]. The DC microgrid architecture has gained popularity and has been implemented in shipboard and aircraft power systems [4]. The *EMS* of the microgrid plays a vital role in the economic dispatch of loads. The generated energy from various *DERs* should be able to meet the load demand alongside two-way communication between grid and local generating units subject to an optimized cost of electricity. In this chapter, the cost of electricity refers to the operating and maintenance costs of the real power flow in the microgrid produced from various generating units.

4.1.2 Prior Work

Earlier research was on AC microgrids, but the focus is now shifting to DC microgrids because of advantages in terms of power quality and lower line losses [5–7]. DC distribution systems are being proposed for connecting *DERs* with loads, the majority of which are DC [8]. In [9, 10], the findings prove that utility involving sole DC distribution network improves overall efficiency, reliability and control. The modern concept of the *EMS* in microgrids was proposed in [11]. The *EMS* takes information from *DERs*, utility and optimizes the economic dispatch in the system. In [12], the authors proposed a stochastic based implementation of matrix-real-coded *GA* on *EMS*. The operational cost minimization was done using Particle Swarm Optimization (*PSO*) [29], Reduced Gradient (*RG*), Cuckoo Search Algorithm (*CSA*) for microgrid in islanded mode [13]. DC Microgrid EMS dispatch with multiple objectives are discussed and studied in [14, 15]. In [8], the O & M costs are optimised but does not allow the BESS to take part in power transaction with the grid.

Reference [26] present a mess adaptive search algorithm to harvest energy and established a high data-rate, cloud-based architecture.

4.1.3 Contributions

The total O & M costs of the microgrid are minimized, which maximizes the profit for the consumer. *GA* and *PSA* are implemented to optimize the overall cost-function of the studied microgrid. It was found that *PSA* gives better optimized results in comparison with the *GA*. The proposed *EMS* for the considered microgrid allows a two-way power exchange, i.e., buying/selling power from/to the utility in order to minimize the overall cost of electricity. Additionally, the *ESS* also takes part in the power exchange with the utility.

4.1.4 Layout of the Chapter

This chapter is organized in the following manner. Section 4.2 discusses the problem statement. Section 4.3 illustrates the mathematical modelling of the microgrid. Section 4.4 provides a brief explanation of the optimization algorithms. A case study and simulation results are discussed in section 4.5. Finally, concluding statements are provided in section 4.6.

4.2 Problem Statement

Though a microgrid provides an environmentally friendly way of generating power, there is a constraint on the economic aspects for its implementation in the real-time scenario. In this chapter, the chosen microgrid consists of renewable energy sources, an energy storage system, diesel generators and residential loads which operate in both the grid-connected and islanded modes. The rated voltage of the microgrid is 380 V. The primary goal of the study is to design a reliable system with good power quality, which has ample rated *DERs* and capable of meeting the loads. The secondary goal is so-called economic scheduling or optimal power dispatch of the available cluster of *DERs*. In simpler words, there's a necessity to optimise the time of use, generator on/off scheduling, maximum penetration of renewable energy, energy storage, etc., aspects related to the *EMS* model.

A typical *EMS* architecture (Figure 4.1) is taken into account as a part of tertiary control in a Smartgrid and regulates overall cost. *EMS* consists of a data-acquisition component whose role is to collect the day ahead load schedule and forecast of renewable energy parameters. Also, based on the load demand profile, the renewable energy sources power generation

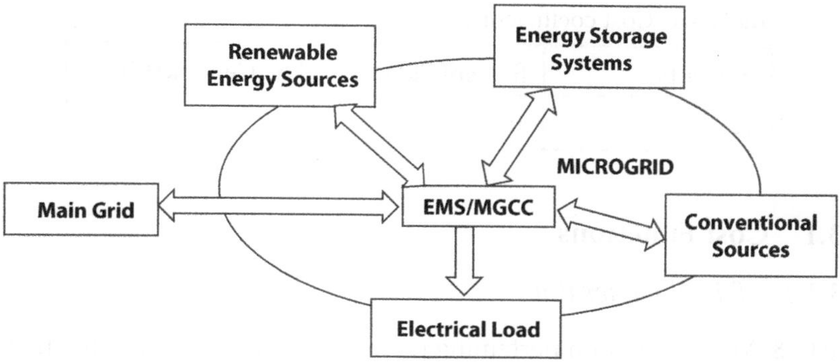

Figure 4.1 Typical EMS architecture.

and real-time price of the electricity allocates the power from each of the *DERs* and grid such that the demanded load is met. As the microgrid has the ability to generate localised power, it may serve as one generating unit altogether. When this whole unit is tied to the main grid there may be a possibility to transact power between the microgid and the main grid at a cheaper price. *EMS* of microgrid monitors the exchange of real-power between MG and the main grid. So *EMS* is used with the optimization model which takes care of the economic aspects by allocating optimum power to the load and to the grid during off-peak hours of operation in such a way that the overall power balance of the system is maintained. A proper conroller design and algorithm for the battery is required to create the interface between utility and the MG. The above-stated problems are configured and different AI techniques are implemented in order to arrive at the optimization model. The following section discusses the algorithms used in detail.

The microgrid has to operate in both grid-connected mode and islanded mode. The *EMS* has to take care of the demand requirement of the loads considering the economics of the cost of power consumed by the micro grid owner.

4.3 Mathematical Modelling of Microgrid

The optimization model for the microgrid requires the objective function formulation subject to the constraints. The objective function in this optimization model is the total operating and maintenance cost of the power consumed by the load of the microgrid. The cost functions of the various generation units of the microgrid are discussed next.

Table 4.1 Cost coefficients.

α (Cents)	β (Cents/kW)	γ (Cents/kW²)
12.2	100	13500

4.3.1 Cost Functions

4.3.1.1 Diesel Generator

The O & M cost function pertaining to the diesel generator(s) is mathematically expressed as a quadratic equation as stated in Equation (4.1) [16, 17]. The cost function is a function of the power output of the diesel generator unit. The cost coefficients are taken from [18] and are tabulated in Table 4.1.

$$C_{dg} = \alpha + \beta P_{dg} + \gamma P_{dg}^2 \tag{4.1}$$

4.3.1.2 Solar Generation

The solar generation cost is a function of the investment cost, O & M costs. There is no fuel cost associated with the solar generation unit. The investment cost is excluded from the cost function considered in this paper. The power output from the solar generation unit is intermittent and to match the demand, storage system and other distributed sources of energy are used in the microgrid. The cost function is given by Equation (4.2).

$$C_{solar} = \mathrm{r}_{solar} P_{solar} \tag{4.2}$$

Where r_{solar} is taken as 1.6 (Cents/kW) [19]. The forecasted power from the solar farm is depicted in Figure 4.2 over a 24-hour span.

4.3.1.3 Wind Generation Unit

The wind generation cost depends on the investment cost, maintenance and operation costs. There is no fuel cost associated with the wind generation unit. The investment cost is excluded from the cost function considered in this work. Like the solar generation unit, the wind generation unit is also intermittent and is operated in coordination with the other *DERs* of the microgrid. The cost function of wind generation unit is given by Equation (4.3).

$$C_{wind} = \mathrm{r}_{wind} P_{wind} \tag{4.3}$$

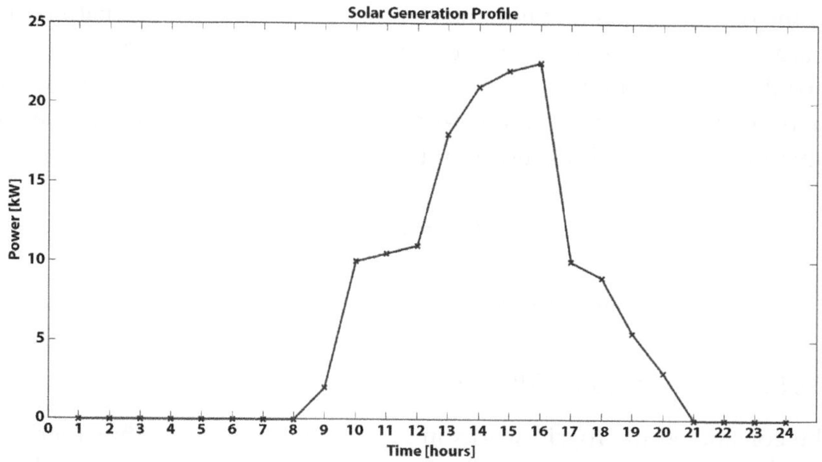

Figure 4.2 Power forecast from the solar farm.

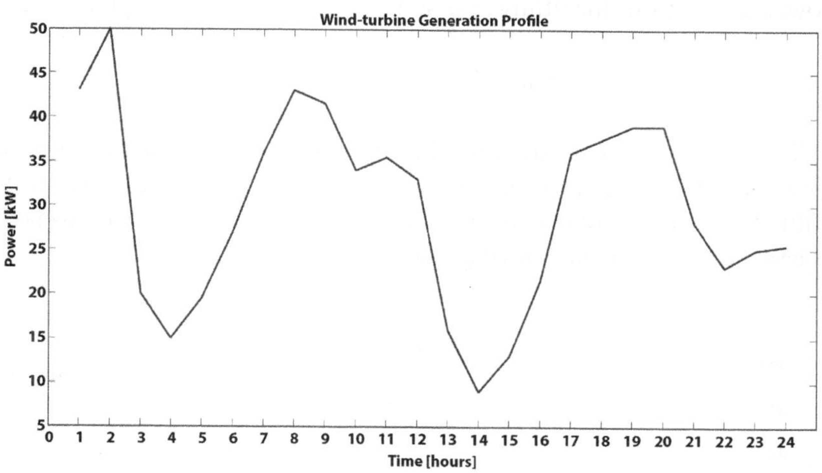

Figure 4.3 Power output forecast from the wind farm.

Where r_{wind} is taken as 1.6 (Cents/kW) [19]. The power forecast from the wind farm is depicted in Figure 4.3 over a 24-hour span.

4.3.1.4 Energy Storage System (ESS)

The *ESS* consists of batteries which store energy and instantly give the power to meet the load requirements in coordination with the other *DERs* in the microgrid. The cost function for the *ESS* considered here depends on the operational and maintenance cost of the batteries. The battery discharges and charges according to the instruction of the *EMS*. The *ESS* is capable of bidirectional power flow. The cost is associated with the discharging of the battery and not with the charging of the battery. The cost function for the *ESS* is as shown in Equation (4.4).

$$C_{ess} = r_{ess} P_{ess} \tag{4.4}$$

Where r_{ess} is taken as 1.6 (Cents/kW) [19].

4.3.1.5 *Transaction with Utility*

The considered microgrid works in grid-connected mode, thus there is a significant amount of exchange of power between the microgrid and utility. The rate associated with the power taken from the utility and that real-time pricing is shown in Figure 4.4. Thus, the overall cost function for the power drawn from the utility is as shown below in Equation (4.5).

$$C_{buy} = \kappa_{buy} P_{transaction}; \ \forall P_{transaction} > 0 \tag{4.5}$$

The *EMS* enables the microgrid to have a two-way exchange of power with the utility. The microgrid is also capable of delivering power to the utility for which a rate is associated. The real-time selling price is always set as less than the real-time buying price.

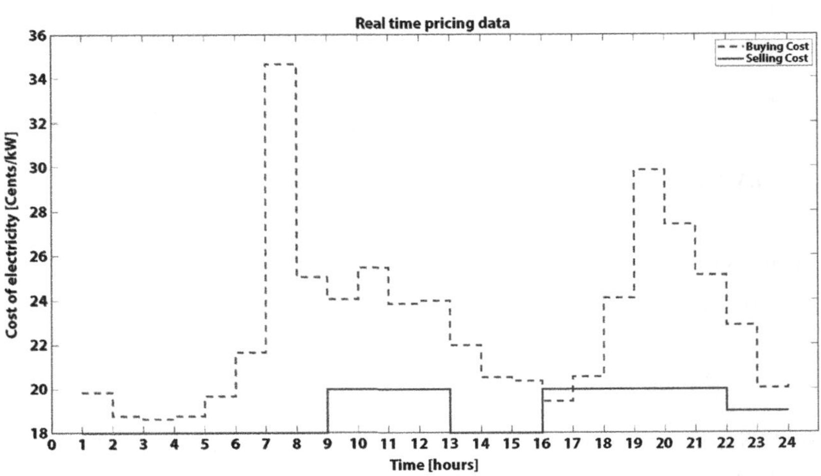

Figure 4.4 Cost of power for transaction with the utility.

The cost function associated with the selling of power to the grid is shown as represented by Equation (4.6).

$$C_{sell} = \kappa_{sell} P_{transaction}; \forall P_{transaction} < 0 \tag{4.6}$$

4.3.2 Objective Function

The microgrid considered in the paper consists of distributed sources of energy that can work in both grid-connected and islanded modes. The motivation is to minimize the total operating and maintenance cost of the considered microgrid over a 24-hour time period. In addition to this, the *DERs* are coordinated efficiently with the grid in such a way that the demand response is achieved. The demand is met from the sources of power in the microgrid so that the overall cost of electric power is minimized and the economic benefits for the microgrid owner are maximized [20].

The total cost function for the microgrid is written as the sum of the individual cost functions of the sources of power by Equation (4.7) i.e.

$$C_{total} = C_{dg} + C_{solar} + C_{wind} + C_{ess} + C_{buy} - C_{sell} \tag{4.7}$$

4.3.3 Constraints

The objective function considered is subjected to the constraints. Major constraints are the overall power balance of the microgrid and the capacity limits of the sources of power in the microgrid (Table 4.2), i.e.

$$P_{dg} + P_{solar} + P_{wind} + P_{ess} + P_{transaction} = P_{demand} \tag{4.8}$$

The constraints related to the generation capacity of the sources of power in the microgrid are as follows:

$$P_{dg}^{min} \leq P_{dg} \leq P_{dg}^{max} \tag{4.9}$$

$$P_{ess}^{min} \leq P_{ess} \leq P_{ess}^{max} \tag{4.10}$$

$$P_{transaction}^{min} \leq P_{transaction} \leq P_{transaction}^{max} \tag{4.11}$$

Table 4.2 Power ratings of DGs.

DER	Maximum power (kW)	Minimum power (kW)
Diesel Generator	35	0
ESS	30	10
Grid	30	–30

4.4 Optimization Algorithm

This section gives a detailed explanation of the optimization algorithms used in solving the economic scheduling problem, that is, *GA* and *PSA*. These algorithms have been applied to various optimization problems.

4.4.1 Heuristic-Based Genetic Algorithm (GA)

The concept of *GA* was introduced during the early 1970s by John Holland (Holland 1975). Encoding and evaluation are the two mechanisms that are involved in *GA*'s biological evolution while solving a problem. The fitness of chromosomes is measured in order to enable reproduction. Eventually, the chromosome population is modified at the end of each iteration. For every run, *GA* picks chromosome pairs at random from the current population and performs genetic operations to result in offspring for the next generation [21]. Over progressive iterations, an ideal/robust optimized solution is developed. The *GA* functions on three standards for generating new populace from the current set which is explained in Algorithm 1 (Figure 4.5) in detail.

```
Algorithm 1: Heuristic based Genetic Algorithm
1.  Begin Procedure
2.  Initialize inputs parameters
    Population size N, Mutation probability p_m, Crossover probability p_c
3.  Randomly generate an initial population using default stochastic uniform distribution
4.  While ((iteration_count<max_iteration_count) & (best_fitness<optimal_fitness)) do
5.        Fitness evaluation for current population
6.        Selection of pair of chromosomes for mating
7.        //Applying genetic operators for new offspring
8.        Crossover
9.        Mutation
10. End while
11. Return: Optimal solution
12. End Procedure
```

Figure 4.5 Step-by-step pseudocode lines of GA.

- *Selection:* involves selecting the parents
- *Crossover:* combines two selected parents to create a pair of offspring for the succeeding generation
- *Mutation:* applies random changes to individual parents to create off-springs

The *GA* begins by creating a random initial chromosome (binary string) population of the $P_{utility}$, P_{dg}, P_{ess} using probabilistic selection. The algorithm then creates a sequence of new population based on the rules. To come up with exploration and exploitation, *GA* uses two genetic operators, namely, crossover and mutation. The genetic operation crossover is performed based on value input parameter crossover probability $p_c \in \{0, 1\}$ which determines a point on the parent chromosome to split. Based on value of mutation probability $p_m \in \{0, 1\}$, the chromosome bits are flipped to result in best set of population. It stops when the stopping criteria is met, i.e., when the number of generations reaches the maximum value of number of iterations. The flowchart for the algorithm 1 is shown in Figure 4.6.

4.4.2 Pattern Search Algorithm (PSA)

The concept of *PSA* (direct method solving) involves minimizing of objective function, say, f_φ pertaining to k trial points/variables. This was developed by Robert Hooke and T. A. Jeeves in 1960 [28]. This adaptive optimization algorithm is carried out for a given number of iterations. For each iteration, two methods, viz., *Searching* and *Polling* are carried out to evolve an optimized function value for considered trail points.

The exploration occurs around the existing trail points. Once the better function value is found, the trial points are updated to a new position. With the updated position, *Searching* begins. When a failure comes across the *Searching* step, then the *Polling* step is called to resume the finding process. When a failure is encountered during the *Polling* step, then that is known as an unsuccessful iteration. To proceed further in finding a better solution, the mesh resolution is improved, and the algorithm is re-run [22, 25, 27].

The steps involved in this optimization Algorithm 2 (Figure 4.7) are explained further.

- *Initialization:* In this step, the initial iteration begins at $x_0 \in \varphi$. Definition of parameters such as *iteration count, mesh size, poll size, and direction set* are done. The trailing points belong to the mesh M_k which is developed based on a given direction

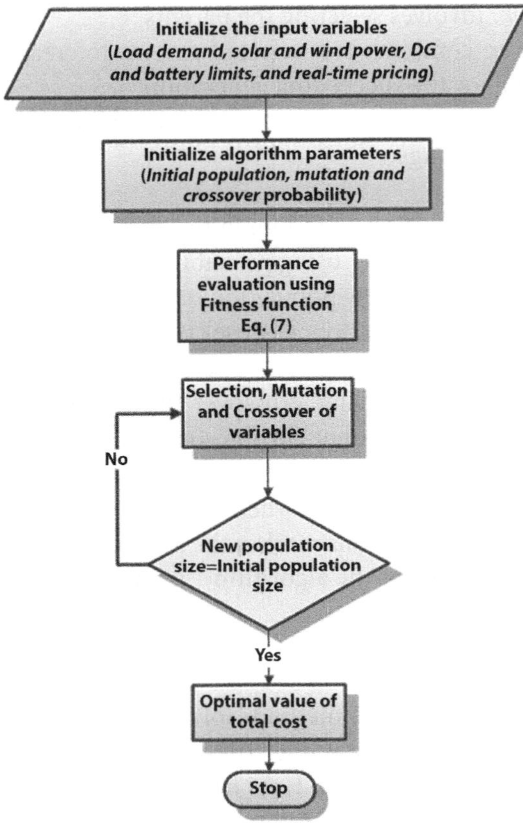

Figure 4.6 Flowchart of heuristic based Genetic Algorithm.

set $D_z \in R$. In the Algorithm 2, S_k consists of points where objective function is evaluated at the beginning of iteration.

- *Searching:* Once the initialization of parameters is done, the *Searching* step is called to find optimal value of function f_φ around the initial set of trail points. Corresponding to the current local variable x_k, the function value of developed set of trail points is compared to that of $f_\varphi(x_k)$. At this point, best objective value is retained, and rest is discarded leading to a new solution set x_{k+1} where the mesh size can increase or be the same.
- *Polling:* In case *Searching* step fails, then the *Polling* step is called to continue till the maximum iteration count is reached. In this step, the exploration is carried around the local variable x_k where the trailing points form a *frame* based

Algorithm 2: Pattern Search Technique [26]

Begin Procedure
Initialization of parameters
    $\{x_0, x_1, \dots, x_k\} \in R,\ \Delta_0 > 0$
    //Mesh definition
    $M_k = \cup_{x \in S_k}\{x + \Delta_k^m D_z\}$
For (iter=1, iter≤max_iter_count, iter++) Do
  SEARCHING step: *Calculate the fitness* f_φ
  Begin If
      $f_\varphi(x_{k+1}) < f_\varphi(x_k)$ ***then*** $x_k = x_{k+1}$
    Else
      POLLING step: $P_k = \{x_k + \Delta_k^m d\} \in M_k$
    Else If
    //Reduce Mesh size
      $\Delta_{k+1} < \Delta_k$
  End If
End For
Return: *Optimal solution*
End Procedure

Figure 4.7 Step-by-step pseudocode lines of PSA.

on parameter called *poll size* $\Delta_k^p \in R^+$. The variable x_k serves as *center of frame* and direction set D_k is developed using two parameters *mesh and poll size* Δ_k^m and Δ_k^p respectively. If an optimal point is found using the *Polling* step then the *center of frame* is updated to x_{k+1}. In the case of *Polling* step failure, the *frame* is renamed as *minimal frame* with x_k as the *minimal center of frame.* Thereby, the mesh size is diminished to improve the resolution for succeeding iterations.

The flowchart for the Algorithm 2 is shown in Figure 4.8.

4.5 Results

This section deals with the simulation results of proposed *EMS* applied to the microgrid as shown in Figure 4.9. The case considered here consists of a single bus with renewable sources of energy, BESS, diesel-generator and load tied to the grid. The load profile, real-time pricing is taken for the span of 24 hours which has been resourced by PMJ [23, 24]. The typical load profile is shown in Figure 4.10. It can be seen that the total load is the sum of loads, that is, Load 1 and Load 2. Load 1 can be considered

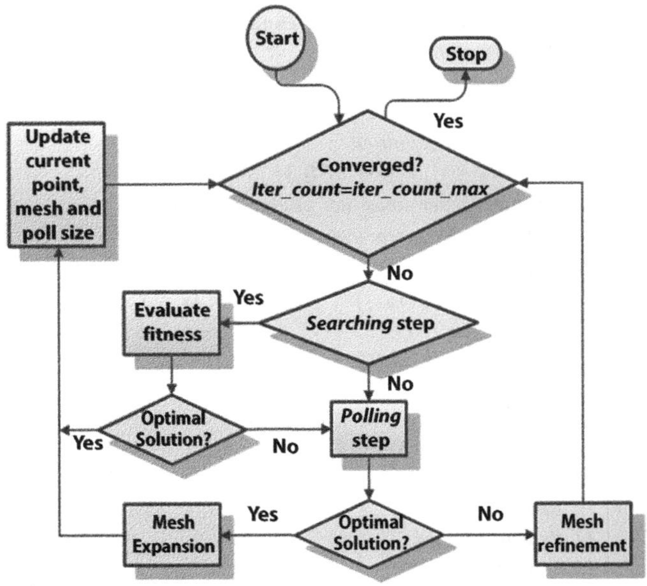

Figure 4.8 Flowchart of heuristic-based Pattern Search Algorithm.

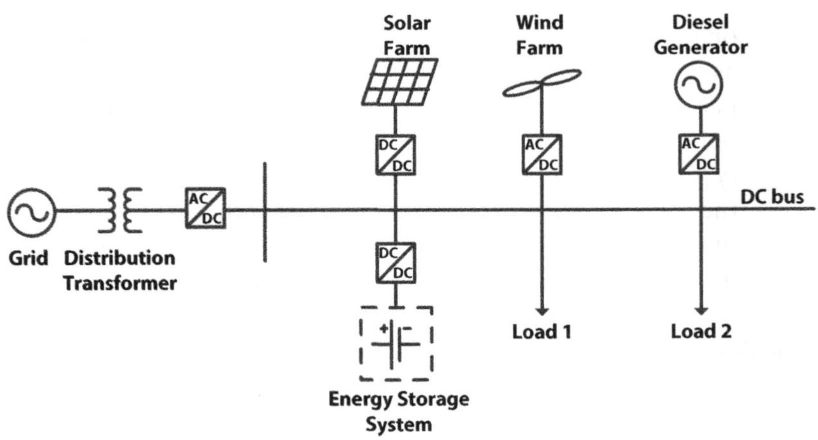

Figure 4.9 Single-line diagram of 380V DC Microgrid.

as group of critical loads while load 2 can be considered as non-critical loads together. In grid-connected mode of operation of MG, the localized *DERs* along with utility cater to the total load demand, that is, critical and non-critical, whereas, in islanded mode of operation, there's a possibility of peak shaving the load. As a result, there is a participation of DERs in supplying power subject to generation equality constraints.

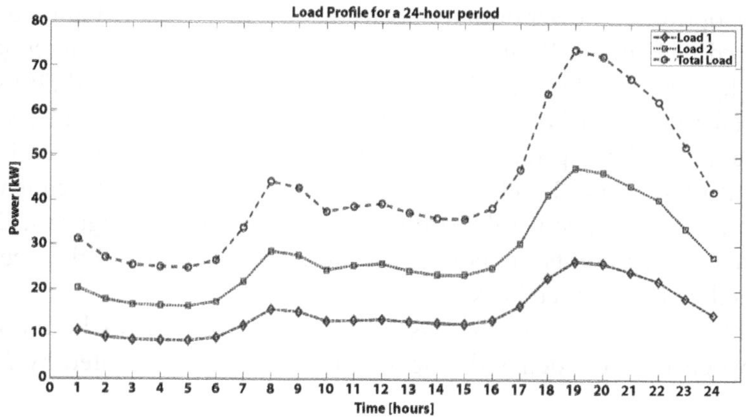

Figure 4.10 Load demand of the microgrid.

The optimization algorithms as discussed in the previous section are effectively applied to solve the *EMS* problem. The optimal dispatch of *DERs* and utility is researched on MATLAB/Simulink® platform based on day-ahead predictions. The evaluation of the fitness function was carried out for as many as 10,000 iterations. The results of the optimization gave the optimal dispatch of power from the diesel generator, energy storage system and utility. The developed algorithms were able to provide dispatch data of when to buy/sell power from/to utility and it is advantageous to the microgrid owner. Figure 4.11 depicts the individual power output for each *DER* as well as grid in an hourly fashion. The positive value of power is the power bought from the utility and the negative value of power is the power sold by the microgrid to the utility.

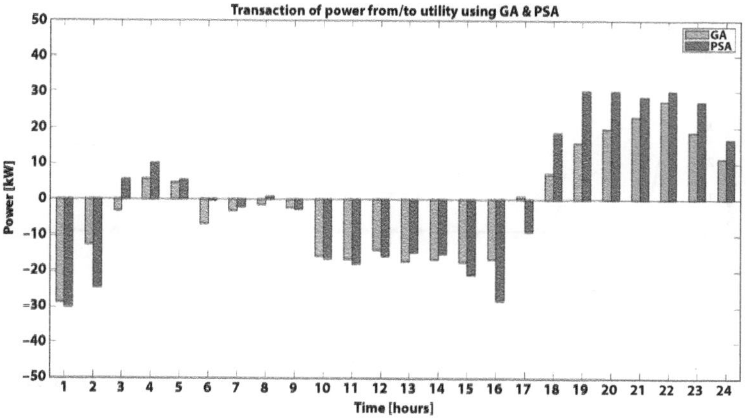

Figure 4.11 Power transaction from/to utility using GA and PSA.

The optimised power output of considered *DERs* using *GA* and *PSA* are pictured in Figure 4.12 and Figure 4.13, respectively. Upon investigating these figures, it is observed that the load is met with considered *DERs* and buy/sell power from/to utility accordingly to reduce total cost. The *ESS* was discharged and charged according to the optimal scheduling given by the optimization results. It followed the State-of-Charge (SOC) principle while iterating in the respective algorithm. The positive power in Figure 4.12 and Figure 4.13 represents the power discharged by the *ESS* and the negative shows the power stored back to *ESS*.

Table 4.3 gives the hourly O & M costs. During off-peak hours of the day, when the load demand is less than the energy generated by *DERs*,

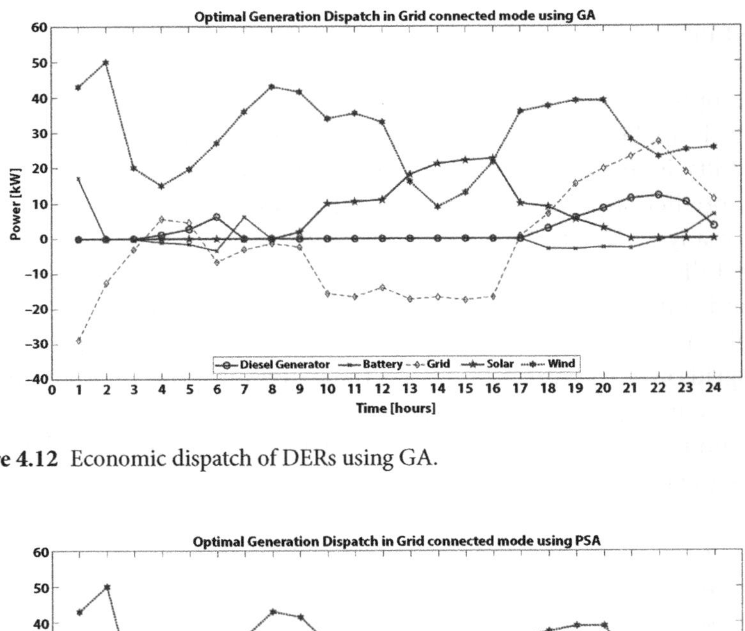

Figure 4.12 Economic dispatch of DERs using GA.

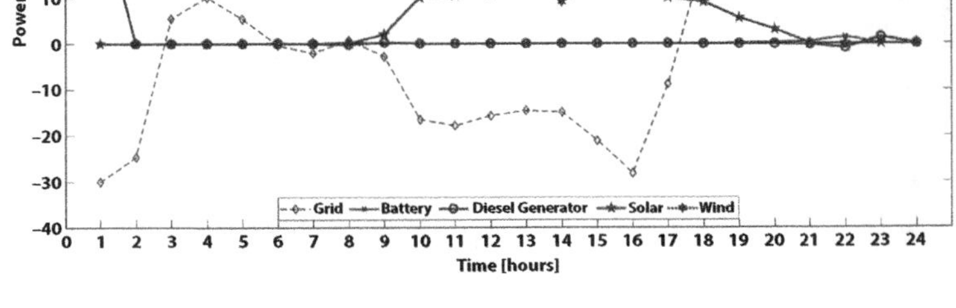

Figure 4.13 Economic dispatch of DERs using PSA.

Table 4.3 Optimization results.

Time (Hour)	Cost ($) by GA	Cost ($) by PSA
1	−4.10	−4.29
2	−1.33	−3.48
3	0.89	1.52
4	185.61	2.35
5	1043.39	1.83
6	5332.51	0.48
7	0.29	0.47
8	1.08	8.58
9	0.33	0.26
10	−2.30	−2.48
11	−2.46	−2.72
12	−1.97	−2.32
13	−2.40	−1.97
14	−2.37	−2.10
15	−2.46	−3.12
16	−2.49	−4.80
17	1.17	−0.91
18	1172.97	5.73
19	4842.44	11.17
20	9777.75	14.26
21	17231.51	31.90
22	19659.13	273.14
23	14119.49	6.31
24	1648.11	4.34

the operating cost value is negative. This negative value attributes to the savings incurred due to power sold to the main grid and the higher the value is, the more beneficiary it is to the microgrid owner. Similarly, when the load demand is more than the energy generated by all the *DERs* (peak period), the optimised operating cost value is positive. This represents the excess energy needed in order to meet the load demand is bought from the utility. Upon comparing the total O & M costs for the complete day, the results evidently show that PSA gave much optimised dispatch values for *DERs* and grid as compared to those of *GA* results.

4.6 Conclusion

In order to benefit the microgrid owner, a cost-effective optimization model was incorporated with the *EMS* of the microgrid. *GA* and *PSA* were implemented for the optimization of the cost function associated with the power flow in the microgrid. The optimization model resulted in the coordinated operation of the *DERs* in the microgrid. The optimization was run for a typical span of a 24-hour profile of the microgrid subjected to the load demand, renewable, energy storage system, diesel generator power output and the optimal operation of the *ESS* was determined. The *EMS* allowed the microgrid to make a two-way transaction of power along with optimizing the operational and maintenance costs of power flow in the microgrid. The *PSA* gave more optimized results in comparison to the *GA*. On the whole, the optimization model gave an economic and reliable aspect to the *EMS* for the microgrid.

Further, the optimization model can be designed for a cluster of microgrids connected with each other and the utility and the overall O & M costs of the power flow through the microgrids can be minimized. The *ESS* can be modelled in such a way that the *ESS* of each microgrid can share the power among them subject to power equality constraints.

References

1. R. Domínguez and A. J. Conejo and M. Carrión. Toward Fully Renewable Electric Energy Systems. *IEEE Transactions on Power Systems*, vol. 30, no. 1 (January 2015): 316–326.
2. L. Che and M. Shahidehpour and A. Alabdulwahab and Y. Al-Turki. Hierarchical Coordination of a Community Microgrid with AC and DC

Microgrids. *IEEE Transactions on Smart Grid*, vol. 6, no. 6 (Nov 2015): 3042–3051.

3. D. Kumar and F. Zare and A. Ghosh. DC Microgrid Technology: System Architectures, AC Grid Interfaces, Grounding Schemes, Power Quality, Communication Networks, Applications, and Standardizations Aspects. *IEEE Access*, vol. 5 (2017): 12 230–12 256.
4. F. Gao and S. Yeoh and S. Bozhko and G. Asher. Coordinated control of a DC electrical power system in the more electric aircraft integrated with energy storage. *2015 IEEE Energy Conversion Congress and Exposition (ECCE)*, (Sept 2015): 5431–5438.
5. H. Kakigano and Y. Miura and T. Ise. Low-Voltage Bipolar-Type DC Microgrid for Super High Quality Distribution. *IEEE Transactions on Power Electronics*, vol. 25, no. 12 (Dec 2010): 3066–3075.
6. B. Wang and M. Sechilariu and F. Locment. Intelligent DC microgrid with smart grid communications: Control strategy consideration and design. *2013 IEEE Power Energy Society General Meeting* (July 2013): 1–1.
7. L. Che and M. Shahidehpour. DC Microgrids: Economic Operation and Enhancement of Resilience by Hierarchical Control. *IEEE Transactions on Smart Grid*, vol. 5, no. 5 (Sept 2014): 2517–2526.
8. C. Li and F. de Bosio and F. Chen and S. K. Chaudhary and J. C. Vasquez and J. M. Guerrero. Operation cost minimization of droop-controlled DC microgrids based on real-time pricing and optimal power flow. *IECON 2015 - 41st Annual Conference of the IEEE Industrial Electronics Society* (Nov 2015): 003 905–003 909.
9. M. Rodriguez and G. Stahl and L. Corradini and D. Maksimovic. Smart DC Power Management System Based on Software-Configurable Power Modules. *IEEE Transactions on Power Electronics*, vol. 28, no. 4 (April 2013): 1571–1586.
10. K. Strunz and E. Abbasi and D. N. Huu. DC Microgrid for Wind and Solar Power Integration. *IEEE Journal of Emerging and Selected Topics in Power Electronics*, vol. 2, no. 1 (March 2014): 115–126.
11. Y. Xin and B. Zhang and M. Zhai and Q. Li and H. Zhou. A Smarter Grid Operation: New Energy Management Systems in China. *IEEE Power and Energy Magazine*, vol. 16, no. 2, (March 2018): 36–45.
12. C. Chen and S. Duan and T. Cai and B. Liu and G. Hu. Smart energy management system for optimal microgrid economic operation. *IET Renewable Power Generation*, vol. 5, no. 3 (May 2011): 258–267.
13. M. Bhoye and S. N. Purohit and I. N. Trivedi and M. H. Pandya and P. Jangir and N. Jangir. Energy management of Renewable Energy Sources in a microgrid using Cuckoo Search Algorithm. *2016 IEEE Students' Conference on Electrical, Electronics and Computer Science (SCEECS)*, (March 2016): 1–6.
14. P. Wang, W. Wang, N. Meng, and D. Xu. Multi-objective energy management system for DC microgrids based on the maximum membership degree

principle. *Journal of Modern Power Systems and Clean Energy*, vol. 6, no. 4 (Jul. 2018): 668–678.

15. A. Parisio and E. Rikos and L. Glielmo. A Model Predictive Control Approach to Microgrid Operation Optimization. *IEEE Transactions on Control Systems Technology*, vol. 22, no. 5 (Sept 2014): 1813–1827.
16. W. Shi and X. Xie and C. Chu and R. Gadh. Distributed Optimal Energy Management in Microgrids. *IEEE Transactions on Smart Grid*, vol. 6, no. 3 (May 2015): 1137–1146.
17. C. A. Hernandez-Aramburo and T. C. Green. Fuel consumption minimisation of a microgrid. *Conference Record of the 2004 IEEE Industry Applications Conference. 39th IAS Annual Meeting*, vol. 3, (Oct 2004): 2063–2068.
18. N. Augustine and S. Suresh and P. Moghe and K. Sheikh. Economic dispatch for a microgrid considering renewable energy cost functions. *2012 IEEE PES Innovative Smart Grid Technologies (ISGT)*, (Jan 2012): 1–7.
19. C. Li and F. de Bosio and F. Chen and S. K. Chaudhary and J. C. Vasquez and J. M. Guerrero. Economic Dispatch for Operating Cost Minimization Under Real-Time Pricing in Droop-Controlled DC Microgrid. *IEEE Journal of Emerging and Selected Topics in Power Electronics*, vol. 5, no. 1 (March 2017): 587–595.
20. Genetic Algorithm. [Online]. https://www.mathworks.com/help/gads/what-is-the-genetic-algorithm.html (accessed May 1, 2019).
21. Pattern Search Algorithm. [Online]. https://www.mathworks.com/help/gads/what-is-direct-search.html (accessed May 1, 2019).
22. Open Energy Information (OpenEI). [Online]. http://en.openei.org (accessed May 1, 2019).
23. Website of the Regional Transmission Organization PMJ. [Online]. http://www.pjm.com (accessed May 20, 2019).
24. C. Audet and J. E. Dennis J. Mesh Adaptive Direct Search Algorithms for Constrained Optimization. *SIAM Journal on Optimization*, vol. 17 (1), (2006): 188–30.
25. V. Torczon. On the Convergence of Pattern Search Algorithms. *SIAM Journal on Optimization*, vol. 7 (1), (1997): 1–25.
26. N. A. Chughtai and M. Ali and S. Qaisar and M. Imran and M. Naeem and F. Qamar. Energy Efficient Resource Allocation for Energy Harvesting Aided H-CRAN. *IEEE Access*, vol. 6 (2018): 43990–44001.
27. Robert Hooke and T. A. Jeeves. 1961. "Direct Search" Solution of Numerical and Statistical Problems. *J. ACM*, vol. 8, no. 2 (April 1961): 212–229 [Online].
28. M. Y. Ali and F. Khan and V. K. Sood. Energy Management System of a Microgrid using Particle Swarm Optimization and Wireless Communication System. *2018 IEEE Electrical Power and Energy Conference (EPEC)*, Toronto, ON, Canada (2018): 1–7.
29. J. H. Holland. 1975. *Adaptation in Natural and Artificial Systems*. University of Michigan Press, Ann Arbor, MI.

5

Energy Management Strategies Involving Energy Storage in DC Microgrid

S. K. Rai[1]*, H. D. Mathur[1] and Sanjeevikumar Padmanaban[2]

[1]*EEE Department, BITS-Pilani, Pilani Campus, India*
[2]*CTiF Global Capsule, Department of Business Development and Technology, Aarhus University, Herning, Denmark*

Abstract

Energy management system (EMS) ensures flexible, economic and optimal cooperation among distributed energy resources of the microgrid and quality power supply to the users. EMS also determines the switching of operating modes like islanded mode (ISM) to grid-connected mode (GCM) and vice versa. It manages the battery storage system's changing and discharging and operation of other dispatchable sources under defined constraints including supply control over normal load, sensitive load and transient load. This chapter presents the literature review of various energy management strategies involving energy storage in a microgrid and a case study is presented by optimizing the involved objective functions using the linear programming (LP) method for a residential microgrid. The optimal operating mode is decided on an hourly basis to get the optimal cost. MATLAB® simulating software is used for the study.

***Keywords*:** EMS, distributed energy sources, optimization, linear programming, MILP

5.1 Introduction

The structure of the traditional electric distribution grid is being modified by the increasing demand of microgrids with high inclusion of distributed electrical energy generators particularly by the renewable energy

**Corresponding author*: p2016503@pilani.bits-pilani.ac.in

Nikita Gupta, Mahajan Sagar Bhaskar, Sanjeevikumar Padmanaban and Dhafer Almakhles (eds.) DC Microgrids: Advances, Challenges, and Applications, (121–158) © 2022 Scrivener Publishing LLC

generating systems. The microgrid can be of ac microgrid, dc microgrid or hybrid microgrid types. Nowadays, dc microgrid is getting attention due to the power generating nature of renewable energy generating system and also due to the nature of power demand by the load. The inclusion of electrical vehicles in the grid system, again, opens the new windows to focus on the dc microgrid due to its ability to behave both as source and load. As per the available literatures, it can be easily concluded that the dc power transmission can increase the system efficiency by about 10% as compared to ac power transmission. But, still, voltage fluctuation at dc node, quality of power, energy management and instability issues related to transition from grid-connected mode to the islanded mode and islanded mode to grid-connected mode are major challenges which need to be handled carefully to have a stable, reliable, efficient and economically viable microgrid. The energy management challenges are enhanced multifold by the stochastic nature of both sources and loads (ex-EVs).

The International Electrotechnical Commission defines the energy management system (EMS) of a microgrid in the standard IEC 61970 as "a computer system comprising a software platform providing basic support services and a set of applications providing the functionality needed for the effective operation of electrical generation and transmission facilities so as to assure adequate security of energy supply at minimum cost".

The management of a microgrid (MG) can be divided into three levels as shown in Figure 5.1. An energy management system is strategic-level management to enhance overall system efficiency based on the data available for hours or days/months.

Figure 5.1 Management in microgrid.

There are two types of approach for EMS, centralized and decentralized. In a centralized energy management system, all the information (DERs Power Generation, Cost Function, Energy Consumption Pattern, etc.) are collected at a single place and then the central EMS controller takes the optimal decision and sends it to all the local controllers to implement the decision. However, in decentralized EMS, the central controller collects the information in real time from local controllers about present and future demand/generation. Based on the received data, the central controller takes the optimal decision to fulfil the global and local needs.

This chapter presents a literature review of energy management techniques proposed by various authors and one example using MATLAB® simulation for the design of EMS of a microgrid based on linear programing (LP) and mixed integer linear programming (MILP) optimization methods.

The chapter is divided into five major sections. Section 5.2 presents the literature review of the various techniques used to achieve optimal energy management goals. The configuration of the microgrid and data for the case study are presented in section 5.3. The objective functions for the various modes of operation are explained in section 5.4, and section 5.5 covers the simulated results of the EMS in both islanded and grid-connected mode using MATLAB®.

5.2 Literature Review

A number of researchers have solved the energy management issue to get optimal operation of MG using a variety of approaches. The different types of approach for EMS are shown in Figure 5.2.

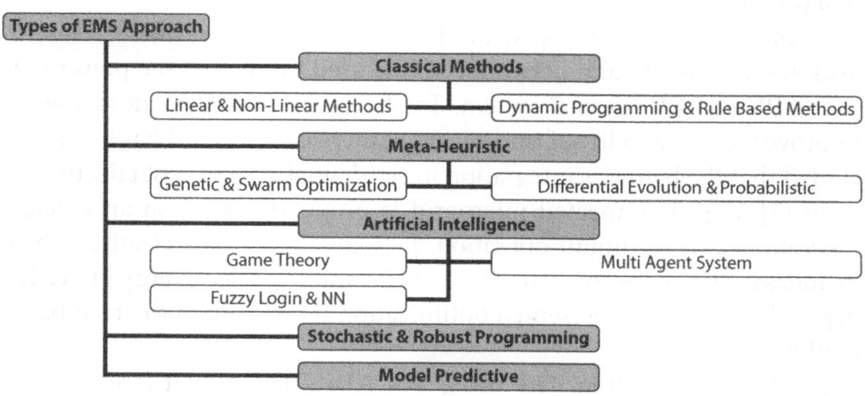

Figure 5.2 Various approaches for EMS.

5.2.1 Classic Approaches of EMS

Tsikalakis and Hatziargyriou [1] developed optimal policies of centralized architecture for energy management using sequential quadratic programming methods. They proposed two market policies to regulate the price bids of grid-connected MG. The aim of the policies is to minimize the operation cost and to maximize the profit.

Kanchev *et al.* [2] developed EMS for microgrid with prosumer (PV system, battery, and ultra-capacitor) and a microturbine. The central energy management system uses the rule-based optimization to optimize the whole microgrid system, whereas prosumer EMS manages the power and frequency regulation of prosumer system.

Sechilariu *et al.* [3] proposed rule-based real-time energy management system for microgrid. It is a grid-connected building integrated microgrid. The rule-based control algorithm is used to switch different operations (load shedding, battery charging and discharging, PV limitation and grid supply modes). The PI controller is used to stabilize the dc bus voltage.

A model to minimize the operational cost for optimal EMS of a residential microgrid is proposed in [4]. The decision is based on the different cost parameters (energy trading cost, penalty cost, EVs batteries wearing cost, the range anxiety term for EVs). The risk level-based analysis is done for operational cost of MG and state of charge (SOC) of EVs battery with three types of load-critical, adjustable, and shiftable loads.

A three-phase unbalanced MG is proposed in [5] based on mixed integer nonlinear energy management. It minimizes the input of conventional sources and optimizes the optional cost of the overall system. The developed MINLP model executes optimization in two sequential stages-MILP-based unit commitment and nonlinear programming-based optimal power flow.

Comodi *et al.* [6] developed an MILP-based EMS for residential microgrid. Radial basis neural network (NN) is used to forecast the power output of PV and PV thermal system. The heat load management is done to improve the profitability of integration of thermal storage. It has been concluded that the batteries integration in residential are not cost effective.

In [7], a grid-connected microgrid is proposed based on an efficient EMS model for optimum operation. It is cost-based optimization which includes both ways – constant energy price and dynamic energy price. The paper shows that the presented optimization is better than the rule-based method.

In [8], MG is simulated using PSCAD/EMTDC software in both islanded and grid-connected mode. A centralized EMS is proposed using rule-based energy management model. The proposed control method ensures the smooth transition of MG modes with stable performance (voltage and frequency).

The optimal EMS for remote military MG is presented in [9] using piecewise linear fuel consumption model of diesel generator as an objective function. The battery rating is decided using Rain-flow Counting method. The two cases are considered for analysis based on the availability of one generator only and one out of two available generators. The proposed system is also verified experimentally.

In [10], a techno-economic MILP-based optimization model is presented. The intermittency and volatility of different parameters are handled using demand response (DR) program. GEMS and Homer software are used for MG sizing optimization.

Sukumar *et al.* [11] proposed mixed mode MG EMS. The controller takes care of power sharing, continuous run, and on/off phenomenon. The power sharing mode considers the power trading. The fuel cell is bound to remain operational in continuous run. Both the issues are solved using linear programming (LP) optimization. The on/off mode of operation is optimized using mixed integer linear programming (MILP).

In [12], EMS for three-phase residential MG is presented. It is a security constrained EMS which uses non-linear optimization method for optimal operation of the system to enhance the reliability of MG. This paper compares the non-linear optimal power flow model with MILP model.

Panwar *et al.* [13] proposed strategic energy management system of a grid-connected microgrid. The EMS is constrained by voltage security and system operational cost. The gradient descent solution method is used for optimization and forward backward sweep algorithm determines power flow.

In [14], online rule-based EMS is proposed to optimize the operation of microgrid in real time. The optimized output is used to switch operations among battery charging-discharging mode, on-off MPPT (Maximum Power Point Tracking) mode of PV system and load-shedding mode depending upon load generation imbalance and state of charge of battery.

Luna *et al.* [15] presented a real-time online MG EMS. It is used to minimize the operational cost and load shedding. Three cases have been analyzed with reference of saving and computational time in the paper-perfect forecast, imperfect forecast and accurate information.

The impact of regulatory framework constraints on energy management system is presented in [16] with a case study. The MILP method of optimization is used. All the reviewed papers of this section have been tabulated in Table 5.1.

Table 5.1 Classical approach of EMS.

Ref.	Published year	Title	Proposed approach
[1]	2011	Centralized control for optimizing microgrids operation. Power and energy society general meeting	Non-linear programming
[2]	2011	Energy management and operational planning of a microgrid with a PV-based active generator for smart grid applications	Rule-based approach
[3]	2013	Building-integrated microgrid: advanced local energy management for forthcoming smart power grid communication	Rule-based approach
[4]	2014	Optimal energy management for a residential microgrid including a vehicle-to-grid system	Mixed integer linear programming
[5]	2014	A centralized energy management system for isolated microgrids	Mixed integer linear programming and nonlinear programming
[6]	2015	Multi-apartment residential microgrid with electrical and thermal storage devices: experimental analysis and simulation of energy management strategies	Mixed integer linear programming
[7]	2015	Optimal energy management for grid-connected microgrid by using dynamic programming method	Dynamic programing

(*Continued*)

Table 5.1 Classical Approach of EMS. (*Continued*)

Ref.	Published year	Title	Proposed approach
[8]	2016	A centralized and heuristic approach for energy management of an ac microgrid	Rule-based approach
[9]	2017	Optimized energy management system to reduce fuel consumption in remote military microgrids	Mixed integer linear programming
[10]	2017	Techno-economic optimization of hybrid photovoltaic/wind generation together with energy storage system in a stand-alone microgrid subjected to demand response	Mixed integer linear programming
[11]	2017	Mix-mode energy management strategy and battery sizing for economic operation of grid-tied microgrid	Linear and mixed integer linear programming
[12]	2017	Security-constrained optimal energy management system for three-phase residential microgrids	Mixed integer linear programming
[13]	2017	Operation window constrained strategic energy management of microgrid with electric vehicle and distributed resources	Nonlinear programming
[14]	2017	Energy management and control system for laboratory scale microgrid-based wind-PV-battery	Rule-based approach

(*Continued*)

Table 5.1 Classical Approach of EMS. (*Continued*)

Ref.	Published year	Title	Proposed approach
[15]	2018	Online energy management systems for microgrids: experimental validation and assessment framework	Mixed integer linear programming
[16]	2019	Regulatory framework embedded energy management system for microgrids: the case study of the Spanish self-consumption scheme	Mixed integer linear programming with regulatory framework
[17]	2020	An MPC-based dual-solver optimization method for DC microgrids with simultaneous consideration of operation cost and power loss	Mixed integer linear programming, nonlinear programming
[18]	2020	Smart meter data to optimize combined roof-top solar and battery systems using a stochastic mixed integer programming model	Mixed integer linear programming
[19]	2020	An energy management system for isolated microgrids with thermal energy resources	Mixed integer linear programming
[20]	2021	Day-ahead optimal power flow for efficient energy management of urban microgrid	Mixed-integer non-linear programming

In [17], a MCP is proposed with dual-solver framework where MILP is used to solve the economic scheduling problem and non-linear programming is used to handle the transmission loss problem.

By using MILP, a least-cost planning model is presented in [18]. This model uses the smart meter data, electrical vehicle's data and data of distributed sources for optimal scheduling among PV, BESS and EV changing and discharging. The two case studies are presented for the validation of the proposed model.

A thermal load-based islanded microgrid with a novel Energy Management System (EMS) model is presented in [19]. The focus of the EMS is to minimize the fuel cost and simulation is done for different winter days. The optimization problem is formulated based on MILP.

An energy management system for a community microgrid is exercised in the paper [20]. The intelligent deep learning is used for the day-ahead prediction. The aim of the proposed model is to reduce the electricity price using Mixed-Integer Non-Linear Programming (MINLP) optimization problem which is validated with experimental results.

5.2.2 Meta-Heuristic Approach of EMS

The Meta-Heuristic approach of energy management includes genetic algorithm (GA), particle swarm optimization (PSO), ant colony optimization, gravitational search optimization, bacterial foraging, artificial bee colony, tabu search, crow search algorithm, artificial fish school algorithm, imperialist competition algorithm etc. All the reviewed papers of this section have been tabulated in Table 5.2.

Chen, C. *et al.* [21] proposed an energy management system for MG based on matrix real coded genetic algorithm. The energy management system generates the control signal based on the algorithm using all the data and forecasting for each generation, storage and loads units.

In [22], an EMS is developed to minimize the operating cost of MG. In this paper, a novel self-adaptive modified θ-PSO is proposed to get optimal control decision and it has been shown that it is better than GA, combinatorial PSO, fuzzy self-adaptive PSO, adaptive modified PSO and θ-PSO.

M. Motevasel *et al.* [23] presented a multi-objective EMS based upon a modified bacterial foraging method. It optimizes the operational cost and functions. In this paper, it has been shown that the proposed algorithm performs better than GA and PSO.

Marzband *et al.* [24] introduced an EMS model for an islanded microgrid using gravitational search algorithm. It minimizes the operating cost with penalty cost on undelivered power. The proposed algorithm is tested through experiment using three different cases - normal operation, sudden demand, and plug and play features. It has been proven that the proposed algorithm performs better than PSO.

K. Huang *et al.* [25] proposed EMS to minimize the overall operational cost of MG which includes fuel cost of conventional sources, operation cost of distributed energy resources and trading cost with main grid. The power

Table 5.2 Meta-Heuristic approach of EMS.

Ref.	Published year	Title	Proposed approach
[21]	2011	Optimized energy management system to reduce fuel consumption in remote military microgrids	Matrix real coded genetic algorithm
[22]	2013	Considering uncertainty in the optimal energy management of renewable micro-grids including storage devices	self-adaptive modified θ-PSO
[23]	2013	Multi-objective energy management of CHP (combined heat and power)-based microgrid	Modified bacterial foraging
[24]	2014	Experimental validation of a real-time energy management system using multi-period gravitational search algorithm for microgrids in islanded mode	Gravitational search algorithm
[25]	2015	Optimal energy management of grid-connected photovoltaic microgrid. Cyber technology in automation, control, and intelligent systems (CYBER).	Modified artificial fish school algorithm
[26]	2015	Optimal probabilistic energy management in a typical microgrid based on robust optimization and point estimate method	Particle swarm optimization
[27]	2015	An efficient two-stage stochastic optimal energy and reserve management in a microgrid	stochastic weight tradeoff PSO

(*Continued*)

Table 5.2 Meta-Heuristic approach of EMS. (*Continued*)

Ref.	Published year	Title	Proposed approach
[28]	2016	Optimal energy management for stand-alone microgrids based on multi-period imperialist competition algorithm considering uncertainties: experimental validation	Imperialist competition algorithm
[29]	2016	Real-time experimental implementation of optimum energy management system in stand-alone microgrid by using multi-layer ant colony optimization	Ant colony optimization
[30]	2017	Optimal energy management for industrial microgrids with high-penetration renewables	Regrouping PSO
[31]	2017	A memory-based genetic algorithm for optimization of power generation in a microgrid	Memory-based genetic algorithm
[32]	2017	A real-time evaluation of energy management systems for smart hybrid home microgrids	Artificial bee colony
[33]	2017	A heuristic method for optimal energy management of dc microgrid. DC microgrids (ICDCM)	Modified crow search algorithm
[34]	2017	Energy management in multi-microgrid systems development and assessment	Tabu search

(*Continued*)

Table 5.2 Meta-Heuristic approach of EMS. (*Continued*)

Ref.	Published year	Title	Proposed approach
[35]	2018	Grey wolf optimizer-based battery energy storage system sizing for economic operation of microgrid	GWO, PSO, ABC, GSA and GA
[36]	2019	Modified PSO algorithm for real-time energy management in grid-connected microgrids	Modified PSO
[37]	2020	Optimal sizing of smart home renewable energy resources and battery under prosumer-based energy management	PSO
[38]	2020	Energy management system for two islanded interconnected microgrids using advanced evolutionary algorithms	Porcellio Scaber algorithm (PSA)
[39]	2021	A multi-objective home energy management system based on internet of things and optimization algorithms	IoT, PSO, BOA

generated by renewable energy resources has been predicted using least square support vector machine.

In [26], an optimal EMS model is developed to optimize the overall operation of a microgrid. The uncertainty of renewable energy generations is modeled using Beta and Weibull probability density functions in point estimated method along with uncertainties in load demand by robust optimization.

Mohan *et al.* [27] proposed an EMS for optimal operation of a microgrid in the point of view of energy and reserve management. The affine arithmetic and stochastic weight tradeoff PSO-based power flow method is used to solve the optimization problem which has been shown to be better than interval arithmetic method.

M. Marzband *et al.* [28] developed an EMS to minimize the operational cost of MG with penalty cost on undelivered power. The forecast of power generation and demand has been obtained by hybrid Neural Network (NN) and Markov chain-based prediction methods. The proposed approach is verified experimentally and it is concluded that it is better than PSO.

In [29], an EMS is developed based on ant colony optimization to optimize operational cost for an islanded MG. The proposed approach has produced better result in reduction of operational cost than the modified conventional EMS and PSO-based EMS by 23% and 5%, respectively.

In [30], an optimal energy management for an industrial MG that can be operated islanded and grid-connected modes is proposed based on regrouping PSO. The operational cost is minimized in the islanded mode whereas the energy trading profit is maximized in grid-connected mode. The paper concludes that the proposed method is better as GA in respect to global optimum solution and computational time.

An EMS using memory-based GA is introduced for optimal power management with the goal of minimizing the operational cost of a grid-connected microgrid in [31]. The proposed approach is proven better that GA, PSO with inertia factor and PSO with constriction factor.

In [32], a multi-period artificial bee colony-based EMS is proposed for residential MG to minimize the operational cost and performance. The developed system works in two layers – day-ahead scheduling and real-time scheduling in 5 min. interval. The paper has concluded that this approach is more productive than mixed integer non-linear programming algorithm.

A modified crow search-based EMS has been presented in [33]. It is used to optimize the daily energy scheduling for a grid-connected microgrid with the final aim to optimize the operational cost and bidding cost. It has been concluded that the proposed approach is more efficient and effective than GA, PSO, fuzzy self-adaptive PSO and crow search algorithm.

In [34], an EMS is proposed for a multi-microgrid system to reduce the total operational cost using Tabu search method. It is a multi-state variable concept-based optimization model. This model considered the uncertainties of renewable energy sources and EVs in the optimization process. The energy loss of the system is determined using the forward-backward-based probabilistic power flow method.

A mix-mode energy management system is proposed in [35]. The LP and MILP is used to solve the EMS problem and grey wolf optimizer (GWO), particle swarm optimization (PSO), artificial bee colony (ABC), gravitational search algorithm (GSA) and genetic algorithm (GA) are used to optimize the battery energy storage system sizing problem. The GWO produces the optimal solution and offers great savings in the operational cost.

A particle swarm optimization (PSO) and dynamic penalty-based real-time energy management system is proposed in [36]. The proposed model optimizes the battery energy.

A multi-objective home energy management system is presented in [37]. It has a two-stage optimization model for operating cost and energy cost. The PSO and MILP techniques are used for optimization purpose. The three different case studies are presented for the effectiveness of the proposed model.

In [38], an energy management system based on a modified version of the basic Porcellio Scaber algorithm (PSA) is presented for the interconnected microgrid. It shows better results than other meta-heuristic techniques.

A home energy management system based on IoT is proposed in [39]. The two-stage optimization process is modeled to optimize the consumption cost and user satisfaction. The effectiveness of multi-objective optimization system is proved by comparing with PSO and BOA-based algorithms.

5.2.3 Artificial Intelligence Approach of EMS

The Artificial Intelligence approach of energy management mainly includes fuzzy logic (FL), neural network (NN), multi-agent system (MAS) and game theory (GT). All the reviewed papers in this section have been tabulated in Table 5.3.

F. A. Mohamed *et al.* [40] proposed a modified game theory-based multi-objective EMS for grid-connected microgrid. It is used to optimize the operation and emission cost. The developed model is compared with the performance of multi-objective GA, Sequential Quadratic Programming and Mesh Adaptive Direct Search.

Kyriakarakos *et al.* [41] introduced a fuzzy logic-based EMS for an isolated microgrid with the main aim of optimizing operational cost. The three categories of load demands are considered for the study – electric load, water load, and transport load. The software platforms of Transys, Genopt, Matlab, and Trnopt are used for simulation and analysis.

In [42], an EMS is proposed for maximization of the use of sources, storages and energy trading from the grid by using reinforcement learning. The Q-learning method in reinforcement learning is used to optimize the operation of microgrid with two hours-ahead forecasting.

Table 5.3 Artificial intelligence approach of EMS.

Ref.	Published year	Title	Proposed approach
[40]	2011	Multi-objective optimization using modified game theory for online management of microgrid	Modified game theory
[41]	2012	A fuzzy logic energy management system for polygeneration microgrids	Fuzzy logic
[42]	2013	Reinforcement learning for microgrid energy management	Reinforcement learning based-NN
[43]	2013	Energy management in microgrids using demand response and distributed storage – a multi-agent approach	Multi-agent system
[44]	2013	Leader-follower strategies for energy management of multi-microgrids	Game theory
[45]	2015	Optimal scheduling of a microgrid with a fuzzy logic-controlled storage system	Fuzzy logic
[46]	2015	Electrical microgrid optimization via a new recurrent neural network	Recurrent NN
[47]	2015	A multi-agent decentralized energy management system based on distributed intelligence for the design and control of autonomous polygeneration microgrids	Multi-agent system
[48]	2016	Intelligent energy management control for independent microgrid	Multi-agent system

(*Continued*)

Table 5.3 Artificial intelligence approach of EMS. (*Continued*)

Ref.	Published year	Title	Proposed approach
[49]	2016	Dynamic energy management system for a smart microgrid	Neural network
[50]	2016	Energy management for joint operation of CHP and PV prosumers inside a grid-connected microgrid: a game theoretic approach	Game theory
[51]	2017	Neural networks for power management optimal strategy in hybrid microgrid	Neural network
[52]	2017	Low complexity energy management strategy for grid profile smoothing of a residential grid-connected microgrid using generation and demand forecasting	Fuzzy logic
[53]	2017	Multi-agent autonomous decision making in smart micro-grids energy management: a decentralized approach. German conference on multi-agent system technologies.	Multi-agent system
[54]	2017	Energy sharing management for microgrids with PV prosumers: a Stackelberg game approach	Game theory
[55]	2018	Energy management with a world-wide adaptive thermostat using fuzzy inference system	Fuzzy logic
[56]	2018	Fuzzy control algorithm for battery storage and demand side power management for economic operation of the smart grid system at Naresuan University, Thailand	Fuzzy logic

(*Continued*)

Table 5.3 Artificial intelligence approach of EMS. (*Continued*)

Ref.	Published year	Title	Proposed approach
[57]	2019	User-centered neuro-fuzzy energy management through semantic-based optimization	Neuro-fuzzy
[58]	2020	Intelligent multi-microgrid energy management based on deep neural network and model free reinforcement learning	Deep neural network and reinforcement learning
[59]	2020	A fuzzy logic-based supervisory control for power management in multibus dc microgrid	Fuzzy logic
[60]	2021	An energy management system design using fuzzy logic control: smoothing the grid power profile of a residential electro-thermal microgrid	Fuzzy logic

An optimal multi-agent system-based EMS for optimal operation of microgrid with the main aim of reducing peak demand and minimizing electricity cost is proposed in [43]. The optimized decision has been achieved by encouraging the customers to participate in a DR program using an index-based incentive mechanism.

G. E. Asimakopoulou *et al.* [44] introduced a bi-level leader follower programming method for energy management of multi-microgrid. Out of two layers the top one is used to minimize the production cost whereas the bottom one is used to maximize the net profit of a microgrid. The LINDOGlobal package in GAMS software is used to analyze the problem.

J. P. Fossati *et al.* [45] induced fuzzy-based EMS for microgrid scheduling and battery power management. It uses two GA algorithms for operational optimization. The energy scheduling and fuzzy rules are determined by first GA algorithm whereas the membership functions of fuzzy are tuned by second GA algorithm.

In [46], EMS is proposed for a grid-connected microgrid based on the recurrent NN approach. The main aim of EMS is to maximize the power output of renewable energy sources along with minimizing the traded power from the grid. The forecasting system is designed using hybrid wavelet functions and extended Kalman filter-based NN approach.

C-S. Karavas *et al.* [47] proposed a fuzzy cognitive maps-based MAS approach for energy management system for an islanded microgrid. The multi-agent system is based on five agents – RER agent, battery agent, desalination agent, fuel cell agent, and electrolyzer agent. Transys, Matlab, Genopt and Trnopt software packages are used for analysis and the results are compared with a fuzzy-based system in terms of economic advantage.

In [48], a MAS-based intelligent EMS is presented for an islanded MG. The following agents are considered in the proposed approach – photovoltaic system, wind turbine, fuel cell, battery, and load. The battery scheduling and load shedding are used for load generation balance and STATCOM is used to improve the voltage profile and to reduce the harmonics.

An intelligent adaptive dynamic EMS is proposed for grid-connected MG in [49]. The optimization strategy is achieved through an active NN. The main aim of the developed system is to maximize the utilization of RERs, DR, and battery life along with reduction of carbon emission. The paper concludes that the proposed approach is better than decision tree approach-based dynamic EMS.

In [50], an EMS is presented for a grid-connected microgrid using a leader follower-based game theory. It is used to maximize the profit while maintaining Stackelberg equilibrium to ensure fair share of profit distribution. The non-linear programming is used by prosumers and differential evolution is used by microgrid operator to reach the Stackelberg equilibrium. The paper concludes that the proposed system produces better results than a centralized optimization model.

In [51], neural network approach-based EMS is presented for overall cost minimization of microgrid. The forecast system is developed using radial basis NN. The paper concludes that the presented approach is better than PSO.

A fuzzy logic-based EMS is introduced for a grid-connected residential microgrid in [52]. The developed system aims to minimize the fluctuations and power peaks in the trading from main grid. The conclusive remark of the paper is that the proposed approach performs better than SOC based EMS.

S. Ghorbani, *et al.* [53] proposed MAS-based decentralized EMS for a grid-connected microgrid. The agents of the system are the consumers, storage units, generation units, and grid. The proposed approach has been proven better than a centralized approach.

In [54], an optimal energy management is proposed for a grid-connected microgrid based on the Stackelberg game theory. It is a leader-follower model where the microgrid is leader and all photovoltaic (PV) are followers. The billing mechanism approach is used to handle the uncertainties. The global profit is maximized by leader whereas utilization is maximized by followers.

A very versatile fuzzy logic-based energy management system for residential buildings is proposed in [55]. The proposed model is applicable to all the regions of the world and is very cost effective with the minimum energy saving of 6.5%.

In [56], a fuzzy logic-based algorithm is developed and presented for BESS power management and DSM. The designed fuzzy-based EMS aims the low cost of operation which reduces the electricity bill by 17.58%.

A cloud-based energy management system for building is included in [57]. The developed model uses neuro-fuzzy system and has user interface, intelligence and interoperability. It was tested in a real environment and was able to save energy by more than 25%.

A deep neural network (DNN) and model-free reinforcement learning (RL) techniques are used to build the EMS for an intelligent multi-microgrid in [58]. This model is designed with the aim of reducing demand side peak to average ratio and optimizing the profit. The effectiveness of the proposed model is proved by providing the comparison with a conventional model.

The power management of a DC microgrid using fuzzy logic is presented in [59]. The proposed design is very simple and does not require intensive mathematical calculations. The model is validated with experiment and shows better results than conventional methods.

In [60], an Energy Management System (EMS) is designed based on Fuzzy Logic Control (FLC) which is used to optimize the performance parameters of a grid-connected electro-thermal microgrid. The developed model has low complexity and reduced number of rules for power trading with the grid and to drive the electrical water heater. It reduced the peak power trading by 11.4% and improves power profile.

5.2.4 Model Predictive, Stochastic and Robust Programming Approach of EMS

The papers based on Model Predictive, Stochastic and Robust Programming approach of EMS are reviewed in this section and are tabulated in Table 5.4.

V. S. Tabar *et al.* [61] introduced a multi-objective stochastic EMS with the aim of minimizing the operational and emission cost of a grid-connected microgrid. The augmented epsilon constraint method is used to solve the optimization problem using GAMS software.

Table 5.4 Model predictive, stochastic and robust programming approach of EMS.

Ref.	Published year	Title	Proposed approach
[61]	2017	Energy management in microgrid based on the multi-objective stochastic programming incorporating portable renewable energy resource as demand response option	Stochastic optimization
[62]	2017	A robust optimization method for energy management of CCHP microgrid	Robust optimization
[63]	2017	Including smart loads for optimal demand response in integrated energy management systems for isolated microgrids	Model predictive
[64]	2017	A two-stage optimization and control for CCHP microgrid energy management	Model predictive
[65]	2018	Towards optimal energy management of microgrids via robust two-stage optimization	Stochastic optimization
[66]	2018	Coordination of pumped-storage unit and irrigation system with intermittent wind generation for intelligent energy management of an agricultural microgrid	Robust optimization
[67]	2019	A single-level, rule-based model predictive control approach for energy management of grid-connected microgrids	Single-level, rule-based model predictive control

(Continued)

Table 5.4 Model predictive, stochastic and robust programming approach of EMS. (*Continued*)

Ref.	Published year	Title	Proposed approach
[68]	2020	Stochastic energy management of microgrid with nodal pricing	Stochastic optimization
[69]	2020	Energy management system for hybrid PV-wind-battery microgrid using convex programming, model predictive and rolling horizon predictive control with experimental validation	Convex programming, model predictive, rolling horizon predictive control
[70]	2020	A Lyapunov optimization-based energy management strategy for energy hub with energy router	Lyapunov optimization
[71]	2021	Multi-agent reinforcement learning for energy management in residential buildings	Reinforcement learning

L. Zhao *et al.* [62] proposed an optimal EMS of a grid-connected microgrid with the objective of minimizing the operational cost of a microgrid. The operational cost of a microgrid includes trading cost, cooling cost of head and power unit and battery degradation cost. The uncertainties of a microgrid are studied in detail.

In [63], a model predictive control approach-based EMS is proposed for islanded microgrid with the facility to estimate the residential controllable load using supervised NN. The main goal of the system is to optimize the operating cost and penalty cost. The paper concludes that the proposed approach performs better than decoupled EMS model which splits up the main problem into two subproblems.

A two-stage coordinated approach for an efficient EMS is proposed for a grid-connected microgrid in [64]. The first stage deals with economic dispatch and the second stage makes adjustment in it. The following objective functions have been considered for the study: operating cost of microturbine, battery degradation cost, and energy trading cost with the main grid.

In [65], a two-stage robust optimization-based EMS is developed for a grid-connected microgrid. The first stage handles the day-ahead unit commitment whereas the second stage handles the trade and economic dispatch. The complexity of the developed system is relaxed by using Lyapunov optimization. The paper concludes that the proposed approach is better than the greedy algorithm.

An energy management system is proposed in a grid-connected community microgrid for agriculture purpose in [66]. A stochastic coordination framework is developed to minimize the overall cost of the system. The uncertainties are handled using the point estimation method.

In [67], a novel, single-level, rule-based model predictive control scheme has been included which converts the standard MILP problem into a linear problem with almost no loss in performance and significant reduction in the computation time.

In [68], minimization of energy cost on the purchase of energy from the grid is proposed through an energy management system based on stochastic framework. Monte Carlo simulations are used to handle the uncertainties.

The two-layer optimization based on convex programming and model predictive control with rolling horizon predictive control is used in [69]. The Energy Management System achieves a daily cost reduction of up to 30%.

A Lyapunov optimization-based energy management system for the energy hub is proposed in the paper [70]. A relaxed form of energy management problem is formulated by penalty. The results are compared with a greedy algorithm.

A multi-agent reinforcement learning-based energy management system is proposed in [71]. The optimum solution of the objective is obtained by Q-learning and lower cost is obtained compared to conventional methods.

5.3 Case Study

An energy management system for a residential dc microgrid shown in Figure 5.3 is presented in the study. The microgrid is able to run in both islanded mode (ISM) and grid-connected mode (GCM) with bidirectional energy trading facility. Details of the parameters of the microgrid are tabulated in Table 5.5.

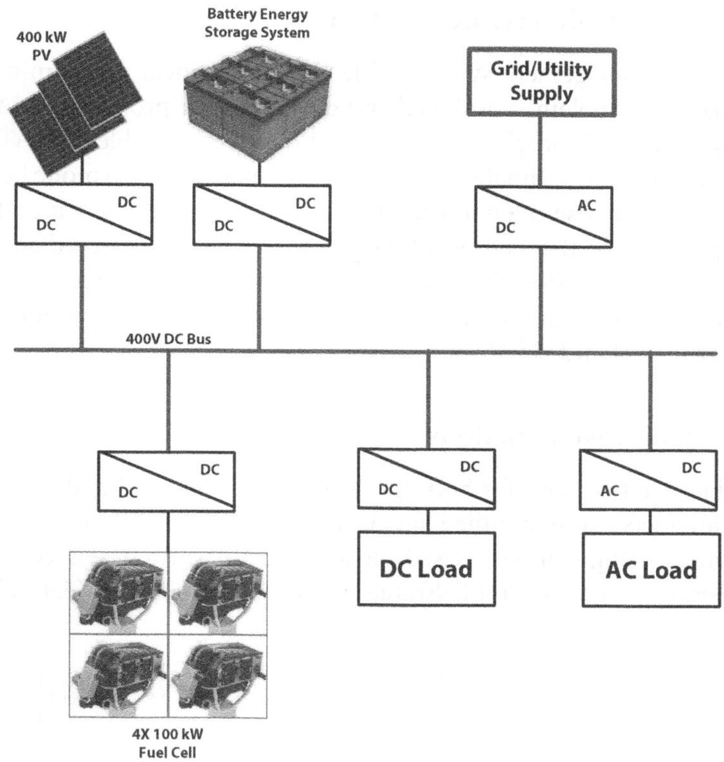

Figure 5.3 Residential microgrid.

Table 5.5 DC microgrid subsystem's parameters.

Parameters/name	Rating/value	Parameters/name	Rating/value
Solar panel	400 kWp	Changing/discharging efficiency of BSS	95%
Fuel cell	4 X 100 kW	SOC range of BSS	20–100%
Operating range of fuel cell/unit	10% to 100%	Battery life time	5 Years
DC grid voltage	400 V	Rate of interest on finance	7%
Grid supply	230 V, 50 Hz	Specific cost of power capacity	$234/KW
Power range of BSS	–200 kW to 200 kW	Specific cost of energy capacity	$167/kW

5.3.1 Energy Management System

The energy management system (EMS) based on linear programing (LP) optimization problem is designed and developed for proposed residential microgrid, and "linprog" function is used which is available in MATLAB® software to get the optimal solutions. The optimal operating mode between ISM and GCM is selected at each hour by EMS to get efficient and optimal operation of the microgrid based on the day-ahead predicted data of generation and load. The fuel cell is a dispatchable source and it provides energy in continuous run mode for both the operating modes of the microgrid: ISM and GCM.

5.3.2 Objective Functions

The objective functions for both the operating modes ISM and GCM have been developed to reduce the daily operating cost {equations (5.1) to (5.8)}. The battery sizing is also exercised and so the per day investment cost of battery energy storage system (BESS) is also formulated in equation (5.9 and 5.10).

$$OF_{ISM} = Minimize \sum_{n=1}^{24} \left\{ \left(Cost_{NGn} \frac{P_{FCn}}{\eta_{FC}} \right) + \alpha \left(P_{BESSn} \right) \right\} \tag{5.1}$$

Constraints:

$$P_{PVn} + P_{FCn} + P_{BESSn} = P_{LOADn}; \text{ n=1, 2, 3, ..., 24} \tag{5.2}$$

$$P_{FC\_min} \le P_{FCn} \le P_{FC\_max}; \text{ n=1, 2, 3, ..., 24} \tag{5.3}$$

$$OF_{GCM} = Minimize \sum_{n=1}^{24} \left\{ \left(Cost_{NGn} \frac{P_{FCn}}{\eta_{FC}} \right) + Cost_{EPn} \left(P_{LOADn} - P_{TOTALn} \right) + \alpha \left(P_{BESSn} \right) \right\} \tag{5.4}$$

Constraints:

$$P_{PVSn} + P_{FCn} + P_{BESSn} + P_{GRIDn} = P_{LOADn}; \text{ n=1, 2, 3, ..., 24} \tag{5.5}$$

$$-\infty \le P_{GRIDn} \le +\infty; \text{ n=1, 2, 3, ..., 24} \tag{5.6}$$

Constraints for BESS:

$$E_{min} \leq E_n \leq E_{max}; \text{ n=1, 2, 3, ..., 24} \tag{5.7}$$

$$P_{Charging\_n} \leq P_n \leq P_{Discharging\_n}; \text{ n=1, 2, 3, ..., 24} \tag{5.8}$$

Investment Cost:

$$\textit{Per Day Investment Cost}_{BESS} = \frac{1}{365}\left\{\frac{r(1+r)^{LT}}{(1+r)^{LT}-1}\right\} * C_{Cap} \tag{5.9}$$

$$C_{Cap} = (C_P * R_P) + (C_E * R_E) \tag{5.10}$$

Where;

$Cost_{NGn}$ = Natural Gas price in dollar at nth hour
P_{FCn} = Power generated by Fuel Cell at nth hour
η_{FC} = Average efficiency of Fuel Cell
$\alpha = 1 \times 10^{-6}$ (Dispatch Solution Constant for BESS)
P_{BESSn} = Power of Battery Storage System at nth hour
$Cost_{EPn}$ = Tariff of energy purchased from grid in \$/kWh at nth hour
P_{LOADn} = Load Demand Power at nth hour
P_{TOTALt} = Power Generated by Distributed Sources at nth hour
E_{min} = The minimum permissible energy level of BSS
E_{max} = The maximum permissible energy level of BSS
E_n = The energy level of BESS at nth hour
r = Interest Rate
L_T = Life Time
C_P = Specific cost of power
C_E = Specific Cost of Energy
R_P = Rated power capacity
R_E = Rated energy capacity

5.3.3 Result and Discussion

The photovoltaic (PV) system and battery energy storage system (BESS) are the prime source of the considered residential microgrid. Firstly, the power from PV system and BESS has been utilized by the load. The dispatchable fuel cell is in continuous run mode and it runs with its 10% capacity if there is no power requirement. In the study, the average efficiency of fuel cell is considered equal to 50%.

Figure 5.4 shows the day-ahead power generation for PV system and load demand. The reference for the range of the operation of fuel cell, utility grid price and natural gas price has been taken from [72]. Figure 5.5 shows the prices of the utility grid and natural gas. In the initial stage of the study, arbitrary, the capacity of the BESS is chosen equal to 3 MWh.

The comparison among the day operating cost (OC) of the proposed energy management system with various in day initial charge of the BESS

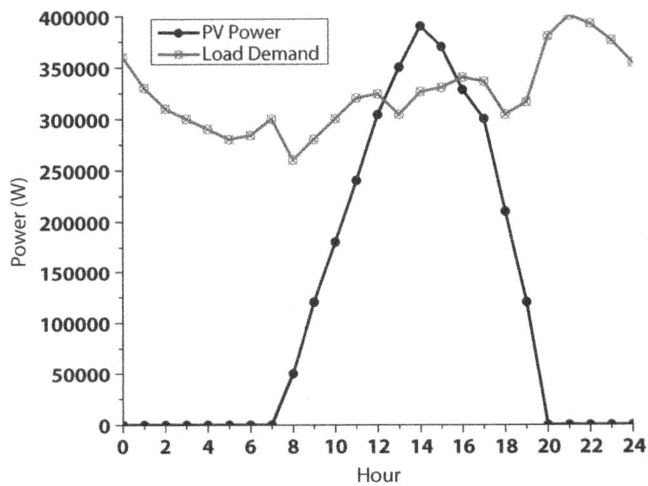

Figure 5.4 Day-ahead PV power generation and load demand.

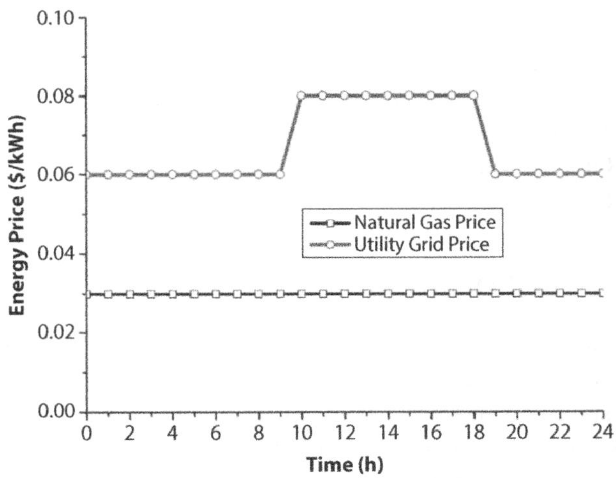

Figure 5.5 Hourly energy price of the grid and natural gas.

is shown in Figure 5.6. The optimal operating mode which is hourly best operating mode between ISM and GCM is the most efficient. The day operating cost of ISM is greater than GCM throughout the day. A sharp decrement in the day operating cost is observed between 90-100% of the day initial charging level of the BESS for optimal operating mode. The hourly operating cost at 20% day initial charging level of the BESS for islanded mode of operation (ISM) and grid-connected mode of operation (GCM) are shown in Figure 5.7 and Figure 5.8, respectively.

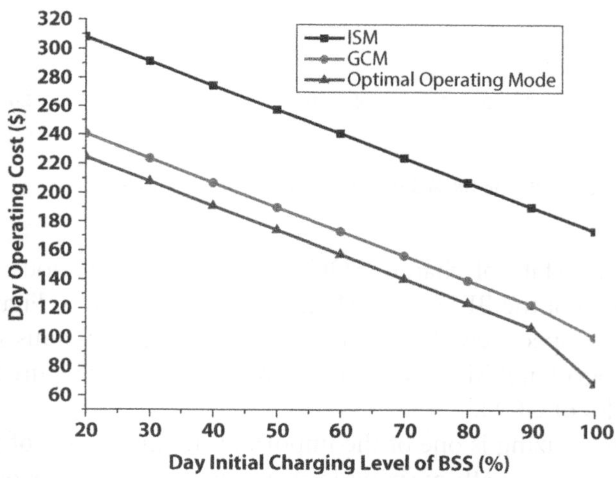

Figure 5.6 Day operating cost with respect to day initial charging level of BESS.

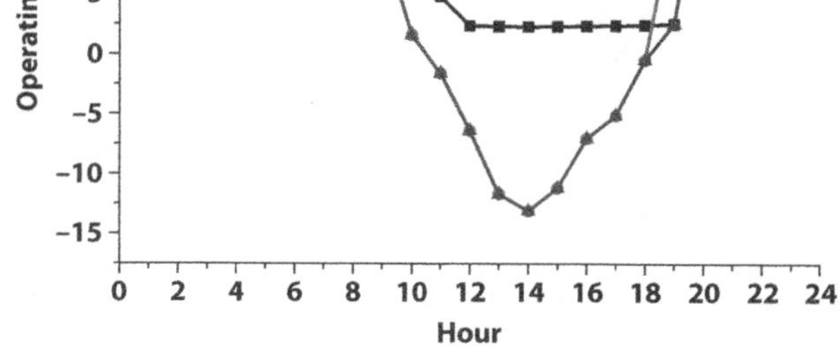

Figure 5.7 Hourly OC for 20% day initial charging level of BESS.

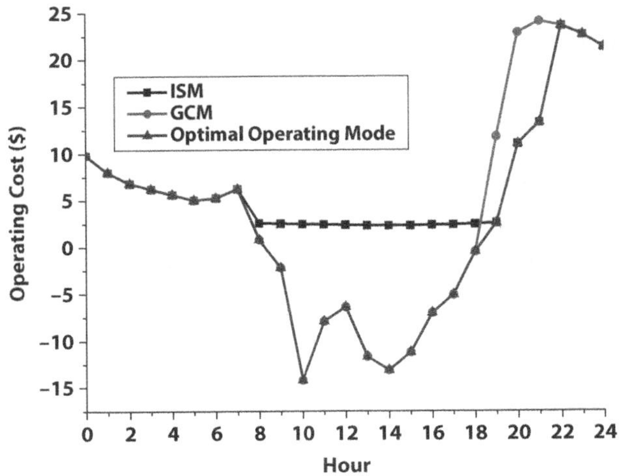

Figure 5.8 Hourly OC for 100% day initial charging level of BESS.

The hourly state of charge (SOC) of the BESS at various day initial charging level of the BESS for ISM and GCM are shown in Figure 5.9 and Figure 5.10, respectively. The hourly power sharing by various subsystems of the microgrid in ISM is shown in Figure 5.11, whereas Figure 5.12 shows the same plot of GCM.

The battery sizing is one of the important design aspects of the microgrid and therefore, battery sizing is also exercised to get the optimal size

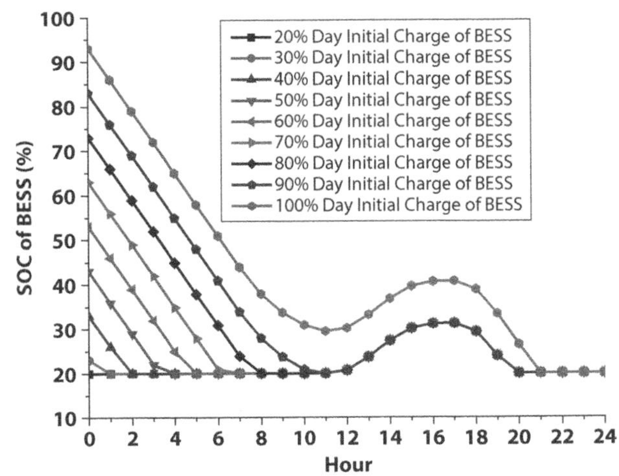

Figure 5.9 Hourly SOC of BESS in ISM for various day initial charging level.

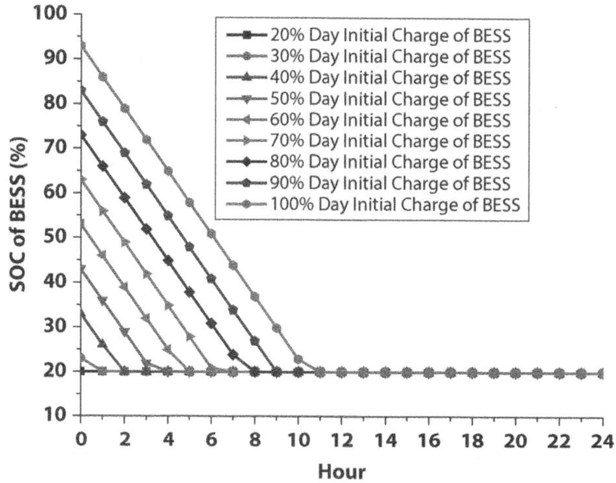

Figure 5.10 Hourly SOC of BESS in GCM for various day initial charging level.

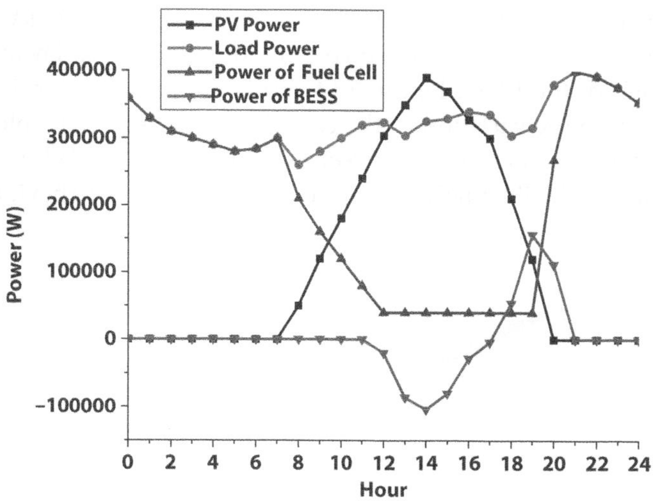

Figure 5.11 Hourly power sharing in ISM for 20% day initial charging level.

of the BESS. The plot of day operating cost with respect to variation in the capacity of BESS is shown in Figure 5.13. The capacity of BESS varies from 500 kWh to 6 MWh. The optimization does not converge in ISM below the mentioned minimum value of the capacity. Since the day operating cost in case of ISM is greater than GCM, ISM operating mode has been as reference for optimal size selection of the BESS. It is very clear from

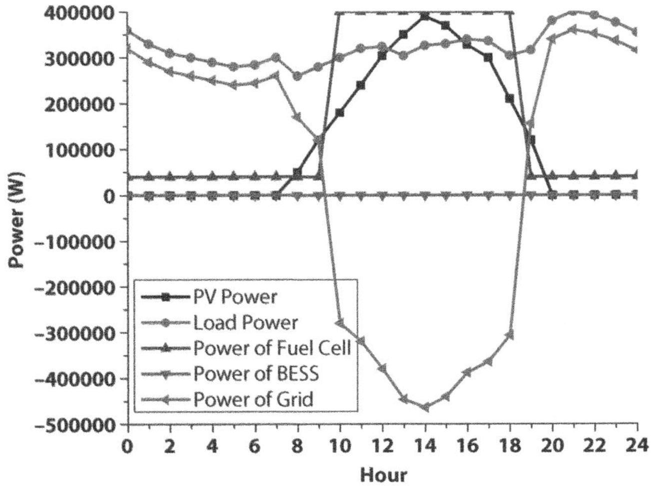

Figure 5.12 Hourly power sharing in GCM for 20% day initial charging level.

Figure 5.13 that the operating cost is decreasing with increase in capacity in ISM operating mode, but it saturates at the capacity of 3825 kWh, which is optimal capacity. This study is further modified with per day investment cost (PDIC) of the BESS. Figure 5.14 shows the modification in the cost with respect to the variation in the capacity of the BESS. The plot reflects that the selection of capacity between 500 kWh to 3825 kWh will be best by considering the fluctuation in the load and stochastic behavior of other subsystems.

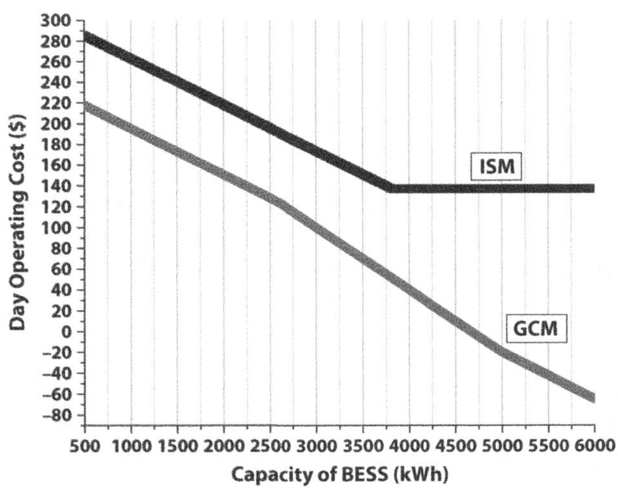

Figure 5.13 Day operating cost with respect to capacity of BESS.

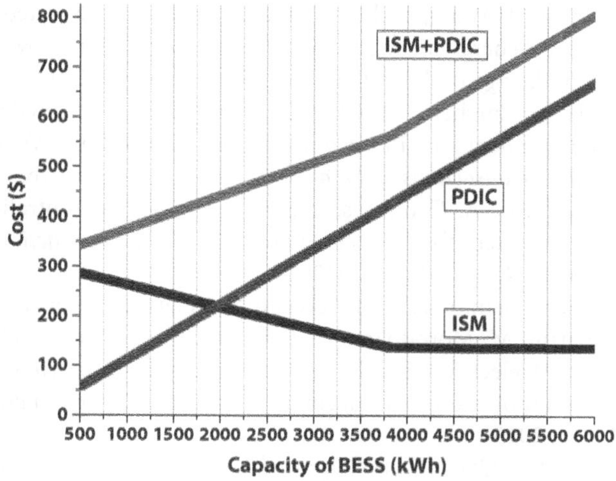

Figure 5.14 Cost with respect to capacity of BESS.

5.4 Conclusion

This chapter presents a literature review for the various approaches of the microgrid - classical, meta-heuristic, artificial intelligence, Model Predictive, Stochastic and Robust Programming. A case study is also included for residential dc microgrid having PV, fuel cell and BESS systems. The microgrid has the capability to work in both islanded and grid-connected mode. On the hourly basis, the best operating mode between ISM and GCM has been selected to get the optimal operating cost of the system. The optimal capacity of the BESS is also exercised with and without per day investment cost and finally, it is found that the optimal capacity range of the BESS for the considered residential dc microgrid is 500 kWh to 3825 kWh.

References

1. G. Tsikalakis and N. D. Hatziargyriou, "Centralized control for optimizing microgrids operation," 2011 IEEE Power and Energy Society General Meeting, 2011.
2. H. Kanchev, D. Lu, F. Colas, V. Lazarov, and B. Francois, "Energy Management and Operational Planning of a Microgrid With a PV-Based Active Generator for Smart Grid Applications," *IEEE Transactions on Industrial Electronics*, vol. 58, no. 10, pp. 4583–4592, 2011.

3. M. Sechilariu, B. Wang, and F. Locment, "Building-integrated microgrid: Advanced local energy management for forthcoming smart power grid communication," *Energy and Buildings*, vol. 59, pp. 236–243, 2013.
4. L. Igualada, C. Corchero, M. Cruz-Zambrano, and F.-J. Heredia, "Optimal Energy Management for a Residential Microgrid Including a Vehicle-to-Grid System," *IEEE Transactions on Smart Grid*, vol. 5, no. 4, pp. 2163–2172, 2014.
5. D. E. Olivares, C. A. Canizares, and M. Kazerani, "A Centralized Energy Management System for Isolated Microgrids," *IEEE Transactions on Smart Grid*, vol. 5, no. 4, pp. 1864–1875, 2014.
6. G. Comodi, A. Giantomassi, M. Severini, S. Squartini, F. Ferracuti, A. Fonti, D. N. Cesarini, M. Morodo, and F. Polonara, "Multi-apartment residential microgrid with electrical and thermal storage devices: Experimental analysis and simulation of energy management strategies," *Applied Energy*, vol. 137, pp. 854–866, 2015.
7. L. N. An and T. Quoc-Tuan, "Optimal energy management for grid connected microgrid by using dynamic programming method," 2015 IEEE Power & Energy Society General Meeting, 2015.
8. J. Almada, R. Leão, R. Sampaio, and G. Barroso, "A centralized and heuristic approach for energy management of an AC microgrid," *Renewable and Sustainable Energy Reviews*, vol. 60, pp. 1396–1404, 2016.
9. N. Anglani, G. Oriti, and M. Colombini, "Optimized energy management system to reduce fuel consumption in remote military microgrids," 2016 IEEE Energy Conversion Congress and Exposition (ECCE), 2016.
10. M. H. Amrollahi and S. M. T. Bathaee, "Techno-economic optimization of hybrid photovoltaic/wind generation together with energy storage system in a stand-alone micro-grid subjected to demand response," *Applied Energy*, vol. 202, pp. 66–77, 2017.
11. S. Sukumar, H. Mokhlis, S. Mekhilef, K. Naidu, and M. Karimi, "Mix-mode energy management strategy and battery sizing for economic operation of grid-tied microgrid," *Energy*, vol. 118, pp. 1322–1333, 2017.
12. P. P. Vergara, J. C. López, L. C. D. Silva, and M. J. Rider, "Security-constrained optimal energy management system for three-phase residential microgrids," *Electric Power Systems Research*, vol. 146, pp. 371–382, 2017.
13. L. K. Panwar, S. R. Konda, A. Verma, B. K. Panigrahi, and R. Kumar, "Operation window constrained strategic energy management of microgrid with electric vehicle and distributed resources," *IET Generation, Transmission & Distribution*, vol. 11, no. 3, pp. 615–626, 2017.
14. A. Merabet, K. T. Ahmed, H. Ibrahim, R. Beguenane, and A. M. Y. M. Ghias, "Energy Management and Control System for Laboratory Scale Microgrid Based Wind-PV-Battery," *IEEE Transactions on Sustainable Energy*, vol. 8, no. 1, pp. 145–154, 2017.
15. A. C. Luna, L. Meng, N. L. Diaz, M. Graells, J. C. Vasquez, and J. M. Guerrero, "Online Energy Management Systems for Microgrids: Experimental

Validation and Assessment Framework," *IEEE Transactions on Power Electronics*, vol. 33, no. 3, pp. 2201–2215, 2018.
16. J. de la Hoz, H. Martín, A. Alonso, A. C. Luna, J. Matas, J. C. Vasquez and J. M. Guerrero, "Regulatory-framework-embedded energy management system for microgrids: The case study of the Spanish self-consumption scheme", *Applied Energy*, Vol. 251, p. 113374, 2019.
17. W. Su, S. Yu, H. Li, H. Iu and T. Fernando, "An MPC-based Dual-solver Optimization Method for DC Microgrids with Simultaneous Consideration of Operation Cost and Power Loss," in *IEEE Transactions on Power Systems*, doi: 10.1109/TPWRS.2020.3011038.
18. E. Chatterji and M. D. Bazilian. "Smart meter data to optimize combined roof-top solar and battery systems using a stochastic mixed integer programming model", *IEEE Access*, vol. 8, pp. 133843-133853, 2020.
19. W. Violante, C. A. Cañizares, M. A. Trovato and G. Forte, "An Energy Management System for Isolated Microgrids with Thermal Energy Resources", *IEEE Transactions on Smart Grid*, vol. 11, no. 4, pp. 2880-2891.
20. J. Arkhangelski, A. T. Mahamadou and G. Lefebvre, "Day-Ahead Optimal Power Flow for Efficient Energy Management of Urban Microgrid," in *IEEE Transactions on Industry Applications*, doi: 10.1109/TIA.2020.3049117.
21. N. Anglani, G. Oriti, and M. Colombini, "Optimized Energy Management System to Reduce Fuel Consumption in Remote Military Microgrids," *IEEE Transactions on Industry Applications*, vol. 53, no. 6, pp. 5777–5785, 2017.
22. A. Baziar and A. Kavousi-Fard, "Considering uncertainty in the optimal energy management of renewable micro-grids including storage devices," *Renewable Energy*, vol. 59, pp. 158–166, 2013.
23. M. Motevasel, A. R. Seifi, and T. Niknam, "Multi-objective energy management of CHP (combined heat and power)-based micro-grid," *Energy*, vol. 51, pp. 123–136, 2013.
24. M. Marzband, M. Ghadimi, A. Sumper, and J. L. Domínguez-García, "Experimental validation of a real-time energy management system using multi-period gravitational search algorithm for microgrids in islanded mode," *Applied Energy*, vol. 128, pp. 164–174, 2014.
25. K. Huang, X. Wang, and L. Wang, "Optimal energy management of grid-connected photovoltaic micro-grid," 2015 IEEE International Conference on Cyber Technology in Automation, Control, and Intelligent Systems (CYBER), 2015.
26. S. A. Alavi, A. Ahmadian, and M. Aliakbar-Golkar, "Optimal probabilistic energy management in a typical micro-grid based-on robust optimization and point estimate method," *Energy Conversion and Management*, vol. 95, pp. 314–325, 2015.
27. V. Mohan, J. G. Singh, and W. Ongsakul, "An efficient two stage stochastic optimal energy and reserve management in a microgrid," *Applied Energy*, vol. 160, pp. 28–38, 2015.

28. M. Marzband, N. Parhizi, and J. Adabi, "Optimal energy management for stand-alone microgrids based on multi-period imperialist competition algorithm considering uncertainties: experimental validation," *International Transactions on Electrical Energy Systems*, vol. 26, no. 6, pp. 1358–1372, 2015.
29. M. Marzband, E. Yousefnejad, A. Sumper, and J. L. Domínguez-García, "Real time experimental implementation of optimum energy management system in standalone Microgrid by using multi-layer ant colony optimization," *International Journal of Electrical Power & Energy Systems*, vol. 75, pp. 265–274, 2016.
30. H. Li, A. T. Eseye, J. Zhang, and D. Zheng, "Optimal energy management for industrial microgrids with high-penetration renewables," *Protection and Control of Modern Power Systems*, vol. 2, no. 1, Aug. 2017.
31. A. Askarzadeh, "A Memory-Based Genetic Algorithm for Optimization of Power Generation in a Microgrid," *IEEE Transactions on Sustainable Energy*, vol. 9, no. 3, pp. 1081–1089, 2018.
32. M. Marzband, S. S. Ghazimirsaeid, H. Uppal, and T. Fernando, "A real-time evaluation of energy management systems for smart hybrid home Microgrids," *Electric Power Systems Research*, vol. 143, pp. 624–633, 2017.
33. B. Papari, C. Edrington, T. V. Vu, and F. Diaz-Franco, "A heuristic method for optimal energy management of DC microgrid," 2017 IEEE Second International Conference on DC Microgrids (ICDCM), 2017.
34. S. A. Arefifar, M. Ordonez, and Y. Mohamed, "Energy Management in Multi-Microgrid Systems - Development and Assessment," *IEEE Transactions on Power Systems*, pp. 1–1, 2016.
35. S. Sukumar, M. Marsadek, A. Ramasamy and H. Mokhlis, "Grey Wolf Optimizer Based Battery Energy Storage System Sizing for Economic Operation of Microgrid," *2018 IEEE International Conference on Environment and Electrical Engineering* and *2018 IEEE Industrial and Commercial Power Systems Europe (EEEIC / I&CPS Europe), Palermo*, pp. 1–5, 2018.
36. M. A. Hossain, H. R. Pota, S. Squartini and A. F. Abdou, "Modified PSO algorithm for real-time energy management in grid-connected microgrids," *Renewable Energy*, vol. 136, pp. 746–757, 2019.
37. L. Bhamidi and S. Sivasubramani, "Optimal Sizing of Smart Home Renewable Energy Resources and Battery Under Prosumer-Based Energy Management," *IEEE Systems Journal*, doi: 10.1109/JSYST.2020.2967351.
38. H. E. Keshta, O. P. Malik, E. M. Saied, F. M. Bendary and A. A. Ali, "Energy management system for two islanded interconnected micro-grids using advanced evolutionary algorithms" *Electric Power Systems Research*, vol. 192, p. 106958, 2020.
39. X. Wang, X. Mao and H. Khodaei, "A multi-objective home energy management system based on internet of things and optimization algorithms," *Journal of Building Engineering*, vol. 33, p. 101603, 2021.

40. F. A. Mohamed and H. N. Koivo, "Multiobjective optimization using modified game theory for online management of microgrid," *European Transactions on Electrical Power*, vol. 21, no. 1, pp. 839–854, May 2010.
41. G. Kyriakarakos, A. I. Dounis, K. G. Arvanitis, and G. Papadakis, "A fuzzy logic energy management system for polygeneration microgrids," *Renewable Energy*, vol. 41, pp. 315–327, 2012.
42. E. Kuznetsova, Y.-F. Li, C. Ruiz, E. Zio, G. Ault, and K. Bell, "Reinforcement learning for microgrid energy management," *Energy*, vol. 59, pp. 133–146, 2013.
43. H. S. V. S. Kumar Nunna and S. Doolla, "Energy Management in Microgrids Using Demand Response and Distributed Storage—A Multiagent Approach," *IEEE Transactions on Power Delivery*, vol. 28, no. 2, pp. 939–947, 2013.
44. G. E. Asimakopoulou, A. L. Dimeas, and N. D. Hatziargyriou, "Leader-Follower Strategies for Energy Management of Multi-Microgrids," *IEEE Transactions on Smart Grid*, vol. 4, no. 4, pp. 1909–1916, 2013.
45. J. P. Fossati, A. Galarza, A. Martín-Villate, J. M. Echeverría, and L. Fontán, "Optimal scheduling of a microgrid with a fuzzy logic controlled storage system," *International Journal of Electrical Power & Energy Systems*, vol. 68, pp. 61–70, 2015.
46. M. E. G. Urias, E. N. Sanchez, and L. J. Ricalde, "Electrical Microgrid Optimization via a New Recurrent Neural Network," *IEEE Systems Journal*, vol. 9, no. 3, pp. 945–953, 2015.
47. C. S. Karavas, G. Kyriakarakos, K. G. Arvanitis, and G. Papadakis, "A multi-agent decentralized energy management system based on distributed intelligence for the design and control of autonomous polygeneration microgrids," *Energy Conversion and Management*, vol. 103, pp. 166–179, 2015.
48. T. Bogaraj, J. Kanakaraj, "Intelligent energy management control for independent microgrid," *Sādhanā*, vol. 41, no. 7, pp. 755–769, 2016.
49. G. K. Venayagamoorthy, R. K. Sharma, P. K. Gautam, and A. Ahmadi, "Dynamic energy management system for a smart microgrid," *IEEE Transaction of Neural Network and Learning System*, vol. 27, no. 8, pp. 1643–1656, 2016.
50. L. Ma, N. Liu, J. Zhang, W. Tushar, and C. Yuen, "Energy Management for Joint Operation of CHP and PV Prosumers Inside a Grid-Connected Microgrid: A Game Theoretic Approach," *IEEE Transactions on Industrial Informatics*, vol. 12, no. 5, pp. 1930–1942, 2016.
51. T. Wang, X. He, and T. Deng, "Neural networks for power management optimal strategy in hybrid microgrid," *Neural Computing and Applications*, 2017.
52. D. Arcos-Aviles, J. Pascual, F. Guinjoan, L. Marroyo, P. Sanchis, and M. P. Marietta, "Low complexity energy management strategy for grid profile smoothing of a residential grid-connected microgrid using generation and demand forecasting," *Applied Energy*, vol. 205, pp. 69–84, 2017.
53. S. Ghorbani, R. Rahmani, and R. Unland, "Multi-agent Autonomous Decision Making in Smart Micro-Grids' Energy Management: A Decentralized

Approach," *Multiagent System Technologies Lecture Notes in Computer Science*, pp. 223–237, 2017.
54. N. Liu, X. Yu, C. Wang, and J. Wang, "Energy Sharing Management for Microgrids With PV Prosumers: A Stackelberg Game Approach," *IEEE Transactions on Industrial Informatics*, vol. 13, no. 3, pp. 1088–1098, 2017.
55. S. Javaid, N. Javaid, S. Iqbal, M. Guizani, A. Al-Mogren and A. Alamri, "Energy Management with a World-Wide Adaptive Thermostat Using Fuzzy Inference System," *IEEE Access*, vol. 6, pp. 33489–33502, 2018.
56. K. Mansiri, S. Sukchai and C. Sirisamphanwong, "Fuzzy Control Algorithm for Battery Storage and Demand Side Power Management for Economic Operation of the Smart Grid System at Naresuan University, Thailand," *IEEE Access*, vol. 6, pp. 32440–32449, 2018.
57. S. K. Howell, H. Wicaksono, B. Yuce, K. McGlinn and Y. Rezgui, "User Centered Neuro-Fuzzy Energy Management Through Semantic-Based Optimization," *IEEE Transactions on Cybernetics*, vol. 49, no. 9, pp. 3278-3292, 2019.
58. Y. Du and F. Li, "Intelligent Multi-Microgrid Energy Management Based on Deep Neural Network and Model-Free Reinforcement Learning," *IEEE Transactions on Smart Grid*, vol. 11, no. 2, pp. 1066-1076, 2020.
59. R. R. Deshmukh, M. S. Ballal and H. M. Suryawanshi, "A Fuzzy Logic Based Supervisory Control for Power Management in Multibus DC Microgrid," *IEEE Transactions on Industry Applications*, vol. 56, no. 6, pp. 6174–6185, 2020.
60. D. Arcos-Aviles *et al.*, "An energy management system design using fuzzy logic control: Smoothing the grid power profile of a residential electro-thermal microgrid," *IEEE Access*, doi: 10.1109/ACCESS.2021.3056454.
61. V. S. Tabar, M. A. Jirdehi, and R. Hemmati, "Energy management in microgrid based on the multi objective stochastic programming incorporating portable renewable energy resource as demand response option," *Energy*, vol. 118, pp. 827–839, 2017.
62. Z. Luo, W. Gu, Z. Wu, Z. Wang, and Y. Tang, "A robust optimization method for energy management of CCHP microgrid," *Journal of Modern Power Systems and Clean Energy*, vol. 6, no. 1, pp. 132–144, 2017.
63. B. V. Solanki, A. Raghurajan, K. Bhattacharya, and C. A. Canizares, "Including Smart Loads for Optimal Demand Response in Integrated Energy Management Systems for Isolated Microgrids," *IEEE Transactions on Smart Grid*, vol. 8, no. 4, pp. 1739–1748, 2017.
64. Z. Luo, Z. Wu, Z. Li, H. Cai, B. Li, and W. Gu, "A two-stage optimization and control for CCHP microgrid energy management," *Applied Thermal Engineering*, vol. 125, pp. 513–522, 2017.
65. W. Hu, P. Wang, and H. B. Gooi, "Toward Optimal Energy Management of Microgrids via Robust Two-Stage Optimization," *IEEE Transactions on Smart Grid*, vol. 9, no. 2, pp. 1161–1174, 2018.

66. A. Ghasemi, "Coordination of pumped-storage unit and irrigation system with intermittent wind generation for intelligent energy management of an agricultural microgrid," *Energy*, vol. 142, pp. 1–13, 2018.
67. T. Pippia, J. Sijs and B. De Schutter, "A single-level rule-based model predictive control approach for energy management of grid-connected microgrids", *IEEE Transactions on Control Systems Technology*, vol. 28, no. 6, pp. 2364–2376, 2019.
68. D. Prudhviraj, P. B. S. Kiran and N. M. Pindoriya, "Stochastic Energy Management of Microgrid with Nodal Pricing," in *Journal of Modern Power Systems and Clean Energy*, vol. 8, no. 1, pp. 102–110, 2020.
69. M. Elkazaz, M. Sumner and D. Thomas, "Energy management system for hybrid PV-wind-battery microgrid using convex programming, model predictive and rolling horizon predictive control with experimental validation" *International Journal of Electrical Power & Energy Systems*, vol. 115, p. 105483, 2020.
70. P. Li, W. Sheng, Q. Duan, Z. Li, C. Zhu and X. Zhang, "A Lyapunov Optimization-Based Energy Management Strategy for Energy Hub with Energy Router," *IEEE Transactions on Smart Grid*, vol. 11, no. 6, pp. 4860–4870, 2020.
71. M. Ahrarinouri, M. Rastegar and A. R. Seifi, "Multiagent Reinforcement Learning for Energy Management in Residential Buildings," *IEEE Transactions on Industrial Informatics*, vol. 17, no. 1, pp. 659–666, 2021.
72. K. A. Pruitt, R. J. Braun and A. M. Newman, "Establishing conditions for the economic viability of fuel cell-based, combined heat and power distributed generation systems", *Applied Energy*. Vol. 111, pp. 904–20, 2013.

6

A Systematic Approach for Solar and Hydro Resource Assessment for DC Microgrid Applications

Sanjay Kumar[1], Nikita Gupta[1]*, Vineet Kumar[2] and Tarlochan Kaur[3]

[1]Electrical Engineering Department, UIT, HPU, Shimla, India
[2]Electrical Engineering Department, NIT Hamirpur, H.P. India, India
[3]Electrical Engineering Department, PEC Deemed to be University Chandigarh, India

Abstract

All the major activities of the development and human growth of a nation depend upon energy. The growth in energy requirement leads oil and mining industries to extract large amounts of fossil fuels. These fossil fuels produce a large amount of CO_2 emission leading to adverse environmental effects such as greenhouse gas (GHG) emission, global warming, climate change, etc. Even the use of these fossil fuels is not able to bridge the gap between demand and supply. Therefore, the government has given great attention to exploring DC microgrids based on renewable energy (RE) resources for power generation in urban as well as rural remote areas. Solar, hydro and wind, etc., are the best renewable alternate options to meet the local demand of electricity as per availability of these resources depending upon geographical conditions in local power generation. The major problems with these energy resources are their intermittent behaviour and dependency on weather conditions. Therefore, the reliability and efficiency of DC microgrids can be improved with the integrating of these resources in an effective manner. The integration of these resources is known as a hybrid renewable energy system (HRES). To meet the main objective, locally available resources like solar need to be investigated for DC microgrid application. Designing a HRES for locality and its execution is a tedious task, as it requires parameters such as available solar radiation, available hydro potential, and demand assessment, etc., for DC microgrid application. The present study will be carried out to develop a systematic approach for solar and hydro potential assessment for DC microgrid application.

**Corresponding author*: guptanikita08@gmail.com

Nikita Gupta, Mahajan Sagar Bhaskar, Sanjeevikumar Padmanaban and Dhafer Almakhles (eds.)
DC Microgrids: Advances, Challenges, and Applications, (159–182) © 2022 Scrivener Publishing LLC

Keywords: Renewable energy, solar PV, hydro, hybrid renewable energy system, DC microgrid

6.1 Introduction

Energy is an indispensable input for human activities that offer a driving force for societal growth. The developing countries, especially less developed countries, are energy poor. The International Energy Agency assessed that 1.2 million people still have no access to electricity. The per capita use of energy has significantly increased with a compound annual growth rate of 3.54% during the years 2011-18 [1]. The energy need is still fulfilled by non-renewable sources like crude oil, coal, natural gas, etc. Figure 6.1 shows that coal is the main energy source, accounting for approximately 70.69% of the total output production during the year 2017-18. Crude oil (10.34%) is in second position, while natural gas (8.7%) is India's third main energy source [2].

However, in the present situation, DC-based microgrids are now trending and becoming a sustainable solution for new electrical systems. A generalised schematic of a DC microgrid is shown in Figure 6.2. DC power systems have inherent advantages of no harmonics, no reactive power, and high efficiency over the conventional AC power networked systems. Hence, DC power systems have become an emerging and promising alternative in various emerging applications, which include distributed energy sources like wind, solar and Energy Storage System (ESS); distribution networks; smart buildings, remote telecom systems; and transport electrification like electric vehicles (EVs). All these applications are designed at different voltages to meet their specific requirements individually because of the lack of standardization. Resource estimation for the sizing of renewable energy sources (RES) like solar and micro hydro-based hybrid renewable energy

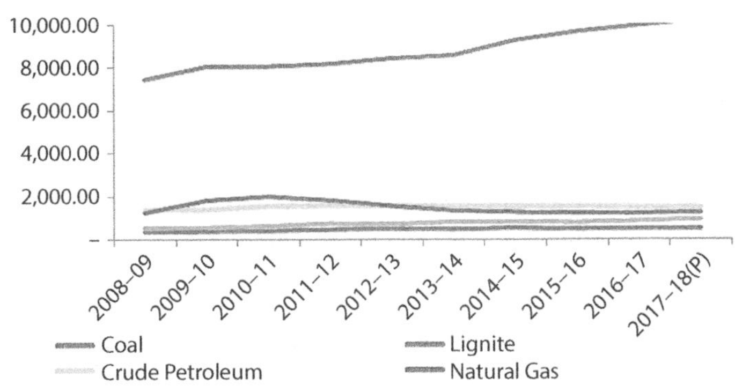

Figure 6.1 Trends in production of energy by non-renewable in India. Source: [2].

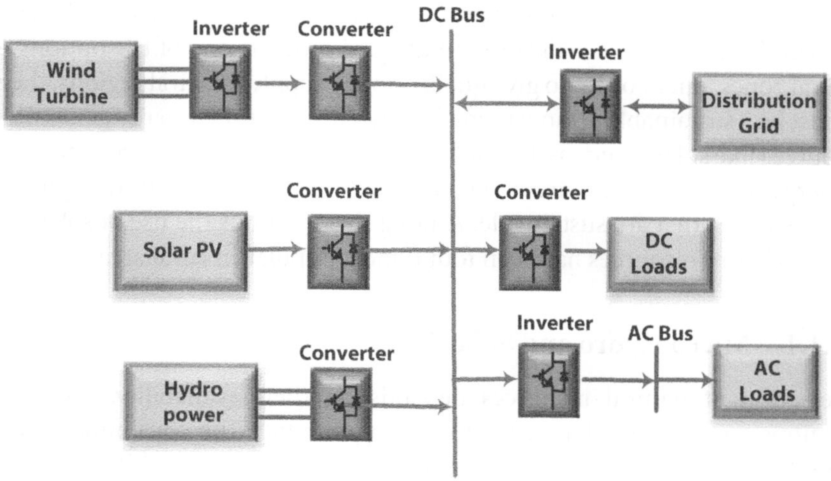

Figure 6.2 Schematic diagram of a DC microgrid.

sources (HRES) is also one of the important criteria for DC microgrids application. In this research a systematic approach for resource estimation has been presented in respect of the application of DC microgrids.

The available solar radiation at a particular location depends upon several geographical and meteorological parameters like pressure, temperature, humidity, rainfall latitude and longitude, etc. [3]. Recently, the use of an Artificial Neural Network (ANN) has been found to be advantageous for solar radiation estimation [4] as these have the capability to solve highly complicated non-convex problems in an easier manner. The rural development of any country is related to access to electricity, but despite that less developed countries are still struggling with continuous electric power availability in rural areas. Electrical energy in a rural community has more significance as activities like education, health, agriculture and local industries are totally dependent on electrical energy for their growth and progress [5]. The absence of electricity in rural areas also leads to urban migration. To stop large migration from rural areas the challenge is to provide electricity with the lowest possible cost. The major barriers for rural electrification are [6]:

(i) Rural population is remote and scattered.
(ii) People living in rural areas have low incomes and are not able to pay for use of electricity.
(iii) Extension of grid is complex and expensive.
(iv) High cost of transmission and distribution.
(v) High fuel transportation cost.

To mitigate these barriers, there is a need to fulfil the electricity demand in rural areas. DC electrification in off-grid mode is one of the options in rural zones, but in order to give quality power supply for rural regions, there must be a sustainable arrangement for power generation. Energy generation sources like solar, wind, hydro, tidal and biogas, biomass, etc., are renewable energy sources [5] and are being considered as alternative power generation sources that are sustainable. Among these options the use of solar and micro hydro resources has been found most popular in rural areas.

6.1.1 Micro Hydro and Solar PV

Selection of natural resources depends on their availability. The hilly regions are generally blessed with hydro potential. The continuous availability of water is known as a perennial in nature. Sun is the ultimate source of energy or electricity generation which directly or indirectly converts light energy into electricity. In its direct form solar energy is utilized using solar thermal heaters, solar collectors [7], etc., and in indirect form converts light energy into electric energy through solar photovoltaic (SPV) system. Figure 6.3 shows a schematic of HRES consisting of a micro hydro power generator and a solar PV power generator.

6.1.2 Renewable Energy for Rural Electrification in Indian Perspective

The government of India is taking various initiatives to encourage the use of renewable energy resources. Use of renewable energy sources is enabled

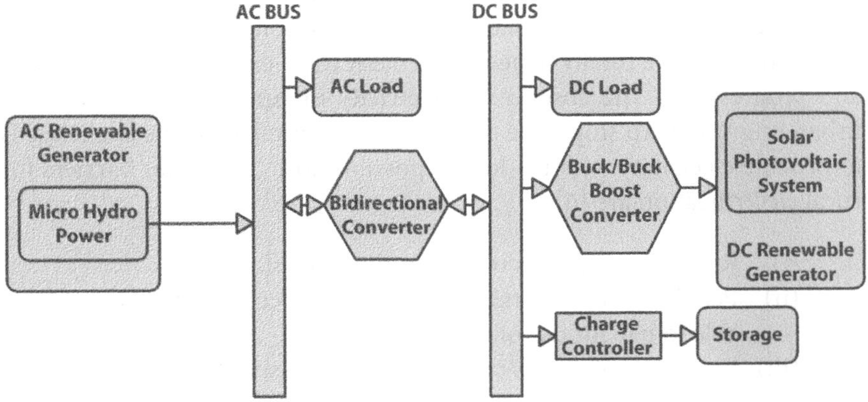

Figure 6.3 Schematic of hybrid renewable energy system.

through various schemes, policy making and implementation of projects. These renewable energy systems are solar lantern [8], water pumping [9], solar cooker [10], off-grid solar electricity [11], biomass cook stoves [12] and gasifier-based systems [13], setting up of biogas plants [14] and manure management program [15] energy from industrial/agricultural wastes [16], development and upgradation of watermills and setting up of micro hydro plant.

The government of India aims to attain 100 GW renewable energy capacity in 2022, of which 73.35 GW has been installed in the country [1]. The government has taken several steps to achieve this goal by setting RE objectives for the states. The renewable purchase obligation (RPO) and solar purchase obligation (SPO) have been set up by each state to meet the demand from renewable energy resources. The main motive of the state governments in promoting these RE polices is to empower the people in rural and remote areas. Therefore, governments are promoting locally available resources, especially solar and micro hydro, through various energy polices. Biomass and wind are also promoted but availability of wind is much lower and is restricted in some areas due to geographical location. To enhance the potential of available RE, technology in state proper planning and techno-economic analysis is required. Resource assessment and sizing optimization for HRES is site-specific and needs to be explored before erecting an energy system [17].

6.1.3 Solar Resource Assessment

Solar energy is among the most abundant energy sources in the world and has the potential to meet the demand of required energy. In water treatment, water heating, industrial process heating, solar cooking besides thermal and photovoltaic-based power generation, solar energy finds applications. The accessibility of solar radiation needs to be confirmed prior to installing solar-based systems at any location. Precise measurement of solar radiation needs instruments like pyrometer and pyrheliometer. However, it is not possible and practical to install such costly instruments at every location. Therefore, it is essential to estimate the solar radiation through other ways where routine measurement using these instruments is not possible. Researchers have suggested multiple models to find associations between different meteorological parameters to deal with this restriction such as temperature, humidity, pressure and the solar radiations [18, 19]. Many empirical methods such as the Angstrom-Prescott model, sunshine model, temperature model, and rainfall model [20–22] have been used by various researchers.

However, in the present era machine learning techniques [23–25] are the most popular. The Artificial Neural Network (ANN) is one of the most common machine learning techniques and widely used for solar radiation estimation. ANN is an efficient and robust solar radiation estimation technique since it can more effectively deal with broad, complex and nonlinear problems [26]. The estimation of daily, monthly and yearly global solar radiation has been estimated by using parameters like ambient temperature, relative humidity, sunshine duration, longitude, latitude, clearness index, total cloud cover, average beam radiation, evaporation, precipitation, pressure, altitude, etc. [27].

The average value of global solar radiation (hourly, daily and monthly) data plays an important role in the growth, dimensioning and performance assessment of the device in many solar energy applications and several scholars have carried out studies using ANN to estimate solar capacity [26–32]. Geographical- and meteorological-based parameters like precipitation, sunshine duration, temperature, relative humidity, clearness index, evaporation, wind speed, longitude, altitude, and latitude, etc., were considered by various researchers for solar potential estimation.

Meteorological parameters were used for the solar potential estimation in Himachal [26]. The authors have tried various possible combinations of the input parameters to find the most suitable combination of the input parameters. Finally, the combination consisting of all input parameters was found more suitable for solar radiation estimation. The authors extended their study to find the importance of input parameter for solar radiation estimation [30]. The estimated solar potential in the proposed study is further used by the authors for designing of the microgrid in rural areas.

Three years' data from 2000 to 2003 were used for training and testing of networks. The scaled conjugate gradient (SCG), and Levenberg-Marquardt (LM) training approaches were used for the solar radiation assessment for 17 cities in regions in Turkey using artificial neural networks. Meteorological and geographical limitations such as latitude, altitude, longitude, month, mean temperature, mean sunshine duration, and relative humidity stayed as input and output in the form of solar radiations in the network. Different combinations of hidden layer and neurons were used for optimal accuracy of system. The mean absolute percentage error (MAPE) was used to validate the model. The value of the MAPE was obtained to be 6.73% [33]. The same input parameters were used by ref. [34] to estimate solar radiation for Cyprus. The back-propagation feed forward network was trained using seven years' data from 1986 to 1992 collected at Athalassa, Cyprus. The one and two hidden layer network architectures were used to estimate the solar radiation by varying hidden layer neuron.

The model with two hidden layers with 23 neurons in the hidden layer was found to be the best. The root mean square error (RMSE) and mean biased error (MBE) were 5.67% and 0.12%, respectively.

Fadre (2009) [35] used an artificial neural network for the estimation of solar energy capacity in Nigeria. Meteorological and geographical parameters like altitude, latitude, longitude, mean sunshine period, mean temperature, month and humidity were used as input to the ANN model and solar radiation as output for network. A total of 11,700 data set observations were used for training and 5850 datasets were used for both testing and validation of the network. Finally, ANN architecture with seven input parameters, five hidden layer neurons and one output (7/5/1) was found more appropriate with a correlation coefficient (R) value of 0.956.

Several input parameters like latitude, longitude, altitude, month of the year, time, temperature, relative humidity, precipitation, wind velocity and long wavelength were used for developing ANN-based models for solar radiation estimation. By taking different combinations of input constraints for each model, 16 combinations were evaluated. The combination of longitude, latitude, altitude, month of year, time, relative humidity, temperature, and long wave length parameters in the winter season showed better performance with nine hidden layer neurons in the ANN architecture with coefficient of correlation (R^2) value of 0.923 [28].

Mishra *et al.* (2008) [36] used mean sunshine per hour, rainfall, month of year, relative humidity, latitude, and longitude to estimate the direct solar radiation for India region. The averaged data of 20 years for the period 1957–1980 were used for training, testing and validations of the six neural networks. Out of six networks, five were taken as feed forward multilayer perceptron network and one network was radial basis function network (RBF). It was concluded that feed forward network performed better than the radial basis function network. For the feed-forward back propagation network, the obtained ranges for root mean square error (RMSE) were found to be between 0.8 and 5.4% and for the radial basis function, RMSE ranges were found between 7 and 29%.

Ref. [37] used day-to-day maximum temperature (T_{max}), difference of daily lower and upper temperature (DT), daily minimum temperature (T_{min}), sunshine hours (S), theoretical sunshine hours (S_o), and extra-terrestrial radiation (H_o) to train the neural network. Six ANN models have been developed by using 32 combinations of input parameters for predicting the monthly mean daily worldwide solar radiation more accurately. The authors concluded that two or three variables are sufficient to predict solar radiation estimation. The combination of DT, H_o and DT, S_o, H_o parameters were found to be the most influential combination for solar radiation estimation. The minimum MAPE obtained for the DT, H_0 combination was evaluated to be 2.61%. From the literature, it can be concluded that ANN can be used for the potential assessment of the solar radiation estimation before designing a grid.

6.1.4 Hydro Resource Assessment

Generally, hydro power plants are classified as very small, small, medium and large. However, different countries have their own variants of classification. Ireland, Portugal, Belgium, Spain and Greece agree to take 10 MW as the higher limit for a small hydro plant, whereas Italy takes this limit up to 3 MW (fixed) and Sweden 1.5 MW. The UK takes 20 MW as the threshold for a small hydro power plant. Thus different countries have different requirements to identify a hydro power plant. In India, general classification of hydro power plants is shown in Table 6.1.

The micro hydro, mini hydro and pico hydro power plant play important roles to strengthen the stand-alone HRES for rural electrification system. The Micro Hydro Power (MHP) plant has several merits as compared to other renewable power generation resources. MHP generation is not so intermittent. It is not strongly affected by weather conditions, and CO_2 emission by these plants are negligible. The additional advantage is that unlike thermal power plants, MHP does not require high starting time and advance techniques to operate [38].

Hydro potential assessment requires flow rate and available head. Generally, discharge can be measured at gauge station; however, it is not possible to install a gauge station at every site due to high cost of instruments. According to the World Meteorological Organization, the flow

Table 6.1 Classification of hydro power plants in India.

Type	Capacity of the hydro power plant
Micro	100 kW & below
Pico	5 kW & below
Small	below 25000 kW
Mini	below 2000 kW
Large	100,000 kW & above
Medium	below 100,000 kW

measurement can be carried out by various methods like statistical, empirical and Rain-fall-Runoff modeling [39]. All these techniques require a huge number of discharge data, which is not available for each site. The studies related to small hydro potential assessment have been already carried out for specific regions [40–42]. However, there is no specific technique available for micro and mini hydropower potential assessment. Flow rate for micro hydro ungauged site can be assessed by the empirical estimation method [17] of Eq. (6.1):

$$Q_{site} = \left[\frac{A_{site}}{A_{guge}}\right] Q_{guge} \tag{6.1}$$

The micro hydro power generation can be assessed using the product of gross head and discharge using equation:

$$P_{\mu H}(t) = \eta_{\mu H}\, Q(t) * g * H_{\mu H}\,(KW) \tag{6.2}$$

Where $P_{\mu H}$ is micro hydro power, $\eta_{\mu H}$ efficiency of turbine, Q flow rate, $H_{\mu H}$ head of micro hydro plant.

6.1.5 Demand Assessment

The main challenge in rural electrification is to precisely estimate the demand of the rural villages. However, it is still a difficult task to create real load projections, mainly because the users have little experience of or exposure to electricity uses. For rural electrification development, essential information regarding load demand is gathered through interview and field survey in desired villages by asking the villagers their desired demand and willingness-to-pay [43, 44]. The cost of energy generated can be reduced by demand side management. Load shifting is the most commonly used demand side management in which the load demand is shifted from peak load period to off-peak load period. The main purpose of a load shifting process is cost saving. The load shifting process is performed by shifting deferrable load where electrical energy is stored and bought during the low prices of electricity [45].

In the present study, a case study has been taken for the resource assessment of solar and micro hydro resource.

6.2 Methodology

6.2.1 Data Collection

The following variety of data from different sources need to be collected and analyzed for potential assessment:

(i) Meteorological data for solar radiation estimation.
(ii) Discharge data for hydro potential estimation.
(iii) Ancillary data for site selection of Solar PV plant.

6.2.1.1 Meteorological and Geographical Data

For the estimation of solar radiation in Himachal Pradesh, meteorological and geographical data from the Chamba district was used.

The average monthly humidity (%), data temperature (°C), wind speed (m/sec.) and pressure (kPa), was obtained from National Aeronautics and Space Administration (NASA) at Langley Research Centre. The global solar radiation data for 23 cities were taken for the period 1986 to 2000 from the solar radiation handbook published by the Indian Metrological Department (IMD) and the Ministry of New and Renewable Energy (MNRE). The data for one district, Hamirpur, was obtained from the Centre for Energy and Environment, NIT, Hamirpur H.P., India. The GSR data for Hamirpur is continuously measured by KIPP and ZONEN CM through four high-temperature pyranometers over a one-minute time step.

6.2.1.2 Discharge Data for Hydro Potential Estimation

Hydro data for the period 1979 to 2008 has been collected from the gauge site in the Chamba district. However, the flow rate data for the selected site of hydro is not available. Therefore, the data of the available gauge site has been collected.

Figure 6.4 provides a diagram of the methods followed in the present analysis. The detailed description of each step has been given at relevant places in the sections to follow. In the following section, a detailed description of the methods adopted in the present study has been given.

Step I - Theoretical Potential Investigation
A field visit was made to see the available option of renewable energy resources for rural electrification in the present study area. Solar and hydro

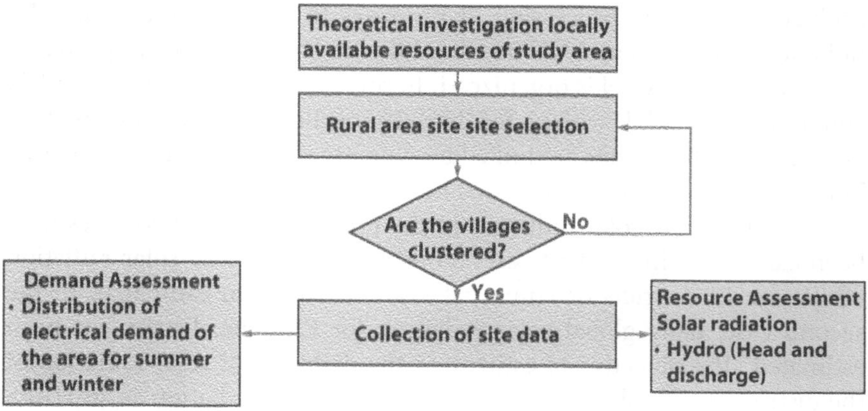

Figure 6.4 A schematic of methodology adopted.

resources are found in a theoretical investigation. The survey of the area quantifies forest cover, agricultural yield, cattle population, availability of solar radiation and hydro discharge, etc., related information. Villages are selected based on the clustering, medium population size and located in distant region. During the field visit geographical details, population size and basic information about general population is collected.

Step II - Demand assessment

Domestic loads (i.e., household), community loads (i.e., schools, panchayat office etc.), commercial loads (i.e., shops) and industrial loads are the major forms of load found in the region. The demand assessment also categorizes the type of appliances people used.

Step III - Resource assessment

Resource assessment includes determining the locally available resources to fulfill the demand. It includes solar and hydro resource potential. An ANN model for solar resource and an empirical model for hydro resource are developed.

Details of Methods or Approaches Used

Solar Radiation Estimation Approach

Renewable resource estimation has attracted the attention of power planners, power companies, governments and others. Several solar resource estimation methods for resource estimation have been used. Here in this work, for solar radiation estimation, the ANN technique has been developed. ANN is a useful technique for pattern recognition, classification problem and image processing. It promises to be appropriate for solar radiation estimation as ANN has the ability to handle nonlinear problems more easily [33]. ANN is organized into three layers, input (I), output (O) and hidden layers (H) with no feedback connections. Each layer consists of neurons and all neurons are linked by synaptic weights to each other (w_{ij}). The ANN architecture is illustrated in Figure 6.5.

To measure the solar radiation in the present analysis, two cases have been examined. In the first case, monthly average global solar radiation (GSR) for 12 districts of Himachal Pradesh has been estimated. In the second case, daily global solar radiation for Chamba district has been estimated. The objective of estimating the daily global solar radiation of the district of Chamba is to use the estimated solar radiation as an HRES design input.

To train different ANN architectures, the training data set is used while the testing data set is used to evaluate the performance of the trained network. When the training process is finished, the validation dataset is used to verify the generalization potential of the trained network (Hagan *et al.*, 1996). Input neuron data is processed to generate output neurons through hidden neurons. In this method, a feed-forward multilayer neural network is commonly used. Each neuron computes the weighted sum of *n* signals h_k for hidden layer $k = 1, 2....n$, w_{jk} is the weight asset of linked neuron j to k and Q_o is the output signal expressed in Eq. (6.3) [33]:

$$Q_o = \sum_{k=0}^{n} w_{jk} h_k \tag{6.3}$$

The network output value for the output neuron (o), Q_0, is obtained in the neural network feed-forward method. The error function (E), is a

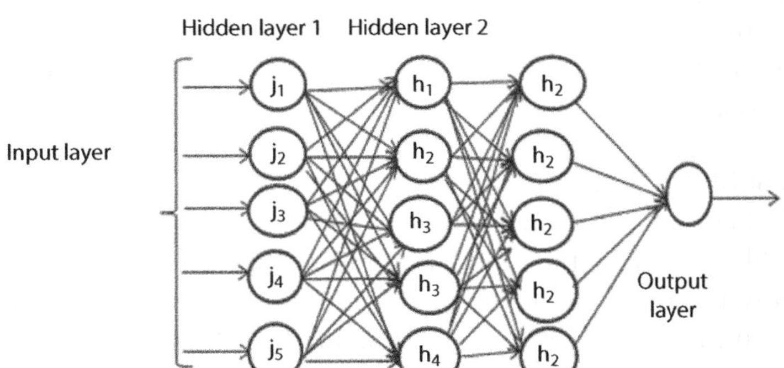

Figure 6.5 Schematic of ANN architecture.

measure of the effectiveness of the network for the output layer processing elements and can be defined as follows [32]:

$$E = \frac{1}{2} \Sigma_{o=1}^{n} (D_0 - Q_0)^2 \tag{6.4}$$

where D_0 is the chosen output vector, Q_0 is the system output vector and n is the quantity of training trials (Arora *et al.*, 2004). The defect is back-propagated through the neural network and reduced by modifying the synaptic weights of different layers between neurons. This can be done based on back propagation learning through certain learning algorithms.

Gradient descent and gradient ascent are the most frequently used back-propagation algorithms with momentum. These are also very sluggish for problem solving. Normal numerical optimizers such as conjugate gradient, quasi-Newton and Levenberg-Marquardt methods are used for faster algorithms [1, 23]. Without having to calculate the Hessian matrix (i.e., the second derivatives of the output index of weights), the Levenberg-Marquardt algorithm was designed to reach the second-order training speed like quasi-Newton techniques. This algorithm uses the following approach to the Hessian matrix.

$$w_{ij+1} = w_{ij} - (J^T J + \mu I)^{-1} J^T E \tag{6.5}$$

where J is the Jacobian matrix that holds first derivatives of the system error regards to connection weights, w_{ij} is a trajectory of current connection weights, μ is a scalar and E is a course of network errors. The Levenberg-Marquardt algorithm does not consider learning rate and momentum factor as its parameters, unlike good gradient algorithms. The relevant training parameters of μ are taken into account. For each effective step (reduction in error), the parameter μ is decreased by multiplying it with mu_dec and has been increased if indeed the value of error is increased. After giving an initial random value, the key scalar parameter μ is updated in an adaptive fashion. In the present analysis, the Levenberg-Marquardt algorithm was used to train the neural network. The back-propagation error is replicated recursively till the error is reduced over the training dataset to an appropriate value. Many neural network architectures have been trained using this algorithm. Depending on the mean absolute percentage error (MAPE) value, the optimal neural network architecture for the dataset is defined.

6.3 Result and Discussion

To estimate the daily global solar radiation of Chamba, a feed forward back-propagation multi-layer artificial neural network with one input layer, two hidden layers and one output layer was considered. The input layer contains five neurons; each represents input limits relative humidity (RH), temperature (T), precipitation (PT), pressure (P) and clearness index (CI) for Chamba.

6.3.1 ANN Architecture

A schematic representation of ANN architecture, adopted in this case, is given in Figure 6.6. Initially, by varying the number of neurons in the secret layers, a few ANN architectures were designed. These networks were tested and trained to compute their performance. The number of networks was subsequently increased and 50 ANN architectures were developed, trained and tested in the present case to determine their generalization capabilities and accuracies.

Data used and Preparation

In the current analysis, the input signal for the ANN control unit separates five distinct process variables, stated as relative humidity (RH), temperature (T), precipitation (PT) pressure (P), and clarity index (CI) and the global daily solar radiation generation from the seven-year NASA site from January 2008 to December 2014.

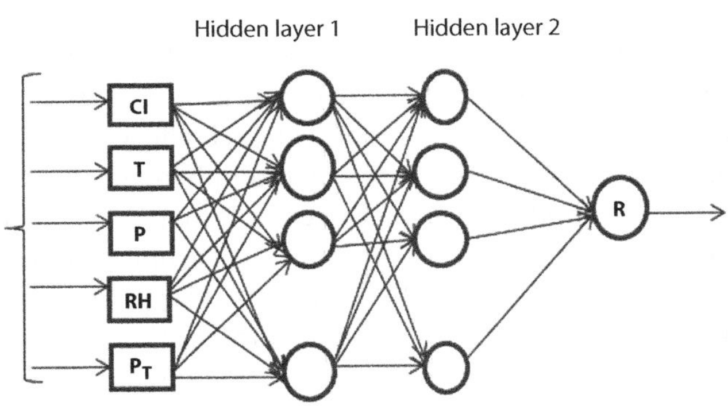

Figure 6.6 The proposed architecture of the artificial neural network.

Training and Testing of ANN

A total of 2,557 data samples were distributed randomly for solar radiation estimation. Of the total data set of 2557, 1,535 (60%) were validated for preparation, 511 (20%) were used for testing purpose and the residual 511 (20 %) were used for validation purpose. When the network output reached an acceptable amount and the mean square error (E) converged to the minimum value taken into account in the present study (0.05), the training process was discontinued.

Result of Case II

Initially, the network consisting of a single hidden layer was used to find the accuracy of the training and testing dataset. One of these network architectures and one secret layer were 0.67 and 0.69, respectively, after conducting multiple experiments with five inputs. For the best performance layer, the MAPE achieved, the finest architecture found was 5/10/1. The correlation coefficient values attained for the 5/10/1 single hidden layer network training and testing were 15.08%. After that, they suggested systems with two secret layers. The performance of the 50 networks was calculated both through training and testing in terms of correlation coefficient and MAPE. The training and testing accuracies in terms of correlation coefficient and MAPE for the 50 networks are represented in Table 6.2.

Figure 6.7 demonstrates the regression scheme (the one that resulted in the best performance) for the 5/48/15/1 network architecture for both training and testing. The regression plot gives the relationship between output and target values of the best network architecture with 5/48/15/1 for potential estimation. The value of regression (R) represents the difference in the outputs and targets. Likewise, the regression plot represents the connection between performance and target values for evaluating 512 datasets. In Figure 6.7, the slope m & intercept c values are 0.62, 1 & 0.54 and 1.2, respectively, for both training and testing datasets, reflecting a good match. Accuracy obtained from this model in terms of R and MAPE shows the model performance. It can also be concluded that as the value of MAPE goes on decreasing, testing and training precisions of the model go on improving. The better the training and testing accuracies, the better the performance of the model.

Table 6.2 Training and testing of ANN architecture data accuracy.

No. of neurons in ANN architectures	Correlation coefficient		MAPE
	Training	Testing	
5-5-1	0.64	0.62	18.52
5-10-1	0.67	0.69	16.47
5-15-1	0.69	0.63	15.40
5-5-5-1	0.70	0.74	17.04
5-7-5-1	0.68	0.72	17.92
5-6-10-1	0.76	0.68	15.31
5-6-11-1	0.74	0.70	15.69
5-7-11-1	0.69	0.72	16.28
5-7-12-1	0.69	0.62	17.9
5-10-12-1	0.68	0.71	16.07
5-13-6-1	0.72	0.69	15.41
5-14-6-1	0.70	0.74	15.02
5-15-6-1	0.73	0.71	15.74
5-15-7-1	0.72	0.73	14.92
5-15-8-1	0.74	0.72	14.27
5-16-5-1	0.71	0.76	14.04
5-16-6-1	0.74	0.74	13.94
5-16-7-1	0.73	0.69	14.65
5-16-10-1	0.63	0.73	16.07
5-18-10-1	0.68	0.66	15.81
5-20-10-1	0.70	0.64	15.13
5-21-1-1	0.72	0.68	15.01
5-22-2-1	0.76	0.65	14.14
5-22-5-1	0.76	0.68	14.96

(*Continued*)

Table 6.2 Training and testing of ANN architecture data accuracy. (*Continued*)

No. of neurons in ANN architectures	Correlation coefficient		MAPE
	Training	Testing	
5-22-10-1	0.72	0.67	15.20
5-25-5-1	0.68	0.64	16.31
5-30-5-1	0.72	0.71	15.70
5-30-7-1	0.69	0.62	16.42
5-32-5-1	0.70	0.64	15.87
5-35-6-1	0.76	0.70	14.02
5-35-10-1	0.73	0.74	14.66
5-37-1-1	0.76	0.72	14.29
5-37-5-1	0.73	0.71	14.83
5-38-3-1	0.73	0.67	15.12
5-40-5-1	0.76	0.71	14.05
5-41-5-1	0.74	0.72	14.31
5-43-5-1	0.76	0.73	14.03
5-44-5-1	0.72	0.74	14.31
5-45-5-1	0.77	0.71	13.96
5-45-6-1	0.75	0.70	14.03
5-46-5-1	0.76	0.69	14.28
5-47-5-1	0.76	0.73	13.82
5-47-10-1	0.75	0.74	13.49
5-48-5-1	0.77	0.72	13.39
5-48-10-1	0.76	0.72	13.62
5-48-15-1	**0.78**	**0.74**	**12.15**
5-48-16-1	0.75	0.74	13.60
5-48-11-1	0.76	0.77	13.96

(*Continued*)

Table 6.2 Training and testing of ANN architecture data accuracy. (*Continued*)

No. of neurons in ANN architectures	Correlation coefficient		MAPE
	Training	Testing	
5-49-5-1	0.75	0.72	14.04
5-50-5-1	0.73	0.72	14.84
5-50-20-1	0.79	0.71	13.62
5-47-6-1	0.73	0.71	14.86
5-52-5-1	0.73	0.63	15.36
5-55-5-1	0.71	0.74	15.91

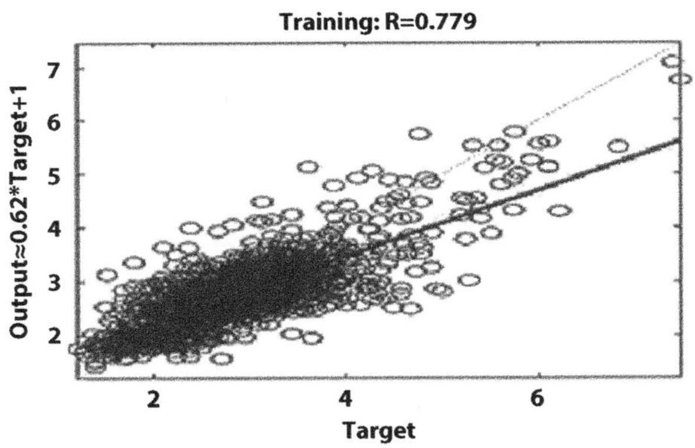

Figure 6.7 Training data regression plot for the superlative architecture 5/48/15/1.

6.3.2 Hydro Resource Estimation

The hydropower potential requires the determination of the available flow rate at the desired location. A discharge sequence of gauge sites has been taken because the flow rate data is not available for the desired location. There is available flow for Mehla Nallah, which is located in the downstream of the Sundri site where a hydro plant has to be erected. For hydro capacity assessment, the Sundri site close to the town of Sundri is taken into account. Consequently, for the Mehla Nallah, monthly average flow rate data was taken from 1979 to 2008. The total volume of the site of Sundri has been calculated using the technique of hydrological estimation. Using the empirical formula given by Eq. (6.6), the calculation of the hydro potential was carried out in ref. [17].

$$Q_{site} = \left[\frac{A_{site}}{A_{guge}} \right] Q_{guge} \tag{6.6}$$

where, Q_{site} the ejection of water in m^3/s. A_{guge} is the catchment area of the gauge (where measurement of flow is accessible) in (m2); A_{site} is the catchment area (where installation of hydro is possible) (m^2); Q_{gauge} is gauge discharge or measurable discharge (m^3/s) and Q_{site} is the expulsion at hydro plant site (m3/s);. The estimated flow for the Sundri site has been shown in Figures (6.8 (a-d)).

After deriving the discharge series, it has been used for the construction of Flow Duration Curve (FDC). By using frequency analysis of derived discharge sequence of flow, the FDC was developed and the stages are as described in the following:

(i) The cumulative number of 26 years for which the sequence of discharges is derived.

(ii) The release series was grouped in descending order and each discharge value was assigned a special ranking

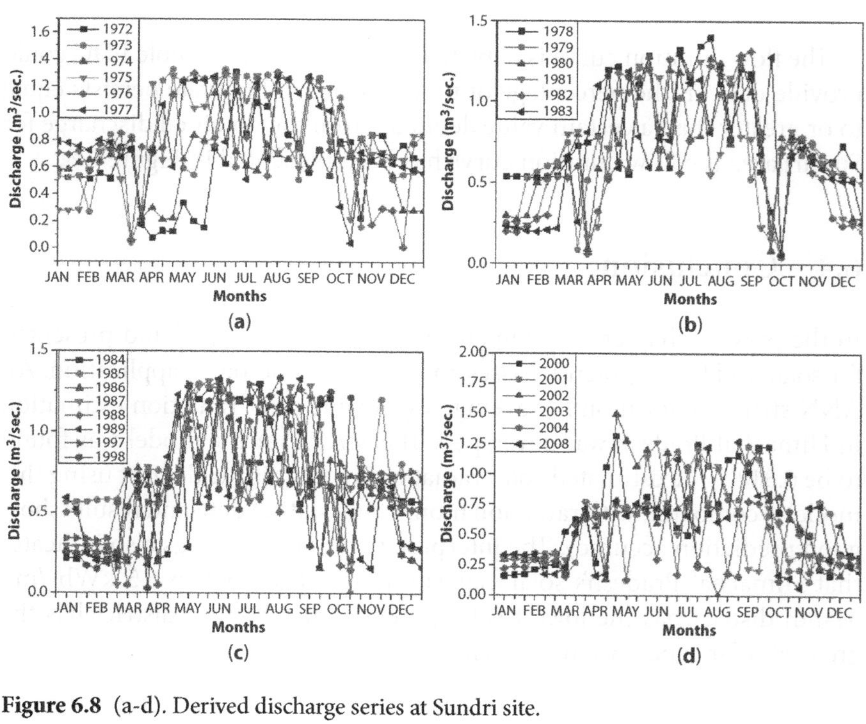

Figure 6.8 (a-d). Derived discharge series at Sundri site.

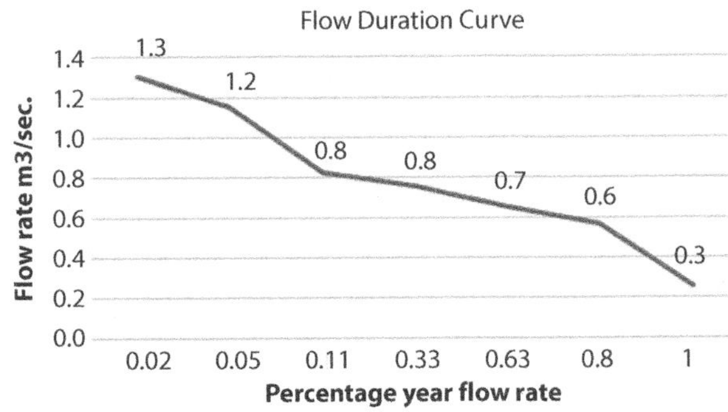

Figure 6.9 Discharge flow curve with percentile dependability.

number, beginning with 1 for the peak flow and 26 for the low discharge value.

(iii) Finally, for each year's flow data, the exceedance probability (P) was calculated using the Weibull method as:

$$P = 100 * \left(\frac{M}{(n+1)} \right) \tag{6.7}$$

The flow duration curve is depicted in Figure 6.9. The whole FDC could provide data on the percentage of time in which a flow rate persists equal to or greater than a certain value defined. From the obtained discharge for Sundri area the flow duration curve has been presented in Figure 6.8.

6.4 Conclusion

In the present chapter, a systematic approach is developed and presented for solar and hydro potential assessment for DC microgrid application. An ANN structure for monthly average worldwide solar radiation estimation in Himachal Pradesh was developed. The MAPE for the model was found to be 6.31%. The estimated solar radiation for Hamirpur district using this model is compared with radiation monitoring station data and results show good prediction accuracy. The interpretation of the findings also indicates that Himachal Pradesh's solar capacity ranges from 3.4 to 6.5 kWh /m^2. Mandi district has the lowest solar potential and Kangra district has the greatest solar capacity amongst all of the divisions of H.P.

For the sizing of HRES in the rural area of Chamba district of H.P, an ANN model for daily global estimation was developed and a weight connection approach was implemented for computing relevant limits. The weighted connection method can be used in various solar estimation studies for ranking the most important parameters. The ranking of the parameters obtained using the weighted connection method has been compared with existing studies and shows that temperature and relative humidity are very crucial variables for solar radiation estimation. The hydro potential for the Sundri site was estimated using empirical relationship, and a flow duration curve was obtained to find the available power generation through micro hydro plant.

References

1. Kumar, S., Kaur, T., Arora, M. K., & Upadhyay, S. (2019). Resource estimation and sizing optimization of PV/micro hydro-based hybrid energy system in rural area of Western Himalayan Himachal Pradesh in India. *Energy Sources, Part a: Recovery, Utilization, and Environmental Effects*, *41*(22), 2795–2807.
2. https://www.iea.org/reports/world-energy-outlook-2019/electricity.
3. Sumithira, T. R., & Kumar, A. N. (2012). Prediction of monthly global solar radiation using adaptive neuro fuzzy inference system (ANFIS) technique over the State of Tamilnadu (India): a comparative study. *Applied Solar Energy*, *48*(2), 140–145.
4. Alsina, E. F., Bortolini, M., Gamberi, M., & Regattieri, A. (2016). Artificial neural network optimisation for monthly average daily global solar radiation prediction. *Energy Conversion and Management*, *120*, 320–329.
5. Sharma, Sumit, and Yog Raj Sood. "Microgrids: A Review of Status, Technologies, Software Tools, and Issues in Indian Power Market." *IETE Technical Review* (2020): 1–22.
6. Ahlborg, H., & Hammar, L. (2014). Drivers and barriers to rural electrification in Tanzania and Mozambique–Grid-extension, off-grid, and renewable energy technologies. *Renewable Energy*, *61*, 117–124.
7. Paliwal, P., Patidar, N. P., & Nema, R. K. (2014). Planning of grid integrated distributed generators: A review of technology, objectives and techniques. *Renewable and Sustainable Energy Reviews*, *40*, 557–570. http://dx.doi.org/10.1016/j.rser.2014.07.200.
8. Velayudhan, S. K. (2003). Dissemination of solar photovoltaics: a study on the government programme to promote solar lantern in India. *Energy Policy*, 31(14), 1509–1518.
9. Beccali, M., Cellura, M., Longo, S., Nocke, B., & Finocchiaro, P. (2012). LCA of a solar heating and cooling system equipped with a small water–ammonia absorption chiller. *Solar Energy*, 86(5), 1491–1503.

10. Joshi, S. B., & Jani, A. R. (2015). Design, development and testing of a small scale hybrid solar cooker. *Solar Energy*, 122, 148–155.
11. Beyer, H. G., & Langer, C. (1996). A method for the identification of configurations of PV/wind hybrid systems for the reliable supply of small loads. *Solar Energy*, *57*(5), 381–391.
12. Kumaravel, S., & Ashok, S. (2012). An optimal stand-alone biomass/solar-PV/pico-hydel hybrid energy system for remote rural area electrification of isolated village in Western-Ghats region of India. *International Journal of Green Energy*, *9*(5), 398–408.
13. Palit, Debajit, Ramit Malhotra, and Atul Kumar. "Sustainable model for financial viability of decentralized biomass gasifier based power projects." *Energy Policy* 39.9 (2011): 4893–4901.
14. Raha, D., Mahanta, P., & Clarke, M. L. (2014). The implementation of decentralised biogas plants in Assam, NE India: The impact and effectiveness of the National Biogas and Manure Management Programme. *Energy Policy*, *68*, 80–91.
15. Chauhan, A., & Saini, R. P. (2014). A review on Integrated Renewable Energy System based power generation for stand-alone applications: Configurations, storage options, sizing methodologies and control. *Renewable and Sustainable Energy Reviews*, *38*, 99–120.
16. Nayal, F. S., Mammadov, A., & Ciliz, N. (2016). Environmental assessment of energy generation from agricultural and farm waste through anaerobic digestion. *Journal of Environmental Management*, *184*, 389–399.
17. Bekele, G., & Tadesse, G. (2012). Feasibility study of small Hydro/PV/Wind hybrid system for off-grid rural electrification in Ethiopia. *Applied Energy*, *97*, 5–15.
18. Bakirci, K. (2009). Correlations for estimation of daily global solar radiation with hours of bright sunshine in Turkey. *Energy*, *34*(4), 485–501.
19. Trabea, A. A., & Shaltout, M. M. (2000). Correlation of global solar radiation with meteorological parameters over Egypt. *Renewable Energy*, *21*(2), 297–308.
20. Boland, J., Ridley, B., & Brown, B. (2008). Models of diffuse solar radiation. *Renewable Energy*, *33*(4), 575–584.
21. Bristow, K. L., & Campbell, G. S. (1984). On the relationship between incoming solar radiation and daily maximum and minimum temperature. *Agricultural and Forest Meteorology*, *31*(2), 159–166.
22. Wan, K. K., Tang, H. L., Yang, L., & Lam, J. C. (2008). An analysis of thermal and solar zone radiation models using an Angstrom–Prescott equation and artificial neural networks. *Energy*, *33*(7), 1115–1127.
23. Jović, S., Aničić, O., Marsenić, M., & Nedić, B. (2016). Solar radiation analyzing by neuro-fuzzy approach. *Energy and Buildings*, *129*, 261–263.
24. Mohammadi, K., Shamshirband, S., Anisi, M. H., Alam, K. A., & Petković, D. (2015). Support vector regression based prediction of global solar radiation on a horizontal surface. *Energy Conversion and Management*, *91*, 433–441.

25. Olatomiwa, L., Mekhilef, S., Shamshirband, S., Mohammadi, K., Petković, D., & Sudheer, C. (2015). A support vector machine–firefly algorithm-based model for global solar radiation prediction. *Solar Energy, 115*, 632–644.
26. Kumar, S., & Kaur, T. (2016). Development of ANN based model for solar potential assessment using various meteorological parameters. *Energy Procedia, 90*, 587–592.
27. Mohandes, M. A., & Rehman, S. (2013). Estimation of sunshine duration in Saudi Arabia. *Journal of Renewable and Sustainable Energy, 5*(3), 033128.
28. Alam, S., Kaushik, S. C., & Garg, S. N. (2009). Assessment of diffuse solar energy under general sky condition using artificial neural network. *Applied Energy, 86*(4), 554–564.
29. Fadare, D. A. (2009). Modelling of solar energy potential in Nigeria using an artificial neural network model. *Applied Energy, 86*(9), 1410–1422.
30. Kumar, S., & Kaur, T. (2020). Efficient solar radiation estimation using cohesive artificial neural network technique with optimal synaptic weights. *Proceedings of the Institution of Mechanical Engineers, Part A: Journal of Power and Energy, 234*(6), 862–873.
31. Zhang, J., Zhao, L., Deng, S., Xu, W., & Zhang, Y. (2017). A critical review of the models used to estimate solar radiation. *Renewable and Sustainable Energy Reviews, 70*, 314–329.
32. Kaur, T., Kumar, S., Kaur, R., & Gera, A. (2018). ANN based global solar radiation prediction: A case study. *Kaav International Journal of Science. Eng. Technol, 5*(4), 54–63.
33. Sözen, A., Arcaklioğlu, E., & Özalp, M. (2004). Estimation of solar potential in Turkey by artificial neural networks using meteorological and geographical data. *Energy Conversion and Management, 45*(18-19), 3033–3052.
34. Tymvios, F. S., Jacovides, C. P., Michaelides, S. C., & Scouteli, C. (2005). Comparative study of Ångström's and artificial neural networks' methodologies in estimating global solar radiation. *Solar Energy, 78*(6), 752–762.
35. Fadare, D. A. (2009). Modelling of solar energy potential in Nigeria using an artificial neural network model. *Applied Energy, 86*(9), 1410–1422.
36. Mishra, A., Kaushika, N. D., Zhang, G., & Zhou, J. (2008). Artificial neural network model for the estimation of direct solar radiation in the Indian zone. *International Journal of Sustainable Energy, 27*(3), 95–103.
37. Premalatha, M., & Naveen, C. (2018). Analysis of different combinations of meteorological parameters in predicting the horizontal global solar radiation with ANN approach: A case study. *Renewable and Sustainable Energy Reviews, 91*, 248–258.
38. Fujii, M., Tanabe, S., Yamada, M., Mishima, T., Sawadate, T., & Ohsawa, S. (2017). Assessment of the potential for developing mini/micro hydropower: A case study in Beppu City, Japan. *Journal of Hydrology: Regional Studies, 11*, 107–116.
39. WMO. 2008. No, 50 Operational Hydrology Report Manual on Low-Flow Estimation and Prediction.

40. Kusre, B. C., Baruah, D. C., Bordoloi, P. K., & Patra, S. C. (2010). Assessment of hydropower potential using GIS and hydrological modeling technique in Kopili River basin in Assam (India). *Applied Energy, 87*(1), 298–309.
41. Girma, Z. (2016). Techno-economic feasibility of small scale hydropower in Ethiopia: The case of the kulfo River, in Southern Ethiopia. *Journal of Renewable Energy, 2016.*
42. Yi, C. S., Lee, J. H., & Shim, M. P. (2010). Site location analysis for small hydropower using geo-spatial information system. *Renewable Energy, 35*(4), 852–861.
43. Bailey, P., Chotimongkol, O., & Isono, S. (2007). Demand analysis and optimization of renewable energy: sustainable rural electrification of Mbanayili, Ghana (Doctoral dissertation).
44. Camblong, H., Sarr, J., Niang, A. T., Curea, O., Alzola, J. A., Sylla, E. H., & Santos, M. (2009). Micro-grids project, Part 1: Analysis of rural electrification with high content of renewable energy sources in Senegal. *Renewable Energy, 34*(10), 2141–2150.
45. Wimmler, C., Hejazi, G., de Oliveira Fernandes, E., Moreira, C., & Connors, S. (2017). Impacts of load shifting on renewable energy integration. *Energy Procedia, 107*, 248–252.

7

Secondary Control Based on the Droop Technique for Power Sharing

Waner W.A.G. Silva[1*], Thiago R. de Oliveira[2], Rhonei P. Santos[3] and Danilo I. Brandao[4]

[1]Federal University of Itajuba (UNIFEI), Itabira, Brazil
[2]Electronic Engineering Department, Federal University of Minas Gerais (UFMG), Belo Horizonte, Brazil
[3]Eletric Engineering Department, Federal University of Ouro Preto (UFOP), João Monlevade, Brazil
[4]Department of Electrical Engineering, Federal University of Minas Gerais (UFMG), Belo Horizonte, Brazil

Abstract

The microgrid energy management usually is based on hierarchical architecture, adapted from the ISA-95 standard, which consists of three control levels: i) the primary control level is responsible for the local and autonomous control of the power converters and ensures power sharing, DC voltage regulation and stability; ii) the secondary control level provides restoration of deviations caused by the primary control level, and iii) the tertiary control level manages the power flow between the microgrid and the mains. The droop control is often used in the primary control level to provide decentralized and autonomous operation for the dispersed parallel power converters without using communication channels and allowing plug-and-play capability. However, the droop technique shows a trade-off between power sharing and voltage regulation, causing power sharing errors and/or voltage deviations, thus, this chapter details the influence of line equivalent resistances in power sharing and in average DC bus voltage errors caused by droop techniques, reviewing the secondary control methods proposed in literature employed to correct such errors. Moreover, two secondary control techniques based on distributed control, proposed by the authors, are highlighted, as well as the design of the communication system.

*Corresponding author: waner@unifei.edu.br

Nikita Gupta, Mahajan Sagar Bhaskar, Sanjeevikumar Padmanaban and Dhafer Almakhles (eds.)
DC Microgrids: Advances, Challenges, and Applications, (183–240) © 2022 Scrivener Publishing LLC

Keywords: DC microgrid, accurate power sharing, voltage restoration, distributed control, consensus-based algorithm, droop-control

7.1 Introduction

Power distribution systems have long been dominated by AC power, since AC-based implementation has historically enabled simple and efficient voltage level transformation, safer fault and overcurrent protection and less costly installation than DC-based systems. Consequently, most research proposals and demonstration sites that address the use of microgrid architectures to couple distributed generation (DG) and energy storage to the utility also consider AC microgrid (μG_{AC}) implementation [1–3]. However, the usage of DC loads, e.g., consumer electronics, smart appliances, plug-in electric vehicles (PHEV), etc., in commercial and residential systems has been increasing significantly. In addition, renewable primary energy sources in DG, as well as, battery energy storage systems (BESS) and fuel cells are intrinsically DC energy sources that provide an interesting environment for DC-based microgrids (μG_{DC}) to be employed. Recently, local area DC power systems are being deployed in *datacenters*, more electric ships, more electric aircrafts and residential/commercial buildings [4].

μG_{DC} provides some advantages in comparison to μG_{AC}, since there are no reactive and harmonic current circulation, skin and proximity effects in cables and the need for phase synchronization [5, 6]. Moreover, a DC bus allows fewer AC/DC power conversion stages in sources and loads, providing an energy efficiency increase of about 2-15% in relation to equivalent AC-based power systems [4].

Figure 7.1 presents a typical residential/commercial μG_{DC} with its main components: a bidirectional utility interface converter (BIC) that interconnects the local DC bus and the main utility at a point of common couple (PCC), renewable resource converters (RRC) that interconnect DG to the DC bus and energy storage converters (ESC) that interface energy storage units (ESU) and the DC bus [7]. The power balance between load and generation in both grid-connected and islanded modes is assured by the energy storage system that consists of ESU and their ESCs. The main DC bus of μG_{DC}, which interlink local power sources and loads, exhibits normally 380 V_{DC} voltage level, although many proposals also consider the existence of secondary busses with lower voltages levels, e.g., 48V_{DC} and 24V_{DC}, to supply low power loads with higher safety than 380 V_{DC}.

The energy management of μG_{DC} is usually based on a hierarchical architecture, adapted from ISA-95 standards, which shows three control levels [8]:

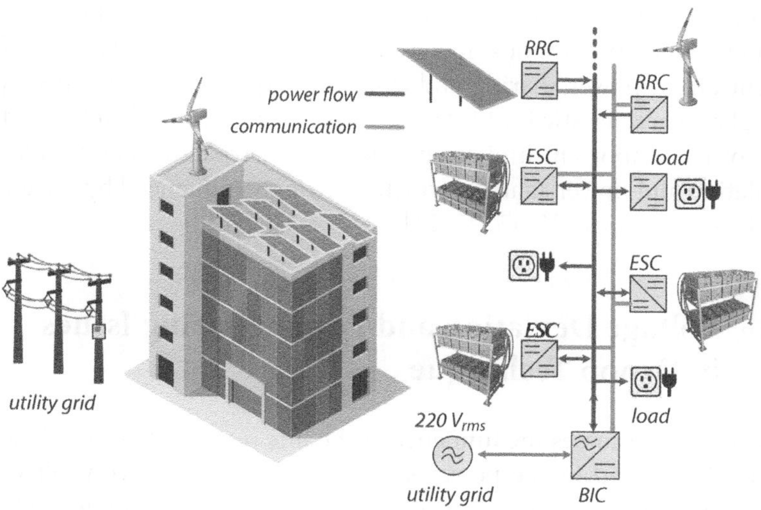

Figure 7.1 Residential/Commercial DC microgrid structure.

- Primary control level: this level refers to the local control of power converters, ensuring power sharing and main DC bus voltage regulation [5, 6, 9]. The control techniques used in this level are based on: i) centralized control based on low bandwidth communication links, where a central element determines the operation mode of all converters. For example, a converter regulates the bus voltage while other ones operate as a current source, in which a central controller cyclically defines the current set-points of each converter in a master-slave approach [10–13]; ii) decentralized strategy based on droop control, in which the output voltage set-point of each converter is dependent on its output power/current [2, 14, 15].
- Secondary control level: it is responsible for ensuring that the electric values of the microgrid (i.e., voltage level) are within predefined ranges, enabling the correction of errors eventually introduced by the primary level [9, 16].
- Tertiary control level: This level manages the power flow between the main grid and the microgrid. It may also provide voltage and current set-points at the PCC and modify primary control parameters, depending on the economic and environmental data to obtain an optimized operation of the μG [8, 9, 17].

In relation to the primary level in μG_{DC}, the droop technique provides reliability and autonomous operation, since it allows the μG_{DC} to achieve decentralized power sharing and expandability, due to its intrinsic plug-and-play characteristic [2, 14, 15, 18–20]. Despite those advantages, droop control has important limitations such as the trade-off between voltage regulation and power sharing, which is strongly influenced by the equivalent line impedances [2, 15, 20, 21].

7.2 Voltage Deviation and Power Sharing Issues in Droop Technique

This section discusses the influence of the line resistances in DC bus voltage deviation and in the power sharing amongst droop-controlled converters. Figure 7.2a shows a steady-state DC bus model of a μG, which is composed of two converters (*Conv-1* and *Conv-2*), represented by their Thevenin equivalent circuits, and two loads (R_{la} and R_{lb}). The converters' rated voltages are V_{o1}^* and V_{o2}^*, respectively, R_{d1} and R_{d2} are the droop coefficients, r_a, r_b and r_c are the line resistances and v_{o1}, v_{o2}, i_{o1}, i_{o2} are the output voltages and currents of the converters, respectively. In Figure 7.2b, the line resistances and loads are lumped into a simplified DC bus model, where $v_{\mu G}$ is the bus voltage, r_1, r_2 and $R_{\mu G}$ are the equivalent line resistances and the total load, respectively, that can be expressed as:

$$
\begin{aligned}
r_1 &= r_a + \frac{r_b R_{la}}{R_{la} + R_{lb} + r_b} \\
r_2 &= r_c + \frac{r_b R_{lb}}{R_{la} + R_{lb} + r_b} \\
R_{\mu G} &= \frac{R_{la} R_{lb}}{R_{la} + R_{lb} + r_b}
\end{aligned}
\tag{7.1}
$$

Considering the simplified model of Figure 7.2b, the output currents of each converter are expressed as:

$$
\begin{aligned}
i_{o1} &= \frac{V_{o1}^*(R_{d2} + r_2) + R_{\mu G}(V_{o1}^* - V_{o2}^*)}{\alpha + \beta R_{\mu G}} \\
i_{o2} &= \frac{V_{o2}^*(R_{d1} + r_1) + R_{\mu G}(V_{o2}^* - V_{o1}^*)}{\alpha + \beta R_{\mu G}}
\end{aligned}
\tag{7.2}
$$

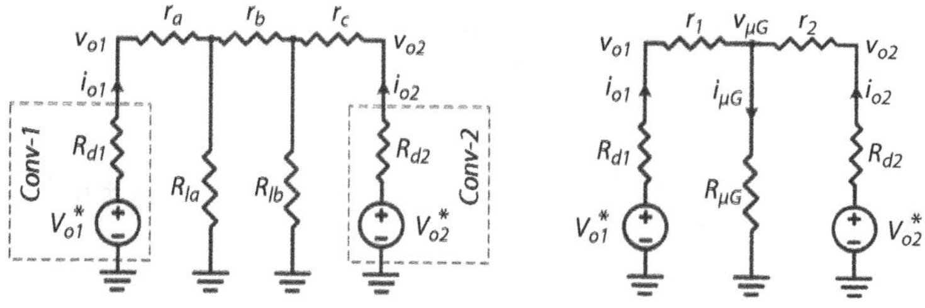

Figure 7.2 DC bus Equivalent Thevenin Circuit (Reprinted with permission from 5052471151140/IEEE.)

where

$$\alpha = (R_{d1} + r_1)(R_{d2} + r_2)$$

$$\beta = R_{d1} + R_{d2} + r_1 + r_2 \tag{7.3}$$

The converter output voltage and power quantities are expressed by (7.4) and (7.5) respectively.

$$\begin{aligned} v_{o1} &= V_{o1}^* - R_{d1} i_{o1} \\ v_{o2} &= V_{o2}^* - R_{d2} i_{o2} \end{aligned} \tag{7.4}$$

$$\begin{aligned} P_{o1} &= v_{\mu G} i_{o1} + r_1 i_{o1}^2 \\ P_{o2} &= v_{\mu G} i_{o2} + r_2 i_{o2}^2 \end{aligned} \tag{7.5}$$

Observing (7.2) and (7.5) and assuming that *Conv-1* and *Conv-2* are identical, i.e., $V_{o1}^* = V_{o2}^* = V_o$ and $R_{d1} = R_{d2} = R_d$, it follows that power ($P_{o1} = P_{o2}$) and current ($i_{o1} = i_{o2}$) sharing can only be achieved concomitantly if, and only if, $r_1 = r_2$. According to (7.1), r_1 and r_2 are directly influenced by the DC bus loads (7.1), therefore, a condition where $r_1 = r_2$ is very improbable in practice. For instance, an increase in R_{la} consequently increases the value of r_1, while decreasing r_2 and leading to a line resistance mismatch. This condition ($r_1 \neq r_2$) enforces power and/or current imbalance.

From (7.2), it is possible to derive (7.6) and (7.7), while (7.5) can be rearranged into (7.8), where m_p is the desired power ratio between *Conv-1* and *Conv-2*.

$$\frac{i_{o1}}{i_{o2}} = \frac{R_{d2} + r_2}{R_{d1} + r_1} \tag{7.6}$$

$$\Delta i_o = i_{o1} - i_{o2} = \frac{2(V_{o1}^* - V_{o2}^*)R_{\alpha G} + (R_{d2} + r_2)V_{o1}^* - (R_{d1} + r_1)V_{o2}^*}{\alpha + \beta R_{\alpha G}} \tag{7.7}$$

$$\frac{P_1}{P_2} = \frac{i_{o1}}{i_{o2}}\left(\frac{v_{\mu G} + r_1 i_{o1}}{v_{\mu G} + r_2 i_{o2}}\right) = m_p \tag{7.8}$$

As $v_{\mu G} = R_{\mu G}(i_{o1} + i_{o2})$, the current ratio can be expressed as:

$$\frac{i_{o1}}{i_{o2}} = -a + \sqrt{b} = m_i \tag{7.9}$$

where m_i is the current ratio required for the system to exhibit the desired m_p, and:

$$a = \frac{(1 - m_p)}{2\left(1 + \frac{r_1}{R_{\mu G}}\right)} \tag{7.10}$$

$$b = m_p\left(\frac{1 + \frac{r_2}{R_{\mu G}}}{1 + \frac{r_1}{R_{\mu G}}}\right) + \left[\frac{1 - m_p}{2\left(1 - \frac{r_1}{R_{\mu G}}\right)}\right]^2 \tag{7.11}$$

Once again, (7.10) and (7.11) show that it is not possible to achieve $m_i = m_p$ when $r_1 \neq r_2$. Therefore, it must be decided whether the μG secondary control layer will promote current sharing (*cs*) or power sharing (*ps*).

Considering (7.6) and (7.7), two approaches can then be used to achieve proportional *cs* or *ps*: i) based on droop coefficient adjustment (*da*), that can either be accomplished by adding a term δR_{dj} (where $j \in [1,2]$) to a converter droop coefficient (7.12), or by applying a gain (g_{Rdj}) to the droop coefficient (7.13); and ii) based on voltage-shifting (*vs*), which shifts a converter reference voltage (V_{oj}^*) by adding a voltage-shifting term (δv_{oj}) to it, as in (7.14).

$$\frac{i_{o1}}{i_{o2}} = \frac{(R_{d2} + \delta R_{d2}) + r_2}{(R_{d1} + \delta R_{d1}) + r_1} \tag{7.12}$$

$$\frac{i_{o1}}{i_{o2}} = \frac{R_{d2} g_{Rd2} + r_2}{R_{d1} g_{Rd1} + r_1} \tag{7.13}$$

$$V_{oj}^* = V_o^* + \delta v_{oj} \tag{7.14}$$

Figure 7.3 shows the static behavior of the converters under the influence of r_1 and r_2. The voltage deviation produced by the droop technique can be expressed as (7.15). The maximum voltage deviation occurs when each converter operates at its maximum output current ratings (i_{ojmax}). Hence, in order to ensure that this deviation does not exceed the desired DC bus voltage maximum limits (ΔV_{omax}), the droop coefficient must be limited, as (7.16) [5].

$$\Delta V_{oj}^* = R_{dj} i_{oj} \tag{7.15}$$

$$R_{dj} \;'' \; \frac{\Delta V_o^* max}{i_{oj} max} \tag{7.16}$$

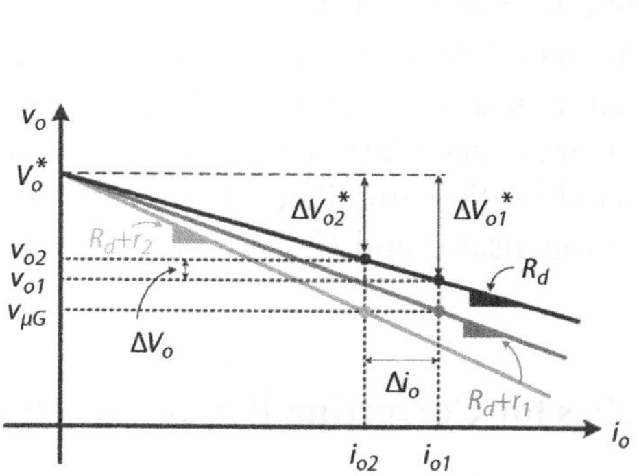

Figure 7.3 Converter static behavior considering the line resistance effect.

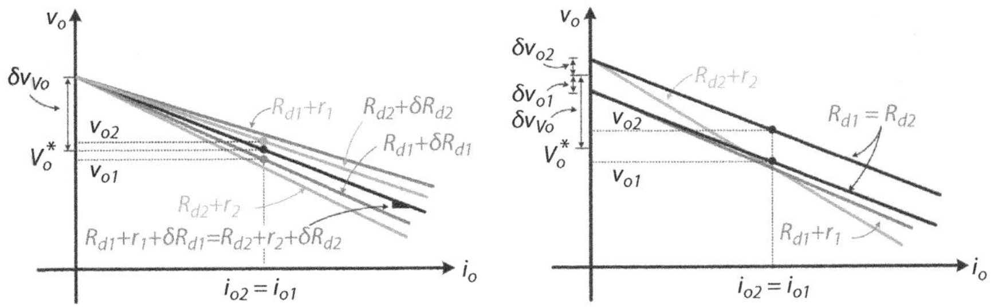

(**a**) Droop Coefficient Adjustment (*da*) (**b**) Voltage-Shifting (*vs*)

Figure 7.4 Approaches for *ps* and *cs*, DC bus voltage restoration.

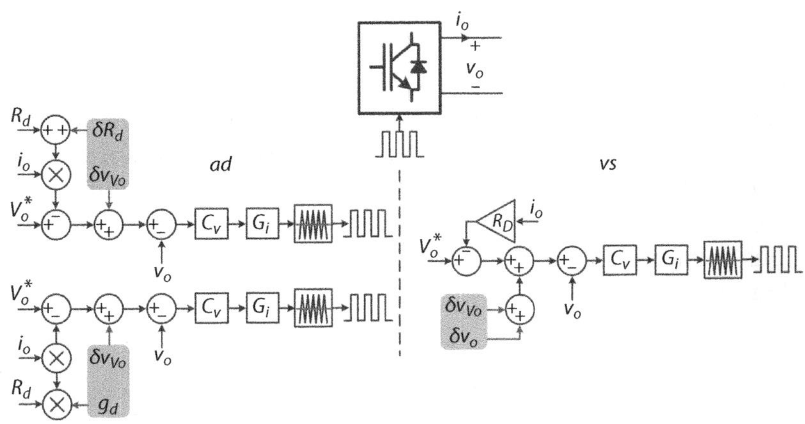

Figure 7.5 Secondary control implementation in the embedded local controller.

The DC bus voltage restoration is usually realized through a voltage-shifting (δv_{VO}) technique, so that, in steady state, $\delta v_{Voj} \approx R_{dj}\, i_{oj}$ and the average of the converter output voltages converge to $v_{oavg} = V_o^*$. Figure 7.4 depicts the system static behavior considering a voltage-shifting strategy for voltage restoration and for *ps* and *cs* both da and *vs* approaches. The implementation of both secondary-level strategies in the converter local controller is depicted by the simplified diagram in Figure 7.5, where C_v is a voltage loop compensator and G_i is the current control loop transfer function (*TF*).

7.2.1 Approaches for Correcting Power and Current Sharing

In terms of power or current sharing, the voltage-shifting approach is not capable of changing the equivalent converter output impedance.

To compensate for the differences between equivalent line resistances, the voltage-shifting term must be constantly updated, even when there is no change in the load bus, since shifting the reference voltage impacts the power balance. In *da* approaches, once the differences between equivalent line resistances have been corrected, the current or power compensation errors will be almost null, even though voltage deviation correction is still running. However, since *da* alters the converter output impedance, system dynamics is affected and the adjustments in the droop coefficient must be restricted in order to prevent instability [7].

To achieve the goals of restoring the voltage level and attaining *ps* or *cs*, the distributed control techniques depend on exchanging at least two converter quantities: output voltage (v_o) and the output power or current (P_o or i_o). This information exchange can occur either among all converters or only among neighboring converters. This last approach comprises techniques based on consensus algorithms.

Lu in [2][2] uses voltage-shifting for voltage restoration and power sharing. Each converter sends v_{oj}, i_{oj} to all other ones and locally calculates the average adjustment terms δv_{Voj} and δv_{oj} that is added in the reference voltage V_o^*. Both goals are achieved simultaneously. In [22][2], each converter sends its output current (i_{oj}) to a common analog communication link and the average current (i_{oavg}) is locally computed as well as the voltage-shifting term δv_{oj}. The voltage deviation is attenuated by employing reduced droop coefficients values. In [23][1] a V/P (voltage/power) droop control is proposed for distributed ESU, where State-of-Charge (SoC) is incorporated in the secondary control goals. In addition to the voltage-shifting terms (δv_{Vo} and δv_o) used to achieve voltage restoration and power sharing, a *da* technique is also performed. The droop adjustment is determined by the deviation between the SoC of an ESU and the system average SoC (SoC_{avg}), therefore a power imbalance is forced to promote SoC equalization, when that occurs, the system is able to achieve power sharing.

Wang in [20][2] employs a *da*-based secondary control. The voltage-shifting term δv_{Vo} corresponds to the compensation error between the average DC bus voltage and the converter output voltage value. The increment δR_d is calculated based on the averages of the droop coefficients and output currents. Such method aims at obtaining the same output impedance for each converter. In [24] [2], the control is *da*-based as well, and uses a quality index ϱ (17), where w_j is the weighting coefficient chosen in function of

[1] In order to maintain a standardized nomenclature in this work regarding the converter output voltage, current and its droop coefficient, the terms used by the authors of these papers have been adapted.

the maximum allowed voltage deviation for the j-th converter. Augustine [25] proposes a proportional droop index (*PDI*), which is a function of the difference between the normalized currents and voltage deviations. In addition to achieving power sharing, voltage deviation is reduced and all converters present equal output voltages.

$$\varrho = (i_{oavg} - i_{oj})^2 + w_j\,(i_{oj}\,R_{dj})^2 \tag{7.17}$$

Zaery in [26][2] employs a distributed control based on a dynamic consensus algorithm in a sparse communication network to solve economic dispatch problems among DGs. Based on local voltage measurements from neighboring, the converters voltage restoration is achieved by reference voltage-shifting. Power sharing is attained by droop adjustment, where the increment δR_{dj} is calculated based on cost functions on the tertiary level. In [19, 27–31], a consensus algorithm is used to achieve dynamic average of the output voltages and currents among neighboring converters by means of voltage-shifting approach. The control structure based on these approaches is similar to that shown in Figure 7.6, with N_j being the total number of converters connected to the j-th converter. In [32], a *da* approach is used, with a structure similar to that shown in Figure 7.6. However, instead of generating a δv_{oj} voltage-shifting term, the PI compensator of the current averaging branch outputs a droop adjustment term. In [33], an adaptive droop approach is proposed based on average voltage and current values among neighboring converters. In this technique, control branch generates increments, δR_{dv} and δR_{di}, that are added to compose the droop gain g_{Rd} that adjusts the droop coefficient. In this way, current sharing and DC bus voltage stability are obtained.

Decentralized secondary level control can also be found in technical literature, using adaptive droop control as a path to improve power or current sharing, while reducing voltage deviation. In [34–36] a decentralized

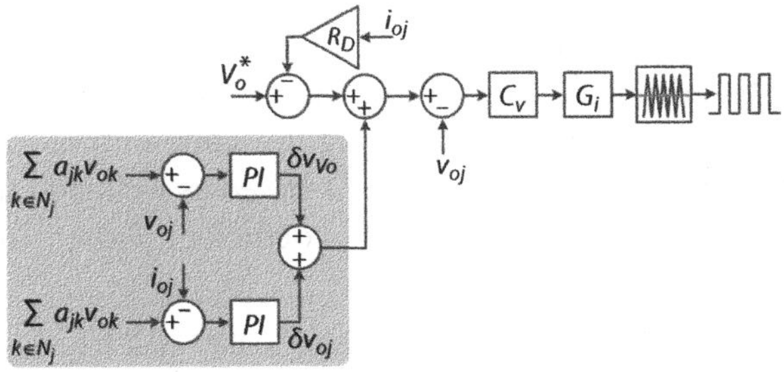

Figure 7.6 Implementation of consensus-based secondary control on local control.

nonlinear adaptive droop control is proposed. It relies on a *da* approach, and the droop coefficient gain (g_{Rd}) calculation employs only local voltage and current measurements, while in [36], only the current measurement is used for computing the droop coefficient gain. Those approaches consist in increasing g_{Rd} along with an increase in the converter output current. Since the equivalent line resistances become much lower than R_d, the adaptive droop control attenuates the output current imbalance and improves the *cs*. As decentralized approaches, those techniques have the advantage of not depending on communication link. However, the implementation of the gain functions is complex, and the secondary control performance, although reducing deviations in electric quantities, varies for each load condition. Peyghami [37] proposes a decentralized adaptive droop control based on sinusoidal signal injection $\tilde{v}$ in V_o^*, where the frequency of $\tilde{v}$ is proportional to the output current of the converter. The disadvantage of such a technique is the introduction of a disturbance in the DC bus voltage and considering the output filters of converters can lead to a point of instability.

Table 7.1 compares the main distributed secondary control techniques found in the literature and the two control approaches that have been proposed by the authors (hybrid [8] and unique *vs* [7]), which is discussed in Section 2.2. The comparison is realized considering: i) the information share between the converters, where '*v*' relates to the converter output voltage, '*i*' to its output current, '*o*' is an information that needs to be calculated before being sent, '*p*' is the output power, '*s*' is the storage unit SoC, *u* is the compensating term and '*$*' relates to an operation cost function; ii) whether the communication should be carried out among all converters or only among neighboring converters; iii) whether the technique employs *vs* or *da* as the secondary level control approach; iv) whether the technique promotes voltage deviation correction and v) whether it performs *cs* or *ps*. It is worth clarifying that the term "indirect voltage restoration" means that the method reduces the deviation indirectly, e.g., decreasing ΔV_o^* by reducing the droop coefficient.

Table 7.1 shows that the techniques used to concomitantly achieve DC bus voltage restoration and *ps* or *cs* require knowledge of the output voltage and current/power measurements of a set of converters in order to compute the system average values for bus voltage and output current/power. In addition, compensators should normally be tuned for each control target, which requires a more complex analysis of the interaction between the control loops to avoid stability issues. Techniques that use a single information exchange usually promote *cs*, while the voltage deviation is attenuated by reducing the droop coefficient value. In this chapter, it is discussed in

Table 7.1 Comparison between distributed secondary control techniques for DC Microgrids.

Literature	Information			Comm.	Approach		Adjustment		Comment
	v	*i*	*o*		*vs*	*da*	ΔV_o^*	*ps/cs*	
[2]	✓	-	-	*all*	✓	-	✓	-	*tertiary control aids in voltage restoration*
[17]	✓	-	-	*all*	✓	-	✓	-	*does not consider line impedances*
[38]	✓	-	*s,u*	*neighbor*	✓	-	✓	*cs*	*uses tertiary control to attain cs*
[19, 27–31, 39–41]	✓	✓	-	*neighbor*	✓	-	✓	*cs*	*consensus-based control*
[20]	✓	✓	δR_d	*all*	✓	✓	✓	*cs*	*controls the droop coefficient average*
[21]	-	-	f_{ml}	*all*	-	✓	-	*ps*	*ps achieved by selecting the maximum load factor* f_{ml}
[22]	-	✓	-	*all*	✓	-	-	*cs*	*indirect voltage restoration*
[5]	✓	✓	-	*all*	✓	-	✓	*cs*	*accurate voltage regulation*
[23, 42]	✓	-	*p,s*	*all*	✓	✓	✓	*ps*	*droop V/P control with accurate voltage regulation*
[24, 43]	-	✓	-	*all*	-	✓	-	*cs*	*indirect voltage restoration*

(*Continued*)

Table 7.1 Comparison between distributed secondary control techniques for DC Microgrids. (*Continued*)

Literature	Information			Comm.	Approach		Adjustment		Comment
	v	*i*	*o*		*vs*	*da*	ΔV_o^*	*ps/cs*	
[25, 44]	✓	✓	-	*all*	-	✓	-	*ps*	*Voltage deviation reduction without vs*
[26]	✓	-	$	*neighbor*	✓	✓	✓	-	*solves economic dispatch problem*
[32]	✓	✓	-	*neighbor*	✓	✓	✓	*cs*	*consensus-based control*
[33]	✓	✓	-	*neighbor*	-	✓	-	*cs*	*significant improvement in voltage deviation*
[6]	✓	-	-	*all*	✓	-	✓	-	*indirect voltage restoration*
[45]	-	✓	-	*all*	✓	-	✓	*cs*	*Employs output current data to attain voltage restoration*
[46]	✓	-	$	*neighbor*	✓	-	✓	-	*solves economic dispatch problem*
[47]	-	✓	$	*neighbor*	✓	-	-	*cs*	*economic regulator for cs*
[48, 49]	✓	-	*dV/dt*	*neighbor*	✓	-	✓	*cs*	consensus-based control
[50]	-	✓	-	*neighbor*	✓	-	-	*cs*	*indirect voltage restoration*
[51]	-	-	-	-	✓	-	✓	-	*does not consider multiple converters*

(*Continued*)

Table 7.1 Comparison between distributed secondary control techniques for DC Microgrids. (*Continued*)

Literature	Information			Comm.	Approach		Adjustment		Comment
	v	*i*	*o*		*vs*	*da*	ΔV_o^*	*ps/cs*	
[52]	-	✓	$1/r_{ij}$	*neighbor*	✓	-	✓	*cs*	*Decentralized voltage restoration achieved in primary control. However, prior knowledge of the μG configuration is required*
Hybrid [7]	-	-	*p*	*neighbor*	✓	✓	✓	*ps*	*Consensus-based ps through da and decentralized voltage restoration through vs*
Unique vs [8]	-	-	λ	*neighbor*	✓	-	✓	*ps*	*Achieves both ps and voltage restoration through the convergence of a single data.*

Section 2 – two distributed control techniques for achieving *ps* and the restoration of the average DC bus voltage, using only information exchanged among neighboring converters. These methods have been proposed by the authors, bringing innovation in comparison to the current literature. The first one introduces a hybrid secondary control strategy where *ps* is attained through a *da*-based consensus algorithm, whereas voltage restoration is performed in a decentralized fashion. The second method uses a factor λ, which lumps together the converter output voltage and power, as well as its power capacity, as the information state to be exchanged between converters in a consensus-based environment. When all converter λ converges to an average value, both *ps* and voltage restoration objectives are achieved.

7.2.2 Hybrid Secondary Control: Distributed Power Sharing and Decentralized Voltage Restoration

The hybrid secondary control strategy distributed control to promote proportional *ps* and a decentralized action to restore the DC bus voltage concomitantly. In this method, a sparse communication network is employed, and each converter exchanges its output power information with its neighbors through a LBC network [8]. The information received is used to compute the droop adjustment term (δR_{dj}) that alters the droop coefficient of each converter, compensating for the equivalent line resistances. Once *ps* is attained, each converter uses only local information to generate a voltage-shifting term (δv_{Voj}), which mitigates the DC bus voltage deviation introduced by the primary level droop control. This strategy reduces information traffic and improves system reliability. In addition, the voltage restoration is robust, being disturbed only by significant changes in the equivalent line resistances, i.e., if the load variations do not change the equivalent line resistances seen by the converters, the voltage regulation is guaranteed even during severe communication failures [8].

Considering (7.8), (7.9) and (7.10), (7.18) can be obtained. It shows that it is possible to achieve *ps* by adjusting the droop coefficients through δR_{dj}. The control diagram implementation is shown in Figure 7.7.

$$\frac{P_1}{P_2} = \frac{(R_{d2} + \delta R_{d2})r_2}{(R_{d1} + \delta R_{d1})r_1} \tag{7.18}$$

It is assumed that N converters share the DC bus, where *Conv-j* (j = 1, 2,..., N) communicates with a set of neighboring converters N_j through a LBC link. In system initialization, Conv-j polls each Conv-k, where $k \in N_j$,

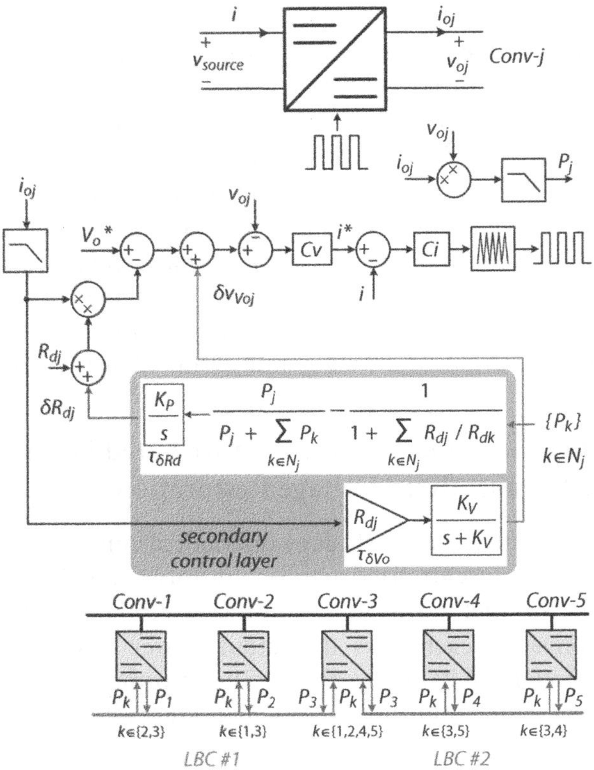

Figure 7.7 Hybrid secondary control diagram (Reprinted with permission from 5052471151140/IEEE.)

and records its nominal droop coefficient R_{dk}. Subsequently, in each control cycle, which is determined by the communication sampling time (τ_{com}), the link is used to exchange output power information between Conv-j and Conv-k synchronously (see Section 7.3 for details on communication synchronization), allowing the j-th converter to calculate its own droop adjustment term (δR_{dj}) as described in (7.19), where P_j and P_k are the output power of *Conv-j* and *Conv-k*, respectively, and K_P is a gain for adjusting the power sharing convergence time.

$$\delta R_{dj}(t) = K_P \int \left(\frac{P_j}{P_j + \sum_{k \in N_j} P_k} - \frac{1}{1 + \sum_{k \in N_j} \frac{R_{dj}}{R_{dk}}} \right) dt \tag{7.19}$$

The dynamic analysis of the control strategy is covered in Section 2.2.1. However, assuming that the algorithm reaches convergence, from (7.19), it can be concluded that the equilibrium leads to proportional *ps* according to:

$$\frac{P_j}{P_j+\sum_{k\in N_j} P_k}=\frac{1}{1+\sum_{k\in N_j}\frac{R_{dj}}{R_{dk}}} \tag{7.20}$$

The output voltage v_{oj} of *Conv-j*, considering δR_{dj}, is expressed in (7.21). Assuming that *ps* correction has already been achieved with the convergence of δR_{dj}, the droop effect ($R_{dj}\,i_{oj}$) can be mitigated by adding a *vs* term (Δv_{Voj}) as in (7.22), where K_V is a gain for adjusting the voltage deviation correction speed.

$$v_{oj}=V_o^*-R_{dj}i_{oj}-\delta R_{dj}i_{oj} \tag{7.21}$$

$$\delta v_{V_oj}(t)=K_V\int(R_{dj}i_{oj}-\delta v_{oj}(t))dt \tag{7.22}$$

Assuming the voltage restoration dynamics to be much slower than that of *ps* correction, changes in i_{oj} along the sampling of v_{oj} can be neglected. Therefore, the solution of (7.22) becomes (7.23). It indicates a first order behavior for the voltage-shifting action with a time constant equal to K_V^{-1}. In steady state, $\delta v_{Voj}=R_{dj}\,i_{oj}$ and adding it in (7.21) results in (7.24), which shows that after a period of $5/K_v$ from the last load perturbation, the converter output voltage exhibits a voltage drop dependent only on its droop adjustment term. Considering that this term compensates the line resistance influence, it can be concluded that in steady state, the system restores the DC bus voltage to its set-point V_O^*.

$$\delta v_{Voj}(t)=R_{dj}i_{oj}(1+e^{-K_Vt}) \tag{7.23}$$

$$\forall t>5/K_V:v_{oj}=V_o^*-\delta R_{dj}i_{oj} \tag{7.24}$$

The correction terms δv_{Voj} and δR_{dj} of *Conv-j* are calculated at each time interval $\tau_{\delta Vo}$ and $\tau_{\delta Rd}$, respectively. Since both actions influence the output current of the converter, to decouple the dynamics between these control loops, the time intervals are defined in relation to τ_{com}, as in (7.25). Considering that the dynamics of *ps* must be faster than the voltage restoration, δR_{dj} is calculated in every communication cycle [8]. Therefore, $\tau_{\delta Vo}$ should be chosen as an integer multiple of τ_{com}, greater than or equal to two. It is noteworthy that, as the term δv_{Voj} is discretely incremented, according to $\Delta\delta v_{Voj} = (R_{dj}\, i_{oj} - \delta v_{Voj})\, K_V\, \tau_{\delta Vo}$, and that the convergence time of the voltage restoration is determined by K_V, an increase in $\tau_{\delta Vo}$ is translated into larger voltage increments. The adjustment gains are proposed to be $K_P = 1/\tau_{\delta Rd}$ and $K_V = 1/5\tau_{\delta Vo}$. Section 3 discusses the design of the communication system and the definition of the time interval τ_{com}.

$$\tau_{\delta Rd} = \tau_{com}$$

$$\tau_{\delta vo} = 2\tau_{com} \tag{7.25}$$

7.2.2.1 *Dynamics and Convergence of the Power Sharing Correction*

Considering (7.19) and the dynamic consensus-based algorithm in discrete time (DCA-DT), which can be represented as [53]:

$$x_j(t+1) = x_j(t) + \epsilon \sum_{k \in N_j} a_{jk}(x_k(t) - x_j(t)) \tag{7.26}$$

where $x_j(t)$ and $x_j(t+1)$ are states of agent j at time t and $t+1$, $\in$ is the weighting constant that adjusts the dynamics of the DCA-DT and a_{jk} is defined according to the communication status between agents j and k: $a_{jk} \neq 0$, if j and k are neighboring agents, $a_{jk} = 0$, otherwise [19]. From the system point of view, (7.26) can be expressed in vector form [53, 54]:

$$X(t+1) = WX(t) \tag{7.27}$$

where $X(t) = [x_1(t), x_2(t), \ldots, x_N(t)]^T$ and W is the weighting matrix of the communication network, defined by [19]:

$$W = I - \epsilon L \tag{7.28}$$

$$L = \begin{bmatrix} \sum_{k \in N_1} a_{1k} & \cdots & -a_{1N} \\ \vdots & \ddots & \vdots \\ -a_{1N} & \cdots & \sum_{k \in N} a_{Nk} \end{bmatrix} \tag{7.29}$$

where I and L are the communication network identity and Laplacian matrices, respectively. Therefore, the states of all agents converge to the consensus value [54]:

$$\lim_{t\to\infty} X(t) = \lim_{t\to\infty} W^t X(0) = \left(\frac{1}{N}1.1^T\right) X(0) \tag{7.30}$$

where 1 is a vector with all elements equal to one and X(0) are the initial states.

As an example, Figure 7.8 represents the Laplacian matrices for a fully connected communication network and a situation where there is a failure in one of the communication links. In both conditions, the secondary control action described in (7.19) is considered and the weighting constant is defined as $\epsilon = \tau_{com}/[(N+1)\,\tau_{ps}]$, with τ_{ps} being the desired time constant for the ps correction. A detailed discussion regarding the selection of a proper weighting constant is presented in [19, 54].

Considering $x_j(t) = R_{dj} + \delta R_{dj}(t) + r_j$, assuming $\delta R_{dj}(0) = 0$ and using (7.27-7.29), it is obtained:

$$\lim_{t\to\infty} X(t) = \frac{1}{\sum_{j=1}^{N} R_{dj}} \left(\begin{bmatrix} R_{d1} & \cdots & 0 \\ \vdots & \ddots & \vdots \\ 0 & \cdots & R_{dN} \end{bmatrix} [1] W \right) X(0) \tag{7.31}$$

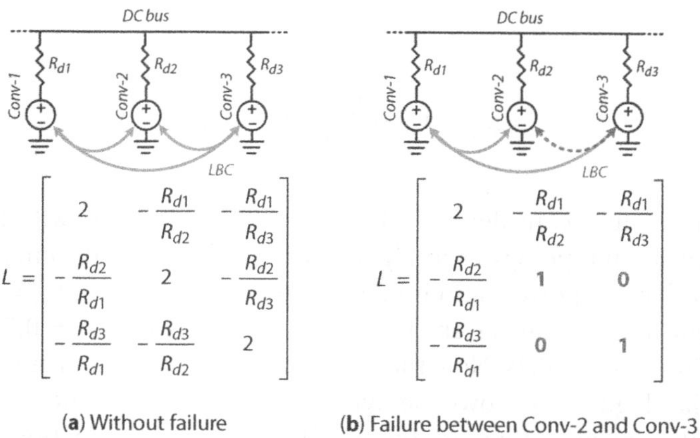

Figure 7.8 Fully connected communication network (Adapted from [8] and reprinted with permission from 5052471151140/IEEE.)

where $[1] = 1.1^T$. Rewriting $x_j(t)$ to $\delta R_{dj}(t) = x_j(t) - R_{dj} - r_j$ and replacing X(t) in (7.31), we obtain:

$$X(t)=\left(\frac{1}{\sum_{j=1}^{N} R_{dj}}\begin{bmatrix} R_{d1} & \cdots & 0 \\ \vdots & \ddots & \vdots \\ 0 & \cdots & R_{dN} \end{bmatrix}[1]W-[I]\right)X(0) \quad (7.32)$$

and

$$\sum_{j=1}^{N} \delta R_{dj}(t)=0 \quad (7.33)$$

Assuming the simplified DC bus model in Figure 7.2b, with two equal converters, then, $R_{d1} = R_{d2}$ and $r_1 \neq r_2$, (7.32) and (7.33) results in:

$$\delta R_{d1} = -\delta R_{d2} = \frac{r_2 - r_1}{2} \quad (7.34)$$

The average of the output voltages between converters is expressed by (7.35). Replacing (7.34) in (7.24) and the result of it in (7.35), then (7.36) is obtained. Assuming $\frac{r_2 - r_1}{4} \ll R_{\mu G}$, then $v_{oavg} \approx V_o^*$.

$$v_{oavg} = \frac{v_{o1} + v_{o2}}{2} \quad (7.35)$$

$$v_{oavg} = V_o^* - \frac{r_2 - r_1}{4}(i_{o1} - i_{o2}) \quad (7.36)$$

As an example, consider a DC bus with three converters with the same communication topology as in Figure 7.8a, the droop coefficients are chosen such that the power ratio is 0.5: 1: 1, then, $R_{d3} = R_{d2} = 2R_{d1} = 2\Omega$. The equivalent line resistances are $r_1 = 0.19\ \Omega$, $r_2 = 0.23\ \Omega$, $r_3 = 0.74\ \Omega$ and τcom= 50 *ms*, τps = 0.2s. Note that, as $r_j \ll R_{dj}$, the power and current ratio are similar. Figure 7.9 shows the dynamic convergence of $\delta R_d(t)$ and $x(t)$ under different operating conditions. The compensation action starts in $t = 0.5$ s and in $t = 2$ s a load variation produces changes in the equivalent line resistances, leading to $r_1 = 0.10\ \Omega$, $r_2 = 0.12\ \Omega$, $r_3 = 0.26\ \Omega$. In both

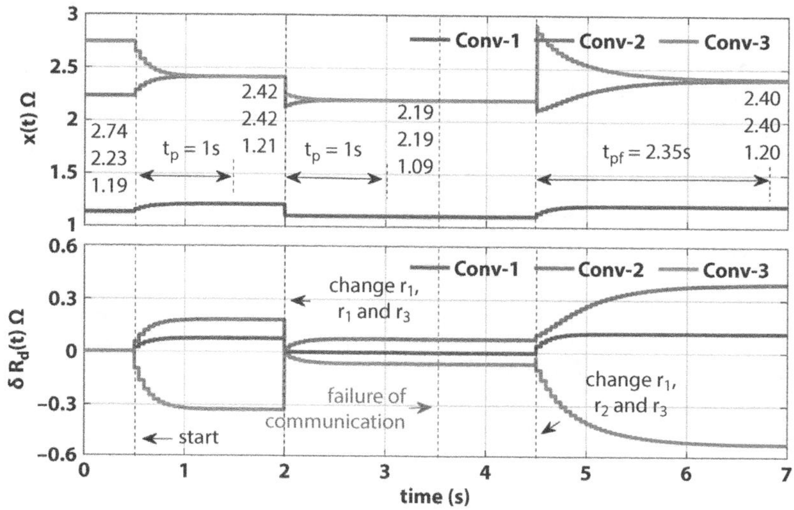

Figure 7.9 $\delta R_d(t)$ Convergence (Reprinted with permission from 5052471151140/IEEE.)

cases, it is observed that $x(t)$ converges to the specified power ratio with the same time interval t_p. At $t = 3.5$ s, a communication failure occurs between Conv-2 and Conv-3, but the power proportion is not affected because no changes in the line resistances had occurred during this event. At $t = 4.5$ s the equivalent line resistances change to $r_1 = 0.10\ \Omega$, $r_2 = 0.10\ \Omega$, $r_3 = 0.9\ \Omega$ and again it is possible to note the convergence being achieved even under a communication failure. However, it is noted that if a communication failure occurs, the time for convergence is longer than no-failure condition.

The error in $x(t)$ decreases exponentially with a rate that is related to the L matrix eigenvalues $\lambda(.)$. Thus, $\lambda(.)$ determines the system global dynamics [55–57]. In this sense, the Laplacian matrix considering a communication failure, Figure 7.8b, presents a set of eigenvalues that results in a convergence time $t_{pf} = 2.35t_p$, i.e., 2.35 times greater than in the initial communication topology.

7.2.2.2 *Communication Delays in Consensus-Based Algorithm*

The dynamic stability of the consensus-based algorithm is sensitive to communication delays. According to [57, 58], considering uniform communication delays, $\tau_{jk} = \tau_d$, the consensus algorithm converges if $\tau_d < \pi/2\lambda_N$, where λ_N is the greatest eigenvalue of the L matrix. Since τ_d is inversely proportional to λ_N, then, conservatively, it can be assumed $\tau_{dmax} < \pi/2_N$ as the communication delay limit, since λ_N will be lower than the order of the

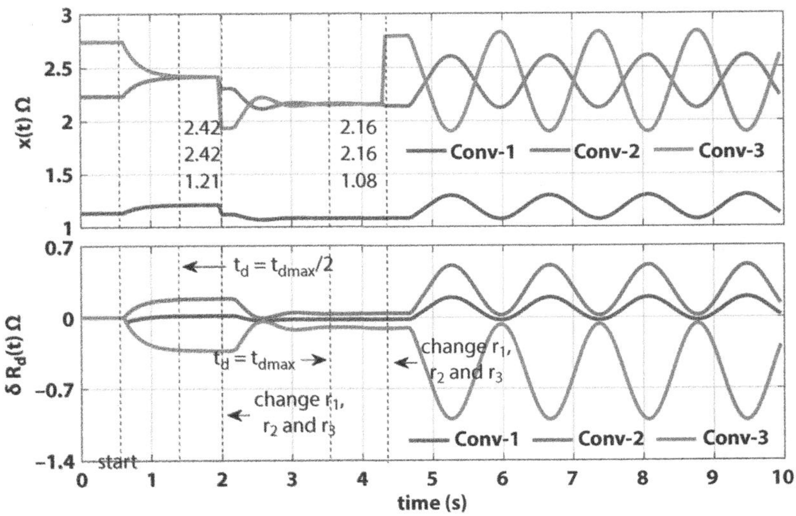

Figure 7.10 $\delta R_d\,(t)$ Convergence under communication delays (Reprinted with permission from 5052471151140/IEEE.)

L matrix. Moreover, for $\tau_d = \pi/2\lambda_N$, the system has a stable oscillatory solution with frequency $\omega = \lambda_N$ [57]. Therefore, the communication in large networks must be carefully designed to minimize communication delay or to increase the maximum tolerable delay limit, for example, by reducing the number of connections per node, which decreases λ_N [8].

To illustrate the influence of long communication delays can have on the μG behavior, we revisit the system described in the previous section, assuming the existence of a communication delay (t_d) between the converters. Figure 7.10 shows the convergence of $\delta R_d\,(t)$, considering the same parameters $R_{d1,2,3}$ and $r_{1,2,4}$ of Figure 7.9. At $t = 0.5$ s, the control is enabled with $t_d = 0$ and at $t = 1.2$ s and at $t = 3.5$ s. The communication delays are included with $t_{dmax}/2$ and t_{dmax}, respectively. It is observed that, for $t_d < t_{dmax}$, the system converges asymptotically; however, the delay decreases the damping of the transient response. Although, when $t_d < t_{dmax}$, the transient response becomes marginally stable, oscillating around the consensus value appears, as highlighted in Figure 7.10.

7.2.2.3 *Secondary Control Modeling*

Assuming the simplified DC μG model of Figure 7.2b, the converter output currents are expressed as:

$$i_{o1} = \gamma_1 v_{o1} - \kappa v_{o2}$$

$$i_{o2} = \gamma_2 v_{o2} - \kappa v_{o1} \tag{7.37}$$

where

$$\gamma_1 = \frac{r_2 + R_{\mu G}}{r_1 r_2 + R_{\mu G}(r_1 + r_2)}$$

$$\gamma_2 = \frac{r_1 + R_{\mu G}}{r_1 r_2 + R_{\mu G}(r_1 + r_2)} \tag{7.38}$$

$$\kappa = \frac{R_{\mu G}}{r_1 r_2 + R_{\mu G}(r_1 + r_2)}$$

The closed-loop control diagram of *Conv-1* is shown in Figure 7.11. Considering that τ_{com} is much slower than the converter response time, δv_{Vo1}, and then δR_{d1} can be seen as disturbances in the voltage reference and the droop coefficient, respectively [8]. The communication delay is represented as $e^{-\tau_d s}$ and a second order Padé approximation is used to model it. The closed-loop transfer function (TF) is expressed by [5, 59]:

$$G_{vo} = \frac{C_v G_i}{1 + C_v G_i} \tag{7.39}$$

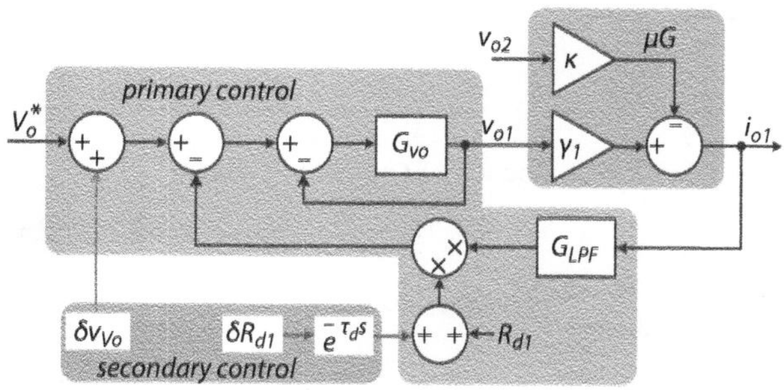

Figure 7.11 Closed-loop secondary control diagram (Reprinted with permission from 5052471151140/IEEE.)

where C_v and G_i are the voltage compensator and the current loop *TF*, respectively. G_i is represented by a unit delay [5]. Therefore, the converters output voltages can be expressed as:

$$\begin{aligned} v_{o1} &= [V_o^* - i_{o1}G_{LPF}(R_{d1} + \delta R_{d1})]G_{vo} \\ v_{o2} &= [V_o^* - i_{o2}G_{LPF}(R_{d2} + \delta R_{d2})]G_{vo} \end{aligned} \tag{7.40}$$

where

$$\begin{aligned} G_{LPF} &= \frac{2\pi fc}{s + 2\pi fc} \\ \delta R_{d1} &= \frac{r_2 - r_1}{2} \\ \delta R_{d2} &= \frac{r_1 - r_2}{2} \end{aligned} \tag{7.41}$$

To evaluate the closed-loop model presented in Figure 7.11, the derived closed-loop model is compared to transient behavior of a complete model considering two converters, with *Conv-1* and *Conv-2*. The simulation results are shown in Figure 7.12 and the parameters used are shown in Table 7.2. At t = 0.2 s (A), the proposed control technique is enabled; at

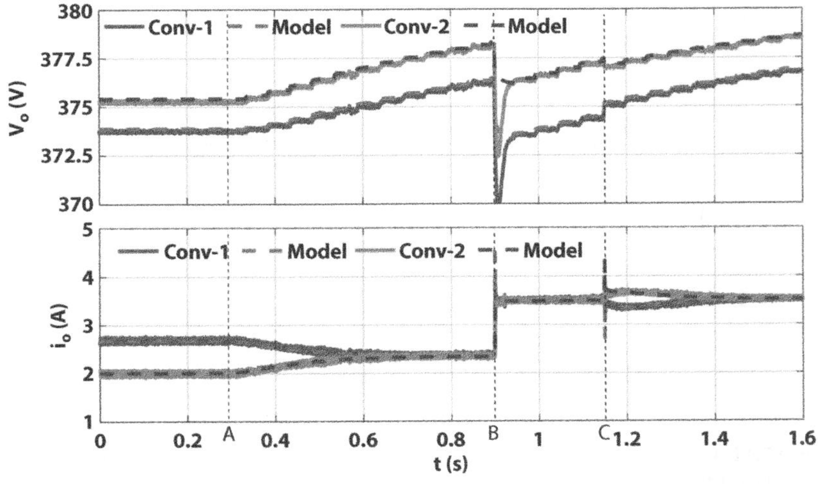

Figure 7.12 Output signals from the complete μG model and the stability analysis model.

$t = 0.8\ s$ (B), a load disturbance of 50% is applied and at $t = 1.025\ s$ (C), r_1 is changed to $r_1 = 0.4\ \Omega$.

Combining (7.37)-(7.41) results in (7.42), which allows the influence of parameter variations on stability to be evaluated.

$$\frac{v_{o1}}{V_o^*} = \frac{G_{vo}}{1 + \left[\alpha_1 G_{LPF}\left(R_{d1} + \frac{r_2 - r_1}{2}\right) + 1\right] G_{vo}}$$
$$\frac{v_{o2}}{V_o^*} = \frac{G_{vo}}{1 + \left[\alpha_2 G_{LPF}\left(R_{d3} + \frac{r_1 - r_2}{2}\right) + 1\right] G_{vo}} \tag{7.42}$$

Figure 7.13 shows the dominant poles root *loci* under variations in r_1, R_{d1}, δR_{d1} and t_d. Table 7.2 describes the parameters used in the analyses. Figure 7.13a shows the influence of the line resistance on the closed-loop system poles. The value of r_2 is kept fixed and r_1 varies from 0.01 Ω to R_{d1}, allowing to evaluate $r_1 < r_2$ and $r_1 > r_2$. Figure 7.13b shows the influence of R_{d1} in the closed-loop poles, where R_{d1} varies from $0.1R_{d1}$ to $1.9R_{d1}$ and the value of R_{d2} is not changed. In Figure 7.13c, it is shown the root locus

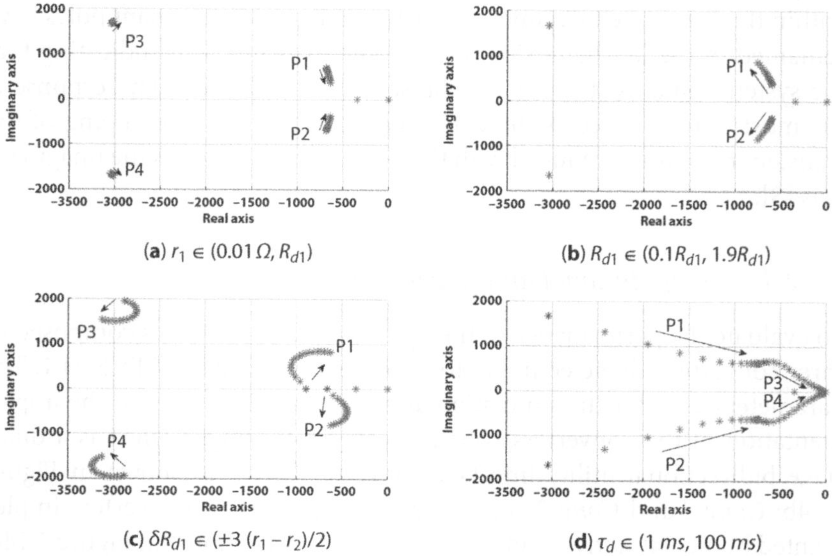

(a) $r_1 \in (0.01\,\Omega, R_{d1})$

(b) $R_{d1} \in (0.1R_{d1}, 1.9R_{d1})$

(c) $\delta R_{d1} \in (\pm 3\,(r_1 - r_2)/2)$

(d) $\tau_d \in (1\ ms, 100\ ms)$

Figure 7.13 Closed-loop root locus under parameter variations (Reprinted with permission from 5052471151140/IEEE.)

Table 7.2 Parameters used in the Stability Analysis of Hybrid Control (Reprinted with permission from 5052471151140/IEEE.)

Item	Symbol	Value
Nominal voltage	V_o^*	380 V
Line resistance	$r_{1,2}$	0.1, 0.9 Ω
Load resistance	$R_{\mu G}$	32,9 Ω
Droop Coefficient	$R_{d1,2}$	2.3, 2.3 Ω
Communication sampling time	τ_{LBC}	50 ms
Communication delay	τ_d	1 ms
LPF cut-off frequency	f_c	100 Hz
Power sharing speed correction gain	K_P	20
Voltage speed correction gain	K_V	2

for variation in δR_{d1}, where δR_{d1} is varied from $-3(r_1 - r_2)/2$ to $3(r_1 - r_2)/2$, and finally, Figure 7.13d shows the variation in τ_d from 1 ms to 100 ms. In the four conditions, only the high frequency complex poles are affected by the parameter variations, with different trajectories. However, in all cases, within the considered parameter variation ranges, all dominant poles have remained in the left-hand side of the complex plane, thus indicating that the system remains stable. It can be seen that the time delay response is the most critical, since as the delay approaches t_{dmax}, the damping of the transient response is reduced with the poles moving towards the imaginary axis [8].

7.2.2.4 *Computational and Experimental Validation*

To evaluate the performance of the hybrid control, a DC bus composed of three converters, depicted in Figure 7.14a, is simulated in PLECS v4.1. The parameters used for the simulation are shown in Table 7.3 and the implementation of the converters considered the following: *Conv-1* is a dual-stage bidirectional utility interface converter (BIC), as shown in Figure 7.14b; *Conv-2* and *Conv-3* are energy storage interface converters implemented by a Dual Active Bridge topology (DAB), as shown in Figure 7.14c. The experimental testbed, shown in Figure 7.15, exhibits the same configuration, where the rated power of the BIC and the two DAB converters are

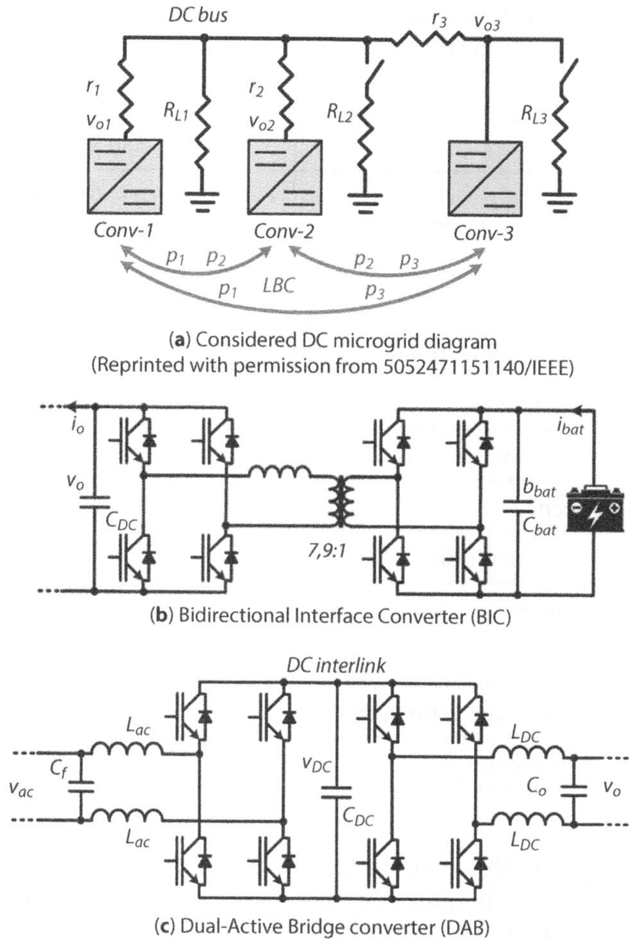

Figure 7.14 Simulated and experimental system used for performance validation.

3.2kW and 1.6 kW, respectively. The experimental communication layer is implemented using a 5125 kbps CAN 2.0 network.

Figure 7.16 shows the simulated results for the hybrid control. At the beginning of the simulation, $t < 1$ *s*, only the primary level droop control is actived, which leads to average voltage of 376.4 V and converter output power distribution of $P_1 = 0.88$ kW, $P_2 = 0.56$ kW and $P_3 = 0.74$ kW, in steady state. In $t = 1$ *s*, the secondary control is enabled, hence, the output power imbalance is corrected with proportional *ps* being achieved in a time interval of $t_{ps} = 1$ *s*. The average output voltage of the converters converges to 380.1 V in $t_{\Delta V_o^*} = 2.5\,s$, as expected. The droop coefficients correction

Table 7.3 Simulation/Experimental Parameters of Hybrid Control (Reprinted with permission from 5052471151140/IEEE.)

Item	Symbol	Value
Nominal voltage	V_o^*	380 V
Line resistances	$r_{1,2,3}$	0.9, 0.9, 0.1 Ω
Load resistances	$R_{L1,2,3}$	65.4, 133 133 Ω
Droop Coefficients	$R_{d1,2,3}$	1.15, 2.3, 2.3 Ω
Communication sampling time	τ_{LBC}	50 ms
Communication delay	τ_d	1 ms
LPF cut-off frequency	f_c	100 Hz
Switching frequency	f_{sw}	15 kHz
DAB transformer turns ratio	N	7.9:1
PI voltage controller	PI	$k_p = 1.8$, $k_i = 276$
PI current controller	PI	$k_p = 0.3$, $k_i = 20$
Power sharing speed correction gain	K_P	20
Voltage speed correction gain	K_V	2

terms converge to $\delta R_{d1} = -0.512\ \Omega$, $\delta_{Rd2} = -0.142\ \Omega$, $\delta_{Rd3} = 0.654\ \Omega$, thus, $\delta R_{d1} + \delta_{Rd2} + \delta_{Rd3} = 0$. In addition, the output power of the converters converges to $P_1 = 1.1\ kW$, $P_2 = 0.55\ kW$, $P_3 = 0.55\ kW$, while the output currents are $i_{o1} = 2.82\ A$, $i_{o2} = 1.46\ A$, $i_{o3} = 1.45\ A$, showing a small difference between the current and power ratios, as expected as well.

At $t = 5.5\ s$ (event A), load R_{L2} is connected to the DC bus, disturbing the output voltages of converters. However, as the system equivalent line resistances are not affected, it does not introduce *ps* deviations. As can be observed in Figure 7.16, all converters proportionally increase their output power, without altering the steady-state values of the droop correction terms, while the decentralized voltage correction action gradually restores the DC bus voltage. At $t = 7.5\ s$ (B), there is a communication failure between *Conv-1* and *Conv-3*; however, there are no load changes, voltage regulation and *ps* were both preserved. At $t = 9\ s$ (C), R_{L3} is connected to the DC bus, disturbing the output voltages and changing in the equivalent line resistances. Once again, the secondary control calculates new values

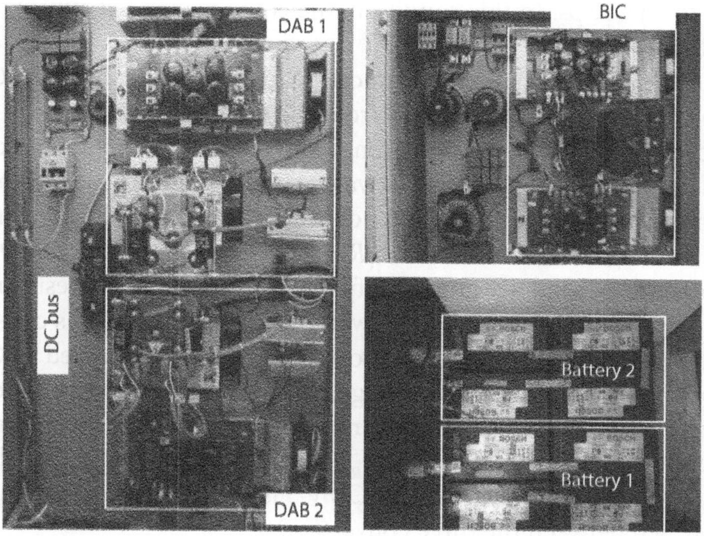

Figure 7.15 Experimental setup.

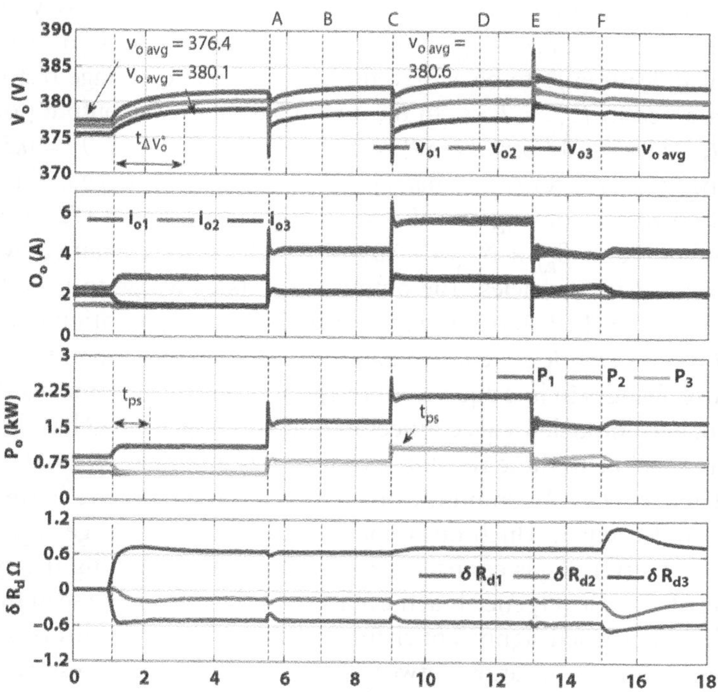

Figure 7.16 Simulation results considering load disturbances and communication failures (Adapted from [8] and reprinted with permission from 5052471151140/IEEE.)

for δR_d to ensure *ps*. In this condition, the convergence last approximately $t_{ps} = 0.1\ s$.

At $t = 11.5\ s$ (D), a new communication failure occurs, between *Conv-2* and *Conv-3*, which isolates *Conv-3* from the other converters. Since it does not receive data from the remaining converters, the secondary control actions are halted, keeping the last values of δv_{Vo3} and δR_{d3}. Since no load variation occurred, this event does not influence *ps* and voltage regulation. At $t = 13\ s$ (E), on the other hand, R_{L2} is disconnected from the DC bus, disturbing the bus voltage and modifying the equivalent line resistances. It can be seen that the power ratio between *Conv-1* and *Conv-2* is conserved, regardless of Conv-3. Voltage deviation correction is maintained by *Conv-1* and *Conv-2*, which reduces the voltage error of the DC bus, but it increases the difference between the output currents i_{o2} and i_{o3}. The droop correction terms converge to $\delta R_{d1} = -0.522\ \Omega$, $\delta_{Rd2} = -0.154\ \Omega$, $\delta R_{d3} = 0.753\ \Omega$. Finally, at $t = 15\ s$ (F), communication with *Conv-3* is restored, and as a result, *ps* is reestablished between the three converters.

Figure 7.17 shows the experimental results for a similar sequence of events as described in Figure 7.16. For $t < 1\ s$, the output power of *Conv-1*, *Conv-2* and *Conv-3* are 0.9 kW, 0.58 kW and 0.77 kW, respectively, and the average DC bus voltage is $v_{oavg} = 376.1\ V$. At $t = 1\ s$, the secondary control is enabled, leading *ps* to be achieved in $t_{ps} \approx 1.2\ s$ and voltage restoration to be attained in $t_{\Delta V_o^*} = 2.5\ s$. In the new steady state, the output power of *Conv-1*, *Conv-2* and *Conv-3* are $P_1 = 1.16\ kW$, $P_2 = 0.58\ kW$, $P_3 = 0.58\ kW$, respectively. It is noteworthy that a proportional *ps* is achieved. The droop correction terms converge to $\delta R_{d1} = -0.469\ \Omega$, $\delta R_{d2} = -0.172\ \Omega$, $\delta R_{d3} = 0.55\ \Omega$, but $\delta R_{d1} + \delta_{Rd2} + \delta_{Rd3} = -0.09\ \Omega$, which can be explained by small measurement inaccuracies. The output currents converge to $i_{o1} = 2.98\ A$, $i_{o2} = 1.52\ A$, $i_{o3} = 1.51\ A$, once again exhibiting a small deviation in *cs*.

At $t = 8.2\ s$ (A), R_{L2} is connected to the DC bus. In theory, this would not cause changes in line equivalent resistances; however, some resistances have not been modeled, e.g., connectors and bonding. *Conv-3* has shown to be more sensible to these resistances which has modified $\delta R_{d3} = 0.55\ \Omega$ to $0.38\ \Omega$ in $t_{ps} = 092\ s$. A communication failure occurs at $t = 10\ s$ (B) between *Conv-1* and *Conv-3*, which does not influence *ps* and voltage regulation. At $t = 11.4\ s$ (C), R_{L3} is connected to the DC bus, disturbing the system. Afterwards, *ps* convergence is achieved in $t_{ps} \approx 0.9\ s$. It can be seen that this event has not caused significant changes in the values of the droop correction terms. Communication between *Conv-1* and *Conv-3* is restored at $t = 13\ s$ (D). At $t = 13.9\ s$ (E) and at $t = 15.3\ s$ (F), R_{L2} and R_{L3} are disconnected, respectively. The hybrid control technique has ensured proportional *ps* and voltage restoration with $v_{oavg} \approx V_o^*$. It is also possible to observe that the

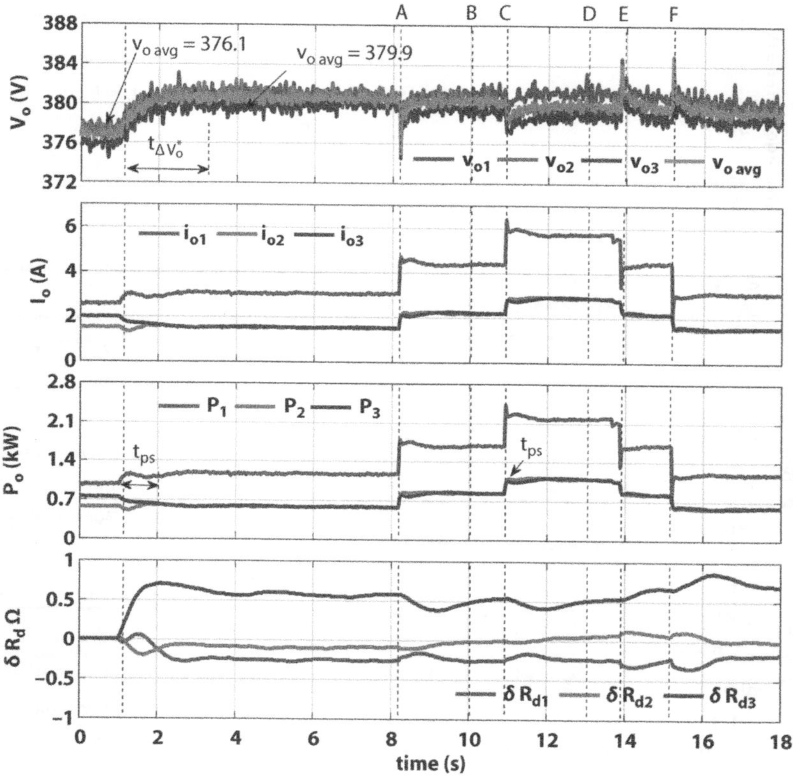

Figure 7.17 Experimental result with load disturbances and communication failures (Adapted from [8] and reprinted with permission from 5052471151140/IEEE.)

intrinsic delay present in the CAN 2.0 communication link has not interfered with the system's behavior or compromised its stability.

To understand the impact of higher values of communication delay in the system behavior, a computational analysis considering three different values of time delays and also the presence of constant power loads (cpl) is carried out. The results are shown in Figure 7.18. At the beginning of the simulation, only R_{L1} is connected to the bus and the system operates with $v_{oavg} = 376\ V$, $P_1 = 0.9\ kW$, $P_2 = 0.76\ kW$, $P_3 = 0.57\ kW$. At $t = 1\ s$, the secondary control is enabled. It can be noticed that, for $t_d \leq t_{dmax}/10$, the communication delay has a negligible impact on system performance and in steady state the obtained output power and current values are: $P_1 = 1.1\ kW$, $P_2 = 0.55\ kW$, $P_3 = 0.55\ kW$, $i_{o1} = 2.9\ A$, $i_{o2} = 1.43\ A$, $i_{o3} = 1.42\ A$. On the other hand, $t_d \approx t_{dmax}$ severely interferes with *ps* inducing oscillations in δR_d, which are reflected in the output voltage, power and current of the

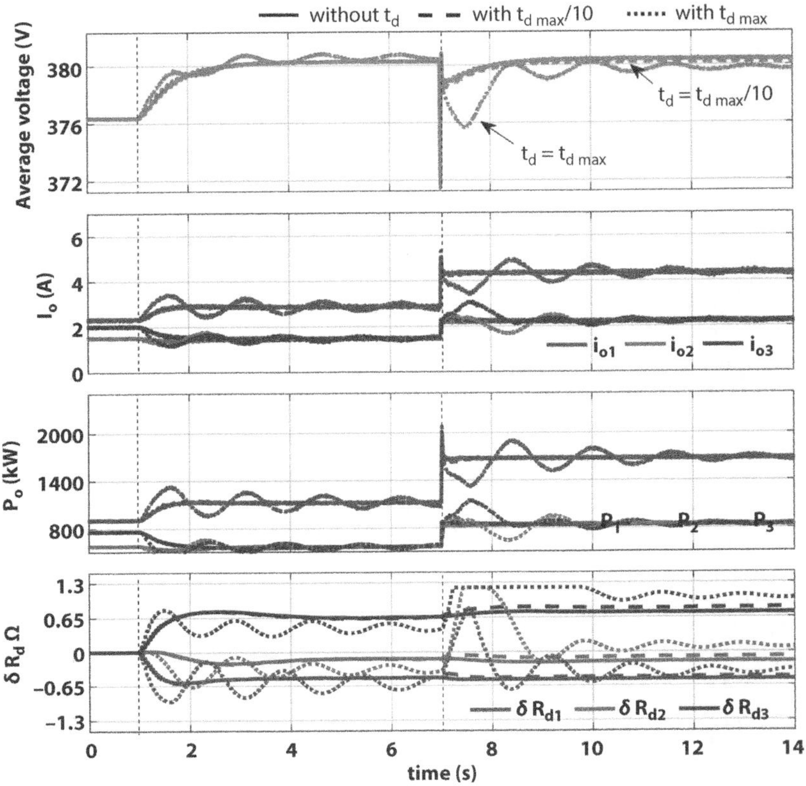

Figure 7.18 Simulation results considering constant power loads and communication delay (Reprinted with permission from 5052471151140/IEEE.)

converters. It is noteworthy that this scenario is extremely unrealistic for a three-node system, since a $t_d = t_{dmax} = 530\ ms$ exceeds the communication delay reported in the literature. Anyhow, the system proved to be tolerant to moderate communication delay ($t_d < t_{dmax}$). Moreover, the average DC bus voltage, even under a very high communication delay, converges with minor steady-state error regarding the 380 V reference value and with small ripple.

At $t = 7\ s$, R_{L3} is connected to the DC bus. It is observed that for $t_d = t_{dmax}$, δR_{d2} and δR_{d3} saturate for a period of time. However, the output power of the converters oscillates around $P_1 = 1.68\ kW$, $P_2 = 0.84\ kW$ and $P_3 = 0.84\ kW$, asymptotically achieving consensus. It is noteworthy the increase of loads displaces the closed-loop poles, moving them away from the right half-plane, as shown in Figure 7.13c.

7.2.3 Secondary Level Control Based on Unique Voltage-Shifting (*vs*)

This control strategy allows the μG to achieve an accurate *ps* with concomitant DC bus voltage restoration through shifting the converter's reference voltage. In contrast to similar approaches, as described in Table 7.1, only one variable per converter is transmitted in the communication link and with unity-gain integrator as a secondary level controller. The information shared among the converters is a λ factor which corresponds to the product between the output voltage of the converter and a quantity related to its output power and rated power [7].

Periodically, each converter calculates its own λ and broadcasts it, afterwards, based on the received data, each converter determines an average λ value, which is then used to perform the voltage-shifting action. The proposed technique reduces the information traffic, allowing the use of ultra-LBC[2] networks, eliminating the need of design several *vs* compensators and providing plug-and-play capability and flexibility to the μG.

Considering again the simplified DC bus model of Figure 7.2b, the average output voltage between *Conv-1* and *Conv-2* can be expressed by (7.43). The intrinsic voltage deviation (ΔV_o^*) produced by droop control can be decreased by δv_{Vo}, which is added to the voltage set-point of each converter ($V_o^* + \delta v_{Vo}$), so that $v_{oavg} = V_o^*$. Replacing (7.2) and (7.4) in (7.43), δv_{Vo} is defined by (7.44), where α and β were already defined in (7.3).

$$v_{oavg} = \frac{v_{o1} + v_{o2}}{2} \tag{7.43}$$

$$\delta v_{V_o} = \frac{2R_{d1}R_{d2} + R_{d1}r_2 + R_{d2}r_1}{2(\beta R_{\mu G} + r_1 r_2) + R_{d1}r_2 + R_{d2}r_1} V_o^* \tag{7.44}$$

Considering (7.7), it indicates that whether appropriate voltage-shifting terms $\delta v_{o1} = \delta v_{Vo} + \delta v_{\Delta i1}$ and $\delta v_{o2} = \delta v_{Vo} + \delta v_{\Delta i2}$ are used to adjust V_{o1}^*

---

[2] Assuming a sampling frequency f_s, then a High Bandwidth communication network (HBC) is defined when the communication frequency (f_{com}) becomes $f_{com} = f_s$ and a Low Bandwidth communication link (LBC) is defined when $f_{com} = f_s/X$, where $X \in Z$ and $X > 1$ [5]. An Ultra-LBC can be defined when X>100.

and V_{o2}^*, respectively, the current sharing error can be then mitigated. Therefore,

$$\begin{aligned} V_{o1}^* &= V_o^* + \delta v_{Vo_o} + \delta v_{\Delta i1} \\ V_{o2}^* &= V_o^* + \delta v_{V_o} + \delta v_{\Delta i2} \end{aligned} \tag{7.45}$$

However, according to (7.7), $r_1 < r_2$ implies that $\delta v_{\Delta i1} < \delta v_{\Delta i2}$, thus, $\Delta i_o = 0$ leads to $v_{oavg} \neq V_o^*$. Otherwise, if $\delta v_{\Delta i2} > 0$ and $\delta v_{\Delta i1} = -\delta v_{\Delta i2}$, implies that $\Delta i_o = 0$ and $v_{oavg} \neq V_o^*$. It is concluded that $r_1 \neq r_2$ precludes achieving concomitant voltage regulation and balanced output currents. One can express the converters' output power imbalance as

$$\Delta p_o = p_{o1} - p_{o1} = v_{o1} i_{o1} - v_{o2} i_{o2} \tag{7.46}$$

Assuming the condition where $\delta v_{\Delta i1} = -\delta v_{\Delta i2} = \delta v_{\Delta i}$ and replacing (7.2), (7.4), (7.44) and (7.45) in (7.46), it is obtained Δp_o as a function of $\delta v_{\Delta i}$ [7]:

$$\Delta p_o (\delta v_{\Delta i}) = a (\delta v_{\Delta i})^2 + b (\delta v_{\Delta i}) + c \tag{7.47}$$

where

$$\begin{aligned} a &= \frac{R_{d2} R_{t1}^2 - R_{d1} R_{t2}^2 + (\alpha + \beta R_{\alpha G})(R_{d2} - R_{d1} + r_2 - r_1)}{(\alpha + \beta R_{\alpha G})^2} \\ b &= \\ & V_o^* \frac{(R_{d2} + r_2)(2R_{d1} R_{t2} - \alpha - \beta R_{\alpha G}) - (R_{d1} + r_1)(\alpha + \beta R_{\alpha G} - 2R_{d2} R_{t1}) - (\beta + 4R_{\alpha G})(\alpha + \beta R_{\alpha G})}{(\alpha + \beta R_{\alpha G})^2} \\ c &= \frac{(\alpha + \beta R_{\alpha G})(R_{d2} - R_{d1} + r_2 - r_1) + R_{d2}(R_{d1} + r_1)^2 - R_{d1}(R_{d2} + r_2)^2}{(\alpha + \beta R_{\alpha G})^2} V_o^{*2} \end{aligned} \tag{7.48}$$

And $R_{tj} = R_{dj} + r_j + 2R_{\mu G}$ for j = 1,2.

The converters' output power are $p_{o1} = v_{o1} i_{o1}$ and $p_{o2} = v_{o2} i_{o2}$, so that, $\Delta i_o = 0$ results in $p_{o1} < p_{o2}$. For the condition where $\delta v_{\Delta i1} = -\delta v_{\Delta i2} = \delta v_{\Delta i}$ and Δp_o $(\delta V_{\Delta i})$. It is noteworthy that, since line resistances are not compensated, a small error between v_{oavg} and $v_{\mu G}$ is expected.

According to [7], the proposed technique aims at computing a single voltage-shifting term (δv_V), for each converter, that promotes accurate and proportional *ps* and DC bus voltage restoration. LBC network is used to exchange information among neighboring converters and a consensus-based algorithm is employed to converge the information state to an average value. However, unlike other proposals, where multiple converter measurements are exchanged in the LBC link, and multiple control loops are required to compose the proper δv_V, this technique defines a factor λ as the information state. It is computed locally by each converter, according to (7.49), where v_{oj} is the j-th converter output voltage and $\bar{P}_{oj}$ is a power quantity defined as (50), where p_{oj} is the converter output power, and P_{MAX} is its rated power. The gain 1/2 is needed to avoid a division by zero in the local controller when $p_{oj} = P_{jmax}$ [7].

$$\lambda_j = \bar{P}_{oj} v_{oj} \tag{7.49}$$

$$\bar{P}_{oj} = 1 - \frac{1}{2}\frac{p_{oj}}{P_{j\max}} \tag{7.50}$$

During a control cycle, each converter sends its λ factor to its N neighbors and receives their respective λ factors, through the LBC network. Afterwards, it locally computes a voltage-shifting term as [7]:

$$\delta v_{V_j} = \int \left(V_o^* - \frac{\lambda_{avg}}{\bar{P}_{oj}} \right) dt \tag{7.51}$$

where, V_o^* is the DC bus voltage reference and λ_{avg} is the average λ, defined as:

$$\lambda_{avg} = \frac{\lambda_1 + \lambda_2 + \cdots + \lambda_N}{N} \tag{7.52}$$

To deal with communication failures and the addition/removal of new agents, the algorithm defines N as the number of neighboring converters that send information to the j-th converter in a control cycle. The proposed local controller diagram is shown in Figure 7.19, where C_v and C_i are the voltage and current loop PI (proportional-integral) compensators, respectively. R_{dj} is the droop coefficient and lowpass filters are used to smooth

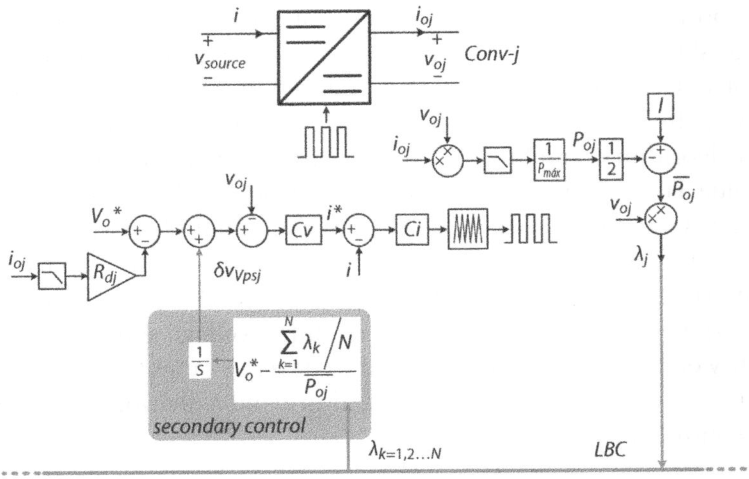

Figure 7.19 Unique *vs* control diagram (Adapted from [7] and reprinted with permission from 5052471360356/IEEE.)

the output current and power measurements. Assuming, for the sake of argument, that the secondary control strategy is able to converge, then $V_o^* = \frac{\lambda_{avg}}{\bar{P}_{o1}} = \frac{\lambda_{avg}}{\bar{P}_{oN}}$, through (7.49), (7.50) and (7.52) it can be shown that a convergence of δv_V will lead to

$$\frac{p_{o1}}{P_{1max}} = \frac{p_{o2}}{P_{2max}} = \cdots = \frac{p_{oN}}{P_{Nmax}}$$
$$V_o^* = \frac{v_{o1} + v_{o2} + \cdots + v_{oN}}{N} \tag{7.53}$$

Although there is a single secondary control loop, *ps* and voltage restoration exhibit different time constants. The convergence analysis of each control objective is addressed separately in Section 7.2.3.1.

7.2.3.1 Power Sharing and Average Voltage Convergence Analysis

Assuming that the control cycle, i.e., the sampling time of the secondary control layer ($t_{\delta v_V}$), is greater than the primary control bandwidth, the dynamic behavior of the proposed strategy can be studied considering the simplified μG model of Figure 7.2b. Applying (7.51) to V_{o1}^* and V_{o2}^* results in a set of nonlinear differential equations, with no explicit analytical solution, thus, the system response can only be assessed by numerical analysis. Evaluating several scenarios with the aid of numerical simulations, it is observed that the power imbalance behavior exhibits a second order system response, which is influenced by the load, droop and line resistances and $t_{\delta v V_{ps}}$ as well. Interestingly, when the converters comply with R_{d1} P_{1MAX}, two different scenarios exhibit the same dependency on the normalized load power and $t_{\delta v V}$ whether both also exhibit the same load type (resistive or constant power load (cpl) and the same k_β, where [7]

$$k_\beta = \frac{R_{d1} + R_{d2} + r_1 + r_2}{R_{d1} + R_{d2}} \tag{7.54}$$

Therefore, the performance of multiple scenarios can be studied by evaluating how changes in k_β impact the system behavior. Notice that the greater the k_β, the more significant are the line resistances in relation to the virtual droop resistances. To quantify the system performance in each scenario, a process estimation function is used to estimate the second order system model from the numerical simulation data. Figure 7.20 presents the root loci and settling time response of the estimated second order model against $t_{\delta v V}$ and the load power for a μG with resistive load and cpl as well.

It can be noticed that, for very low $t_{\delta v V}$, the system shows a dominant pole that grants it a first order behavior. As $t_{\delta v V}$ increases, system damping is decreased, reducing the *ps* settling time (Δt_{ps}), but as the poles become complex and move towards the RHS of the s-plane, Δt_{ps} also increases. Eventually, for large $t_{\delta v V}$, the system becomes unstable, e.g., for $k_\beta \approx 1$, critical stability is reached at $t_{\delta v V} \approx 65\text{ ms}$. An increase in k_β as well as a decrease in the load power raises system damping, reducing Δt_{ps} for complex poles and the contrary for real poles. It is also noticed that in the load variation range, Δt_{ps} varies roughly 6dB, whereas in the $t_{\delta v V}$ range, it varies up to 35 dB. Hence, proper selection of $t_{\delta v V}$ is crucial for the successful operation of the μG. It is noteworthy that, although small differences are observed between the resistive and cpl behaviors, both types have performed similarly. Since the proposed strategy does not alter the output impedance of the converters, once the primary level is designed to ensure a stable operation in the presence of CPL, the secondary layer does not degenerate the load dynamics [7].

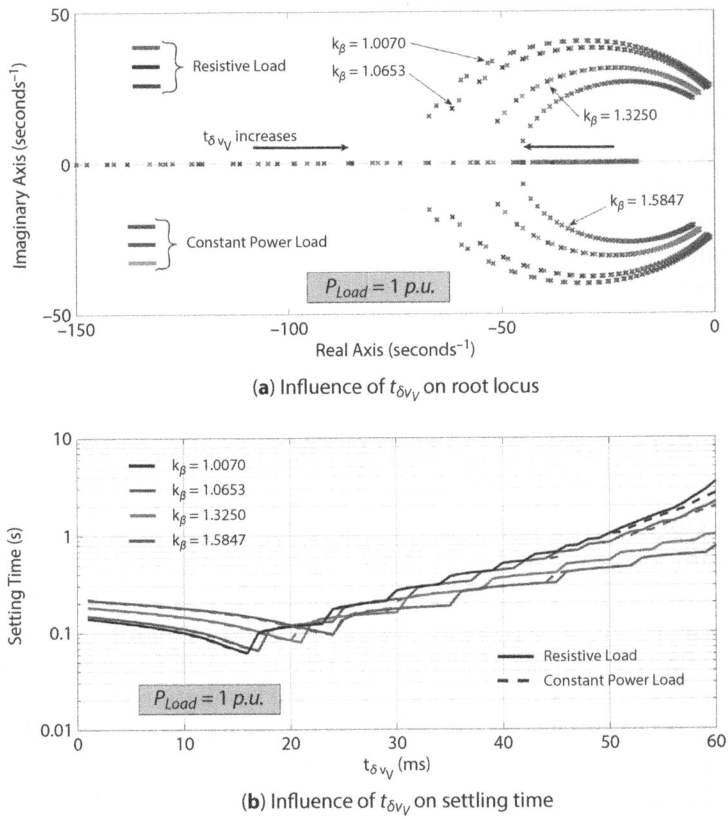

Figure 7.20 Assessment of the power sharing behavior. In (a) and (b), $t_{\delta v_V}$ varies from 1 ms to 60 ms. In (c) and (d), P_{Load} varies from 1 p.u. to 0.1 p.u. (Reprinted with permission from 5052471360356/IEEE.) *(Continued)*

Once power balancing is achieved, the secondary-layer control feedback becomes

$$\frac{\overline{P}_o v_{o1} + \overline{P}_o v_{o2}}{2\overline{P}_o} = \frac{v_{o1} + v_{o2}}{2} = v_{oavg} \tag{7.55}$$

Consequently, the secondary control law can be rewritten as

$$\delta v_{Vj}(t) \approx \int (V_o^* - v_{oavg}(t))dt \tag{7.56}$$

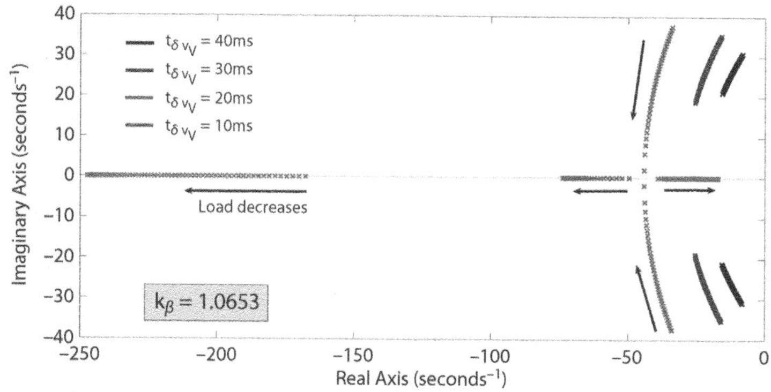

(c) Influence of the Load on the Root Locus.

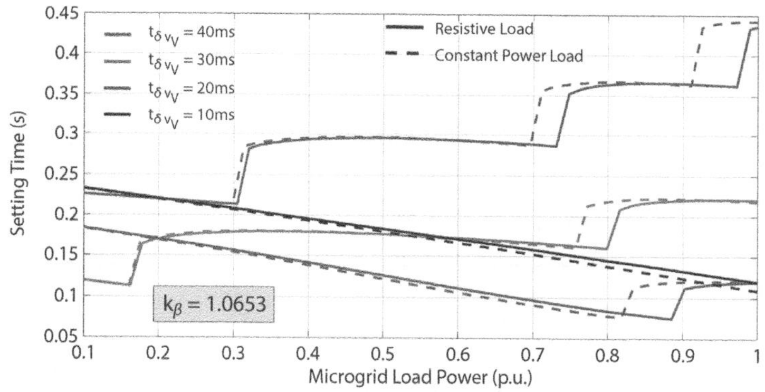

(d) Influence of the load on the Settling Time

Figure 7.20 **(Continued)** Assessment of the power sharing behavior. In (c) and (d), P_{Load} varies from 1 p.u. to 0.1 p.u. (Reprinted with permission from 5052471360356/IEEE.)

Assuming that Δt_{ps} is negligible in relation to the voltage restoration response, replacing (7.56) in (7.2)-(7.4) and (7.41), yields

$$v_{oavg}(t) = V_o^*(1-k_{vo})e^{-(1-k_{vo})t}$$
$$k_{vo} = \frac{R_{d1}(R_{d2}+r_2)+R_{d2}(R_{d1}+r_1)}{2(\alpha+\beta R_{\mu G})} \tag{7.57}$$

Therefore, the DC bus voltage restoration shows a first order response behavior with time constant of $\tau_v = (1 - k_{vo})^{-1}$. Obviously, $\Delta t_{ps} \ll \tau_v$ is required for stability purpose. Figure 7.21 illustrates the transient response

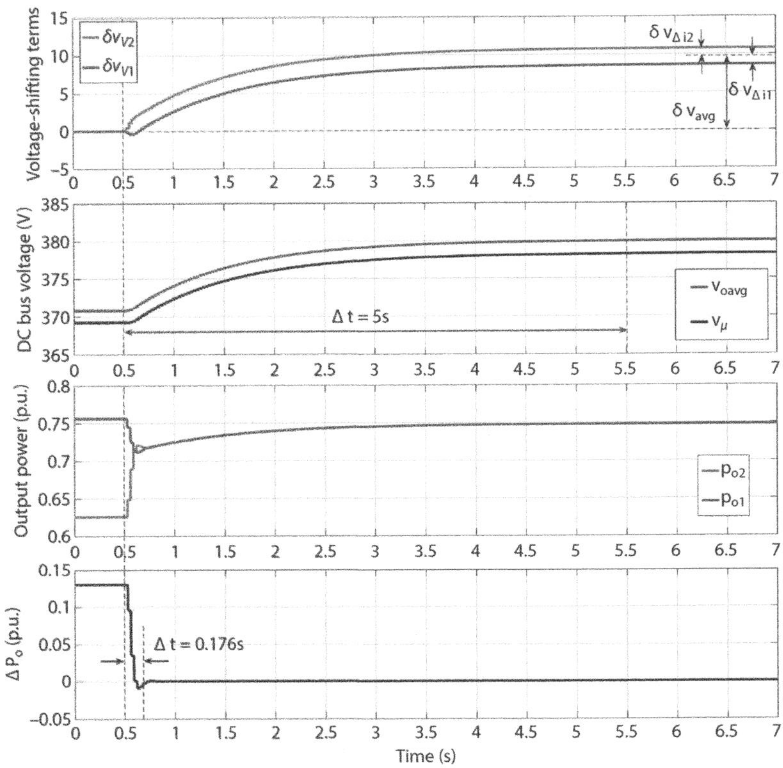

Figure 7.21 Evolution of δv_V and its effect on bus voltage and power imbalance (Reprinted with permission from 5052471360356/IEEE.)

of the proposed technique, for a simplified DC μG model with parameters given in Table 7.4 and $R_{\mu G} = 40\ \Omega$, which results in $\tau_v \approx 1\ s$ and $\Delta t_{ps} \approx 0.176\ s$.

It is noted that before the secondary control strategy is enabled, the μG operates with $v_{oavg} \approx 370$ V, a load voltage of 369 V and a power mismatch of 0.13 p.u. In t = 0.5 s the secondary control is activated. The voltage-shifting terms δ_{vV1} and δ_{vV2} are predominantly defined by $\delta_{v\Delta i}$, hence they diverge from each other, forcing the output imbalance to be mitigated in $\Delta t_{ps} = 0.176$ s. Once $p_{o1} = p_{o2}$, the argument of the secondary control integral action of both converters is the same, therefore the voltage-shifting terms evolve defining a δv_{avg} that leads the DC bus voltage to converge to $v_{oavg} = 380$ V in $\Delta t = 5\tau_v \approx 5$ s. In steady state, $\delta_{v\Delta i2} = -\delta_{v\Delta i1} = 1.035$ V and $\delta v_{avg} = 9.665$ V.

Since the electrical and communication systems interact with each other, they are affected by the transmission rate and communication

Table 7.4 Parameters used for the stability analysis on unique *vs* control (Reprinted with permission from 5052471360356/IEEE.)

Item	**Symbol**	**Value**
Nominal voltage	V_o^*	380 V
Line impedances	$r_{1,2}$	0.1, 0.9 Ω
Load resistances	$R_{\mu G}$	80 Ω
Droop Coefficients	$R_{d1,2}$	1.54, 3.08 Ω
Integrator time interval	$\tau_{\delta vV}$	30 ms
Calculation delay	τ_d	1...100 ms
LPF cut-off frequency	f_c	1.5 k Hz

network configuration [19]. Therefore, it is important to consider the discrete nature of the communication network, hence the secondary layer integral controller is implemented in discrete time, where the time interval between control cycles ($t_{\delta v_V}$) should be enough for all converters to send and receive the λ factors and update δ_{vV}.

7.2.3.2 *Secondary Control Level Modeling*

From the μG model in Figure 7.2b, the relationship between the output voltage and current of *Conv-1* can be expressed as:

$$i_{o1} = \gamma v_{o1} - \kappa v_{o2} \tag{7.58}$$

where

$$\gamma = \frac{r_2 + R_\mu}{r_1 r_2 + r_1 R_{\mu G} + r_2 R_{\mu G}}$$
$$\kappa = \frac{R_{\mu G}}{r_1 r_2 + r_1 R_{\mu G} + r_2 R_{\mu G}} \tag{7.59}$$

replacing $\delta v_{avg} + \delta v_{\Delta i}$ for δv_{V1} in (7.45) and in (7.4) yields to

$$v_{o1} = V_o^* + \delta v_{V1} - R_{d1} i_{o1} \tag{7.60}$$

Considering (7.51) and (7.58)–(7.60), the closed-loop control diagram of *Conv-1* is shown in Figure 7.22. Assuming that $t_{\delta v_V}$ is much larger than the converter local control response time, then the converter can be represented by its Thevenin equivalent circuit. The communication delay is represented by $e^{-\tau_d S}$ and the closed-loop transfer function is expressed as [5, 59]:

$$G_{vo} = \frac{G_{PI}G_C}{1+G_{PI}G_C} \tag{7.61}$$

Such that G_{PI} and G_C are the voltage loop PI compensator and the current loop transfer function, respectively. G_C can be represented as unit delay [5]. Therefore, from Figure 7.22, *Conv-1* output voltage is expressed as:

$$v_{o1} = \frac{\left[V_o^*(1+1/s)+v_{o2}\left(\kappa R_{d1}G_{LPF} - \frac{e_{o2}^{-\tau_d s\underline{P}}}{2\underline{P}_{o1}s}\right)\right]G_{vo}}{1+\left(\gamma R_{d1}G_{LPF} + \frac{e^{-\tau_d s}}{2s}\right)G_{vo}} \tag{7.62}$$

where $G_{LPF} = \frac{2\pi fc}{s+2\pi fc}$.

To assess the stability analysis model accuracy, Figure 7.23 shows the simulated behavior of *Conv-1* and *Conv-2* in different operating conditions for both the evaluated model and the complete converter switching model, considering the parameters shown in Table 7.4. At t = 0.1 s (A) the

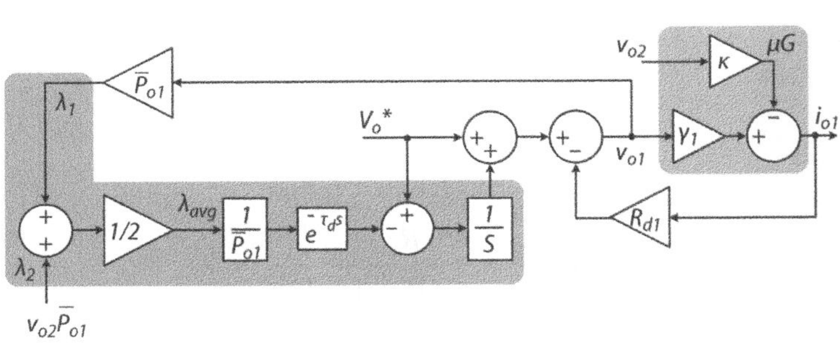

Figure 7.22 Closed-loop control diagram for stability analysis (Reprinted with permission from 5052471360356/IEEE.)

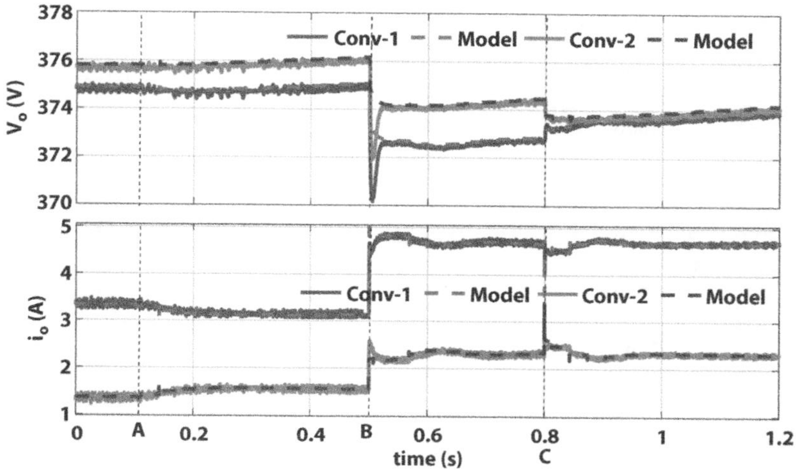

Figure 7.23 Output voltage and current behavior of the switched and stability analysis models of a μG (Reprinted with permission from 5052471360356/IEEE.)

proposed technique is enabled and at t = 0.5 s (B), $R_{\mu G}$ is reduced to 40 Ω, and finally, at t = 0.8 s (C), r_1 changes to 0.4 Ω. It is observed that the models hold high correlation.

Applying small-signal linearization on the variables in (7.60), the small signal model, thus, can be represented as [7]:

$$\frac{\widehat{v_{o1}}}{\widehat{V_o^*}}\Big|_{\widehat{v_{o2}}=0} = \frac{V_o^*(1+1/s)G_{vo}}{1+\left(\gamma R_{d1}G_{LPF}+\dfrac{e^{-\tau_d s}}{2s}\right)G_{vo}} \tag{7.63}$$

The dominant poles root *loci* under variations on r_1, R_{d1} and τ_d are shown in Figure 7.24, also for the parameters described in Table 7.4. To verify the influence of line impedance on the closed-loop system poles, r_1 varies from 0.05 Ω to 1 Ω and r_2 is fixed. R_{d1} varies from 0.5 R_{d1} to 2 R_{d1} with R_{d2} fixed and $R_{\mu G}$ is varied from 0.5 $R_{\mu G}$ to 1000 $R_{\mu G}$, which allows to evaluate the conditions from heavy μG load up to light load. The effect of λ_{avg} calculation delay (τ_d) is assessed considering $t_{\delta v_V}$ as a reference, therefore, τ_d is varied from 1 ms to 100 ms. In all four analyses, the poles *loci* are affected. However, in the considered variation ranges no small-signal instability is foreseen.

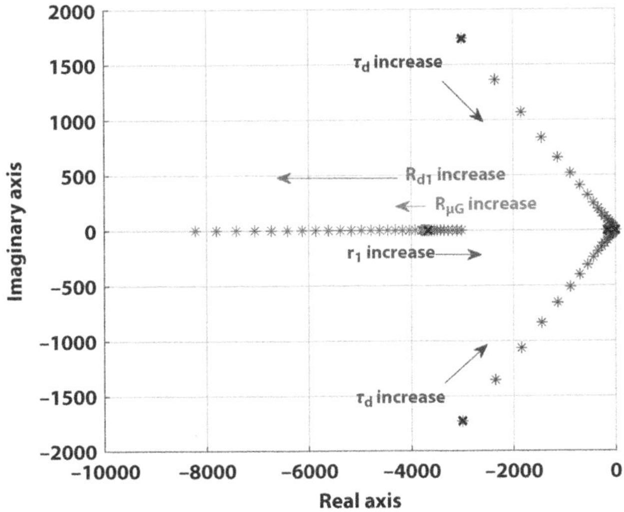

Figure 7.24 Stability analysis of two converters (Reprinted with permission from 5052471360356/IEEE.)

7.2.3.3 Computational and Experimental Validation

To evaluate the enhanced voltage-shifting (*vs*) strategy, a PLECS simulation of a DC bus as shown in Figure 7.25 is performed with *Conv-1*, *Conv-2* and *Conv-3* being DAB converters. The simulation parameters are shown in Table 7.5.

The simulation initial condition is v_{oavg} = 376.2 *V*, P_{o1} = 0.43 p.u., P_{o2} = 0.41 p.u., P_{o3} = 0.31 p.u. with load R_{L1} connected to the DC bus. The dynamic behavior of the enhanced *vs* control under load variations, communication

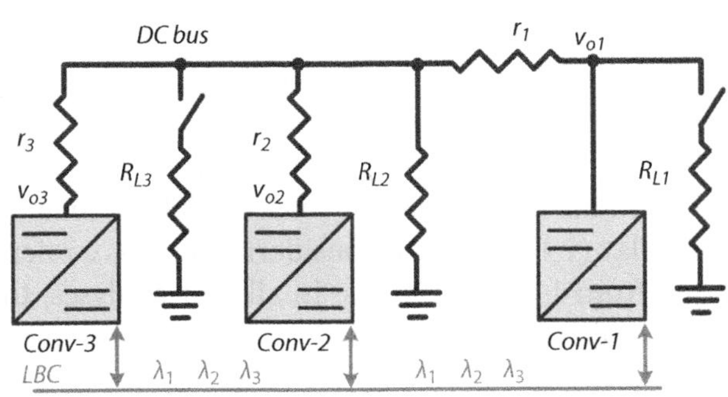

Figure 7.25 Simulated DC bus diagram. (Reprinted with permission from 5052471360356/IEEE.)

Table 7.5 Simulation/experimental parameters used in unique *vs* control. (Reprinted with permission from 5052471360356/IEEE].)

Item	Symbol	Value
Nominal voltage	V_o^*	380 V
Line impedances	$r_{1,2,3}$	0.1, 0.1, 0.9 Ω
Loads resistances	$R_{L1,2,3}$	133, 170 133 Ω
Droop Coefficients	$R_{d1,2,3}$	1. 54, 3.08, 3.08 Ω
Power ratio	$P_{1,2,3}$	3.2, 1.6, 1.6 kW
Integrator time interval	$\tau_{\delta vV}$	30 ms
LPF cut-off frequency	f_c	1.5 kHz
Switching frequency	f_{sw}	15 kHz
DAB transformer turns ratio	N	7.9:1
PI voltage controller	PI	$k_p = 1.8$, $k_i = 276$
PI current controller	PI	$k_p = 0.3$, $k_i = 20$
Power sharing speed correction gain	K_P	20
Current and voltage gains	$H_{i,v}$	0.1, 0.01

failures and converter connection/disconnection is evaluated and the results are shown in Figure 7.26. At $t = 1$ *s*, the secondary control is activated, initiating the *ps* and voltage restoration control processes. It can be seen that proportional *ps* is achieved after $t_{ps} \approx 295$ ms, resulting in $P_{o1} = P_{o2} = P_{o3} = 0.41$ p.u., leading to output currents, in steady state, of $i_{o1} = 2.46$ A, $i_{o2} = 1.25$ *A*, $i_{o3} = 1.27$ *A*. As mentioned in Section 2.3.1, proportional *cs* is not achieved, i.e., $0.5i_{o1} \neq i_{o2}$ and $i_{o2} \neq i_{o3}$, however, the deviation decreases from $\Delta i_{o23} = 0.49$ *A* (initial condition t < 1 s) to $\Delta i_{o23} = 0.02$ *A*. The DC bus voltage converges to 380 V after $t_{\Delta V_o^*} = 5$ *s*, as in (7.57). At $t = 6.2$ *s* (A), R_{L3} is connected to the DC bus causing a voltage drop of about 2 V and power imbalance between the converters. The secondary control restores *ps* at $t_{ps} = 140$ ms, converging to $P_{o1} = P_{o2} = P_{o3} = 0.63$ p.u., and the DC bus voltage the regulated. At $t = 8.5$ (B), R_{L1} and R_{L3} are simultaneously disconnected from the DC bus, increasing the average voltage in approximately 3 V. Is observed that, after $t_{ps} = 135$ ms, *ps* is achieved with $P_{o1} = P_{o2} = P_{o3} = 0.18$ p.u.. Note that the *ps* corrections in (A) and (B) is faster than the

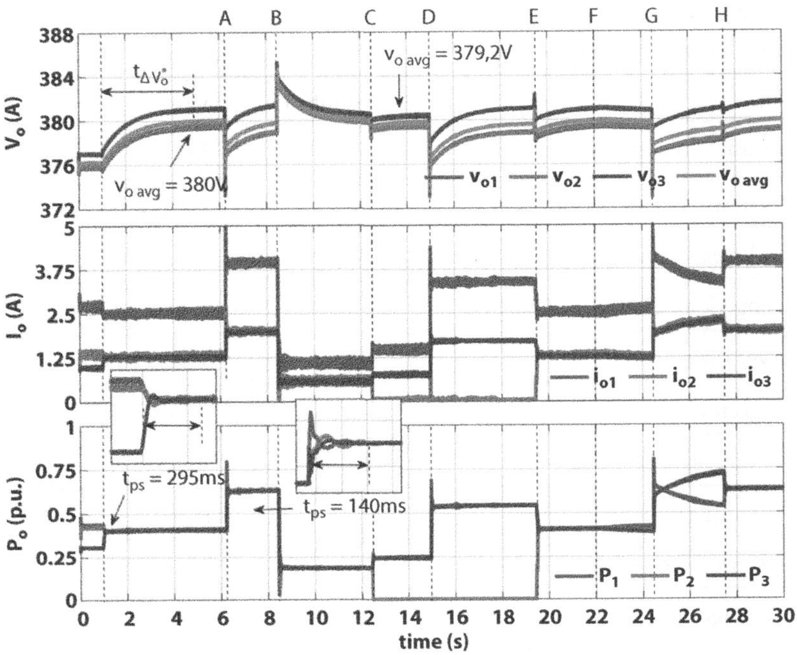

Figure 7.26 Simulation of control unique *vs* secondary control performance. (Adapted from [7] and reprinted with permission from 5052471360356/IEEE.)

beginning of the simulation ($t = 1\ s$), since the integrator argument, i.e., the secondary control error has a smaller magnitude the startup process.

The sequence of events that occur in the simulation after t > 10 s is designed to analyze plug and play capability and robustness against communication failures. At t = 12.5 s (C), *Conv-2* is disconnected from the DC bus and as a result, the DC bus voltage reduces in about 1 V. A new *ps* configuration is achieved, where *Conv-1* and *Conv-3* assume *Conv-2* load portion, where $P_{o1} = P_{o3} = 0.245$p.u. At t = 15 s (D), R_{L1} is reconnected to the DC bus, disturbing the bus. The control strategy reacts in order to adjust the power balance and keep the average voltage regulated. At t = 19.5 s (E), *Conv-2* is reconnected to the DC bus and, after the reconnection, the strategy reaches a new *ps* configuration. During the disconnection and reconnection of a converter, the secondary control is able to maintain stability with regulated DC bus voltage and proportional *ps*.

A communication failure in *Conv-1* occurs at t = 22 s (F), leaving it to operate isolated from the other converters. The secondary control still regulates the average bus voltage and imposes proportional *ps* between *Conv-2* and *Conv-3*, leading to $P_{o2} = P_{o3} = 0.39$ p.u. with $\Delta i_{o23} = 0.03$ A.

However, as *Conv-1* operates only with droop control, the average DC bus voltage between all three converters becomes v_{oavg} = 379.7 V. At t = 24.5 s, R_{L3} is connected to DC bus disturbing the system voltage and its *ps*. As the control action tends to regulate the average voltage between *Conv-2* and *Conv-3*, these converters assume a larger portion of the load, while *Conv-1* detects an increase in voltage and reduces its output power, therefore, *ps* is no longer proportional between all three converters. It is worth noting that although the isolation of a converter precludes accurate *ps* and voltage regulation, it does not prevent the DC bus from maintaining operation. At t = 27.5 s (H), communication with *Conv-1* is reestablished and proportional *ps* is restored.

The control technique is implemented experimentally with a DC bus configuration according to Figure 7.25, where *Conv-1* corresponds to a 3.2 kW BIC, *Conv-2* and *Conv-3* correspond to 1.6 kW DAB converters. These topologies were previously illustrated in Figure 7.15. The control diagrams are built in Texas Instruments TMS320F28335 DSP-platforms and the communication link is implemented with an 125 kbps CAN 2.0 network. The parameters of the converters are shown in Table 7.5.

Figure 7.27 shows the performance of the control strategy for some load conditions. At the beginning of the experiment, for $t < 1$ s, the system operates in a steady state in which v_{oavg} = 375.4 V, P_{o1} = 0.44 p.u., P_{o2} = 0.44 p.u. and P_{o3} = 0.26 p.u., Δi_{o23} = 0.61 A with R_{L1} connected to the DC bus. At t = 1 s, the secondary control is activated. Power sharing is achieved at t_{ps} = 340 ms, where the output power of the converters converge to 0.40 p.u. The average DC bus voltage converges to 380 V in $t_{\Delta V_o^*} \approx 5\,\text{s}$. At t = 6 s (A), R_{L3} is connected to the DC bus forcing the bus voltage to sag 3.6 V and unbalancing the converters' output power. A new proportional power sharing is attained at t_{ps} = 150 ms. At t = 9 s (B), R_{L1} and R_{L3} are simultaneously disconnected from the DC bus, producing a DC bus voltage swell of 4 V. Promptly the *ps* is restored.

The results described after event (B) correspond to the assessment of the µG plug and play capability and resilience to communication failures. At t = 11.5 s (C), *Conv-2* is disconnected from the DC bus, which causes an average DC bus voltage drop of about 1 V, where *Conv-1* and *Conv-3* proportionally assume the load power portion previously supplied by Conv-2. At t = 15.5 s (D), R_{L1} is connected to the bus and the output power of the active converters converge to 0.56 p.u., with i_{o1} = 3.46 A and i_{o1} = 1.69 A. At t = 18.4 s (E), *Conv-2* is reconnected to the bus and the *ps* among the three converters is once again achieved within t_{ps} = 200 ms. At t = 21 s (F), a communication failure occurs, isolating *Conv-1* from the other converters. As the correction of the average voltage deviation is still being performed,

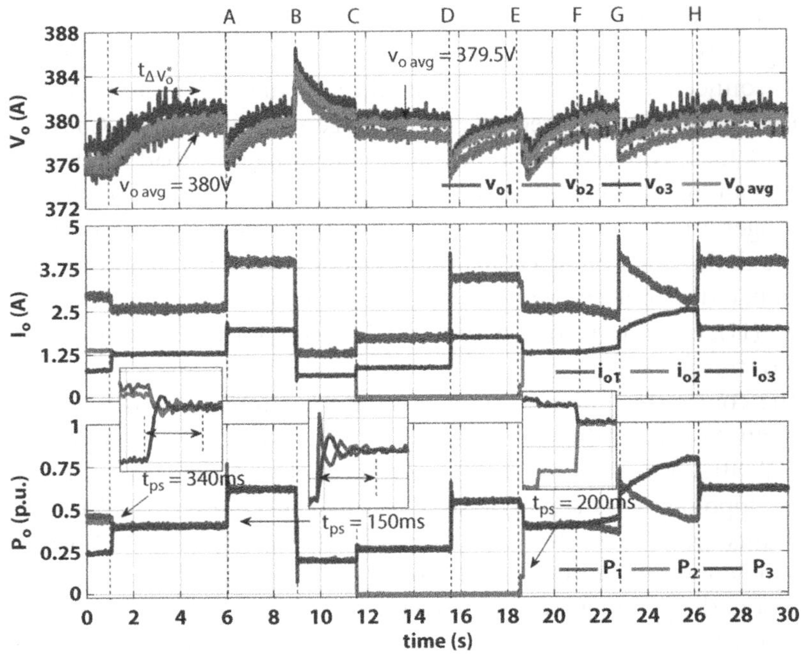

Figure 7.27 Experimental performance of unique *vs* secondary control. (Adapted from [7] and reprinted with permission from 5052471360356/IEEE.)

the output power of *Conv-2* and *Conv-3* increase, while that of *Conv-1*, due to droop effect, is reduced. At $t = 22.7\ s$ (G) the load R_{L3} is reconnected increasing the power imbalance of *Conv-1* in relation to *Conv-2* and *Conv-3*. Finally, at $t = 26.1\ s$ (H), communication with *Conv-1* is restored, which enables the system to regulated DC bus voltage of $v_{oavg} = 380$ V and proportional *ps* with $P_{o1} = P_{o2} = P_{o3} = 0.61$ p.u.

7.3 Design and Implementation of the Communication System

Electrical and communication systems interact with each other, being affected by both the transmission rate and communication network configuration [19]. Therefore, it is important to consider the discrete nature of the communication network in secondary level control system design and implementation. The communication system employed in both the hybrid secondary control and the improved voltage-shifting control detailed in this chapter is based on the following concept: at the beginning of a control cycle ($t = t_0$), each converter samples and holds (S&H) local quantities (i.e., voltage, current or power), which is later transmitted during a time interval $\Delta t_{s\&h}$, and while data is also received from other converters. In addition, a time interval $\Delta t_{c\&u}$ is reserved for each converter to process the received information, calculate and use as new control references before starting new sampling, nt_0; where $n \in \mathbb{Z}^*$. Thus, a control cycle communication time interval t_{com} can be established as:

$$t_{com} = \Delta t_{s\&h} + \Delta t_{c\&u} \tag{7.64}$$

Assuming that a communication node can have up to N_{MAX} converters, the time for a converter to sample, send and receive information ($\Delta t_{s\&h}$) can be estimated based on the LBC message frame. Taking CAN 2.0A as an example, where the data frame consists of 51 control bits and up to 64 data bits [60], it is defined that:

$$\Delta t_{s\&h} = \frac{51+64+3}{bps}(1+t_e)N_{ma'x} \tag{7.65}$$

where additional three idle bits between messages are considered, *bps* is the bit rate per second and t_e is the acceptable lost message rate. Arbitrating $\Delta t_{c\&u}$ as $\Delta t_{c\&u} \geq 4\Delta t_{s\&h}$, one can estimate $t_{com} \geq 5\Delta t_{s\&h}$. Thus, enough time is ensured to compute calculations related to the secondary control, update the parameters of the primary control and leave the communication bus inactive by an adequate period. So, such idle period could be employed to perform other functions of the secondary and/or tertiary control layers. As an example, Figure 7.28 shows the control cycle time frame for the hybrid secondary level control technique described in Section 7.2.2.

To evaluate whether the control cycle time frame is adequately defined, an experiment is performed with two TMS320F28335 microcontrollers (μ_{C1} and μ_{C2}) interlinked by an 125 kbps CAN network. Each microcontroller

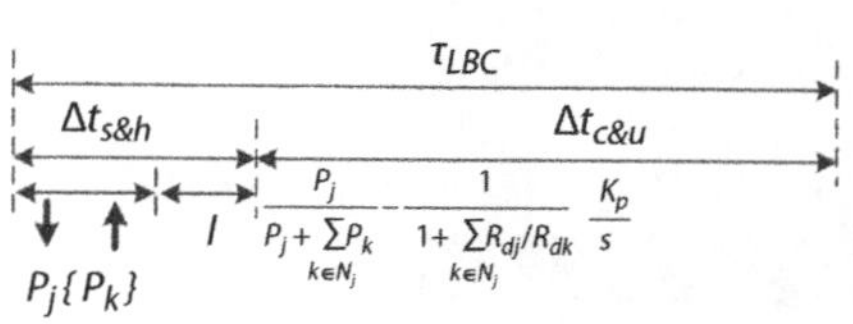

Figure 7.28 Secondary control cycle time frame for the hybrid control technique (Adapted from [8]).

sends a message (msn_1 and msn_2) using the 64 data bits. Once msn_2 is received, μ_{C1} calculates the average between the values contained in msn_1 and $msn_{2,}$ and thereupon returns a message (msn_3) to μ_{C2}.

Assuming (7.65) and adopting $t_e = 0$, thus $\Delta t_{s\&h} = 1.9$ *ms*. Figure 7.29 shows the experimental results, where CH1 presents the logical level of a μ_{C1} pin, which changes to 1 when msn_1 starts sending and returns to 0 after sending msn_3, whereas CH2 shows the voltage level of the CAN bus. The message duration time intervals (t_{msn1}, t_{msn2}, t_{msn3}) are measured as 944 µs, $\Delta t_{c\&u}$ =360 µs and the total control cycle period t_{msn1}, t_{msn2}, $t_{msn3} + t_{c\&u}$ = 3,1 *ms*. In practice, it is usually observed that $t_{c\&u} < t_{s\&h}$, hence considering $t_{com} = 5\Delta t_{s\&h}$ ensures sufficient time for sampling local electric quantities, sending, receiving and processing that information to be performed in one control cycle.

Once t_{com} is defined, the timers used in the communication modules of all microcontrollers that are interlinked by the communication network must be synchronized in order to guarantee that all data sampling and sending occurs at the same time t_0. It is important to mention that, since all microcontrollers are configured to exhibit a total control cycle equal to t_{com}, this synchronization actually relates to locking the phase of the communication module timers. This procedure can be accomplished through the sending of a sync message by one converter assigned to this task, where the remaining converters, after receiving that message, would reset their communication timers. The synchronization method employed in both control techniques described in this chapter (Section 2.2 and 2.3) consists in an auto-synchronization based on priority, where all converter communication timers achieve the same phase without the need of a sync message.

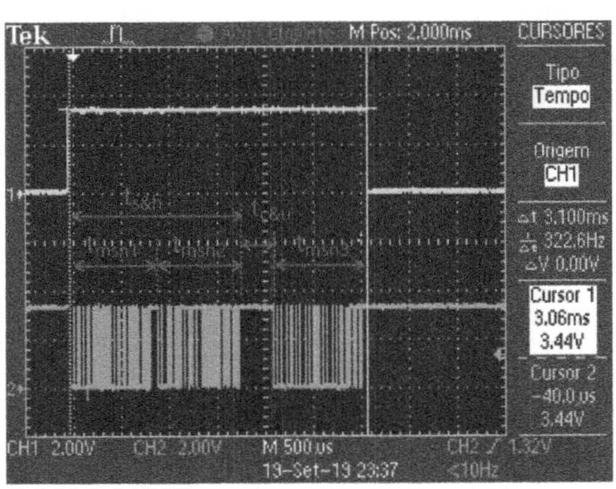

Figure 7.29 Experimental analysis of one communication cycle.

As each microcontroller (μC) is programmed to send messages every $t = t_{com}$, then message sending can be represented in phasor form as shown in Figure 7.30, where the angular frequency is $\omega_{com} = 2\pi/t_{com}$. Such that, θ_j represents the μC timer lag and the t_0 axis represents the instant when sampling/sending is performed. If $\theta_1 = \theta_2 = \theta_N$, all timers are in the same phase and sampling/sending is performed at the same time. In the priority-based auto-synchronization technique, considering the phasor representation, when a vector j passes through axis t_0, each μC_k verifies the received message ID (unique code that identifies each μC). If the received ID is lower than the μC_k own ID, then μC_k resets its communication timer, which in phasor form corresponds to $\theta_k = \theta_j$. Considering Figure 7.30, when phasor 2 passes through axis t_0, it forces $\theta_3 = \theta_2$ and when phasor 1 passes through the axis, $\theta_3 = \theta_2 = \theta_1$. Therefore, after a period equal to t_{com}, all μC will be synchronized.

Figure 7.31 shows the synchronization process for three microcontrollers timers (μC_1, μC_2, μC_3), where CH1 shows μC_1 timer output signal, CH2 shows the same for μC_2, CH3 for μC_3 and CH4 shows the voltage level of the CAN bus. The control cycle is defined as and between timers 1 and 2 and, between timers 2 and 3 there is a 33 ms lag. Figure 7.31a shows that μC_3 is the first to sample and send a message, as its ID is higher than the others ones. Note it has no impact on the timers. After 33 ms, μC_2 sends its message and μC_3, upon receiving it, verifies that the received ID is lower than their values and reset their timers. After another 33 ms, μC_1 sends its message, thereafter, μC_2 and μC_3 upon receiving it, verify that the message ID is lower than theirs values and reset their timers. Therefore, Figure 7.31a proves that timers 2 and 3 achieve the same phase of timer 1 within a period lower or equal to t_{com}. Figure 7.31b shows the three synchronized timers and Figure 7.31c shows in detail the three messages on the CAN bus, although they are commanded to be sent at the same time, the CAN protocol arbitration process defines which message will be sent first.

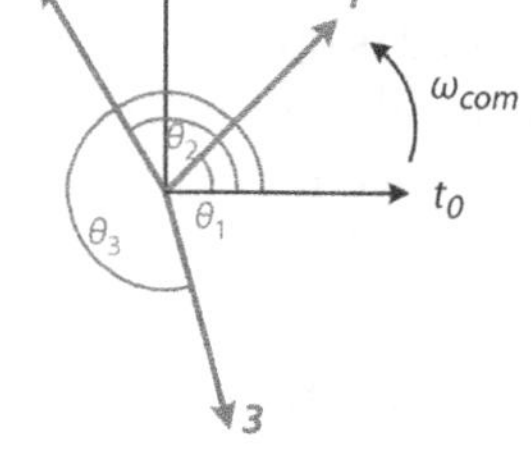

Figure 7.30 Phasor representation of information sent by CAN converters.

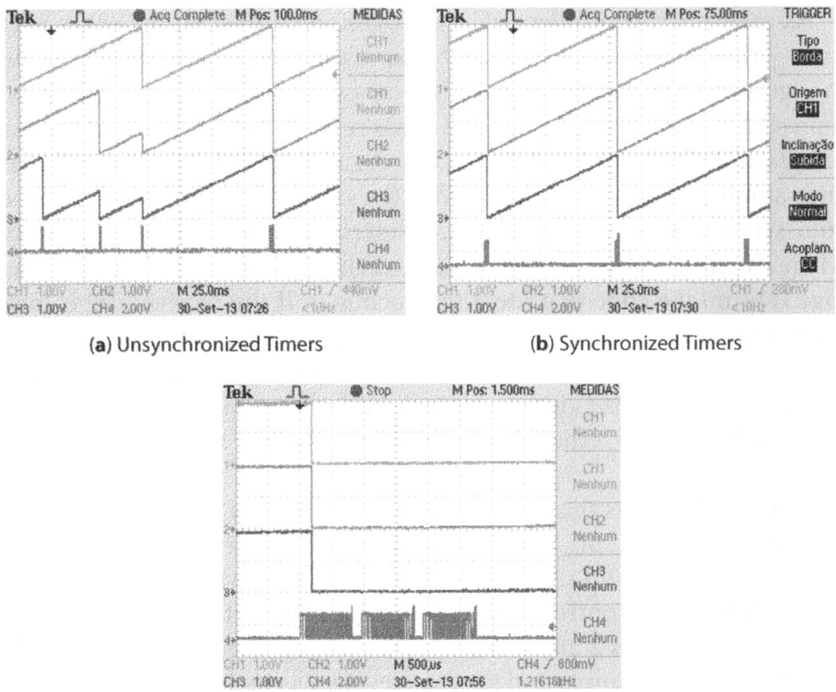

(a) Unsynchronized Timers

(b) Synchronized Timers

(c) Synchronized messaging

Figure 7.31 Auto-synchronization of timers.

7.4 Conclusions

The present chapter presented the main control techniques employed at the secondary layer of DC microgrids. These techniques aim to correct the voltage deviation caused by the primary layer droop technique, and the proportional power/load sharing deviation, produced by the line impedances. The effect of these line impedances on the converters output power/current was presented and discussed. The proposed techniques allow both power sharing and current deviation correction, where the criterion of which approach will be loomed is up to the designer. In UAEs applications, the choice of power sharing deviation correction, instead the current, may be more appropriated, since it eliminates one of the possible unbalanced sources of SoC among the UAEs. The approaches used to correct the power/load sharing are based on the voltage shifting or the adjustment of the droop coefficient, where in some literature is presented with adaptative droop.

Techniques based on voltage shifting are not able to change the converters output impedance and compensate for the difference between the equivalent line resistances, the term voltage-shifting is constantly updated, even if there is no load change in the DC bus. In approaches based on droop coefficient adjustment, once the differences between equivalent line resistances are corrected, the current or power sharing compensation errors will be almost null; however, the system dynamic is affected and the coefficients adjustments must be restricted, in order to prevent instability.

The techniques developed by the authors were based on a different approach with regard to power sharing correction. The use of consensus-based algorithm allowed the verification of the convergence of power sharing for both techniques. Among the contributions of these techniques, compared with the literature proposed approaches, is the ability to simultaneously achieve the voltage deviation and power/load sharing correction with only one information per converter in a sparse communication network. The presented results allow the verification of the performance of the proposed methods as well as the plug-and-play characteristic and robustness under communication failures, that are desirable requirements in microgrid.

References

1. R. P. S. Chandrasena, F. Shahnia, S. Rajakaruna, and A. Ghosh, "Dynamic operation and control of a hybrid nanogrid system for future community houses," *IET Generation, Transmission Distribution*, vol. 9, no. 11, pp. 1168–1178, 2015.
2. X. Lu, J. M. Guerrero, K. Sun, J. C. Vasquez, R. Teodorescu, and L. Huang, "Hierarchical control of parallel ac-dc converter interfaces for hybrid microgrids," *IEEE Transactions on Smart Grid*, vol. 5, no. 2, pp. 683–692, March 2014.
3. E. Unamuno and J. A. Barrena, "Hybrid ac/dc microgrids—part i: Review and classification of topologies," *Renewable and Sustainable Energy Reviews*, vol. 52, pp. 1251 - 1259, 2015. Online]. Available: http://www.sciencedirect.com/science/article/pii/S1364032115008412
4. N. Yang, D. Paire, F. Gao, A. Miraoui, and W. Liu, "Compensation of droop control using common load condition in dc microgrids to improve voltage regulation and load sharing," *International Journal of Electrical Power & Energy Systems*, vol. 64, pp. 752 - 760, 2015. [Online]. Available: http://www.sciencedirect.com/science/article/pii/S0142061514005237
5. X. Lu, J. M. Guerrero, K. Sun, and J. C. Vasquez, "An improved droop control method for dc microgrids based on low bandwidth communication with

dc bus voltage restoration and enhanced current sharing accuracy," *IEEE Transactions on Power Electronics*, vol. 29, no. 4, pp. 1800–1812, April 2014.
6. P. Huang, P. Liu, W. Xiao, and M. S. El Moursi, "A novel droop-based average voltage sharing control strategy for dc microgrids," *IEEE Transactions on Smart Grid*, vol. 6, no. 3, pp. 1096–1106, May 2015.
7. W. W. A. G. Silva, T. R. Oliveira, and P. F. Donoso-Garcia, "An improved voltage-shifting strategy to attain concomitant accurate power sharing and voltage restoration in droop-controlled dc microgrids," *IEEE Transactions on Power Electronics*, vol. 36, no. 2, pp. 2396–2406, 2021.
8. ——, "Hybrid distributed and decentralized secondary control strategy to attain accurate power sharing and improved voltage restoration in dc microgrids," *IEEE Transactions on Power Electronics*, vol. 35, no. 6, pp. 6458–6469, 2020.
9. J. M. Guerrero, J. C. Vasquez, J. Matas, M. Castilla, and L. G. de Vicuna, "Control strategy for flexible microgrid based on parallel line-interactive ups systems," *IEEE Transactions on Industrial Electronics*, vol. 56, no. 3, pp. 726–736, March 2009.
10. J. Banda and K. Siri, "Improved central-limit control for parallel-operation of dc-dc power converters," in *Power Electronics Specialists Conference*, 1995. PESC '95 Record., 26th Annual IEEE, vol. 2, Jun. 1995, pp. 1104–1110 vol. 2.
11. J. Rajagopalan, K. Xing, Y. Guo, F. C. Lee, and B. Manners, "Modeling and dynamic analysis of paralleled dc/dc converters with master-slave current sharing control," in *Proceedings of Applied Power Electronics Conference*. APEC '96, vol. 2, March 1996, pp. 678–684 vol. 2.
12. X. Sun, Y.-S. Lee, and D. Xu, "Modeling, analysis, and implementation of parallel multi-inverter systems with instantaneous average-current-sharing scheme," *IEEE Transactions on Power Electronics*, vol. 18, no. 3, pp. 844–856, May 2003.
13. R. Pradhan, M. Chirayath, and S. Thale, "Coordinated control strategy for a dc microgrid with low bandwidth communication," in *2016 IEEE International Conference on Power Electronics, Drives and Energy Systems (PEDES)*, Dec. 2016, pp. 1–6.
14. J. M. Guerrero, J. C. Vasquez, J. Matas, J. L. Sosa, and L. G. de Vicuna, "Parallel operation of uninterruptible power supply systems in microgrids," in *2007 European Conference on Power Electronics and Applications*, Sept. 2007, pp. 1–9.
15. C. Jin, J. Wang, and P. Wang, "Coordinated secondary control for autonomous hybrid three-port ac/dc/ds microgrid," *CSEE Journal of Power and Energy Systems*, vol. 4, no. 1, pp. 1–10, March 2018.
16. Q. Shafiee, J. M. Guerrero, and J. C. Vasquez, "Distributed secondary control for islanded microgrids—a novel approach," *IEEE Transactions on Power Electronics*, vol. 29, no. 2, pp. 1018–1031, Feb. 2014.
17. J. M. Guerrero, J. C. Vasquez, J. Matas, L. G. de Vicuna, and M. Castilla, "Hierarchical control of droop-controlled ac and dc microgrids - a general

approach toward standardization," *IEEE Transactions on Industrial Electronics*, vol. 58, no. 1, pp. 158–172, Jan. 2011.
18. T. Dragičević, J. M. Guerrero, J. C. Vasquez, and D. Škrlec, "Supervisory control of an adaptive-droop regulated dc microgrid with battery management capability," *IEEE Transactions on Power Electronics*, vol. 29, no. 2, pp. 695–706, Feb. 2014.
19. L. Meng, T. Dragicevic, J. Roldán-Pérez, J. C. Vasquez, and J. M. Guerrero, "Modeling and sensitivity study of consensus algorithm-based distributed hierarchical control for dc microgrids," *IEEE Transactions on Smart Grid*, vol. 7, no. 3, pp. 1504–1515, May 2016.
20. P. Wang, X. Lu, X. Yang, W. Wang, and D. Xu, "An improved distributed secondary control method for dc microgrids with enhanced dynamic current sharing performance," *IEEE Transactions on Power Electronics*, vol. 31, no. 9, pp. 6658–6673, Sep. 2016.
21. J. Fang, Z. Shuai, X. Zhang, X. Shen, and Z. J. Shen, "Secondary power sharing regulation strategy for a dc microgrid via maximum loading factor," *IEEE Transactions on Power Electronics*, pp. 1–1, 2019.
22. S. Anand, B. G. Fernandes, and J. Guerrero, "Distributed control to ensure proportional load sharing and improve voltage regulation in low-voltage dc microgrids," *IEEE Transactions on Power Electronics*, vol. 28, no. 4, pp. 1900–1913, April 2013.
23. D. Dam and H. Lee, "A power distributed control method for proportional load power sharing and bus voltage restoration in a dc microgrid," *IEEE Transactions on Industry Applications*, vol. 54, no. 4, pp. 3616–3625, July 2018.
24. A. Ingle, A. B. Shyam, S. R. Sahoo, and S. Anand, "Quality-index based distributed secondary controller for a low-voltage dc microgrid," *IEEE Transactions on Industrial Electronics*, vol. 65, no. 9, pp. 7004–7014, Sep. 2018.
25. S. Augustine, N. Lakshminarasamma, and M. K. Mishra, "Control of photovoltaic-based low-voltage dc microgrid system for power sharing with modified droop algorithm," *IET Power Electronics*, vol. 9, no. 6, pp. 1132–1143, 2016.
26. M. Zaery, E. M. Ahmed, M. Orabi, and M. Youssef, "Operational cost reduction based on distributed adaptive droop control technique in dc microgrids," in *2017 IEEE Energy Conversion Congress and Exposition (ECCE)*, Oct 2017, pp. 2638–2644.
27. L. Guo, Z. Guo, X. Li, C. Wang, C. Hong, and Y. Zhang, "Consensus-based distributed coordinated control for islanded dc microgrids," in *2017 IEEE Power Energy Society General Meeting*, July 2017, pp. 1–5.
28. V. Nasirian, S. Moayedi, A. Davoudi, and F. L. Lewis, "Distributed cooperative control of dc microgrids," *IEEE Transactions on Power Electronics*, vol. 30, no. 4, pp. 2288–2303, April 2015.
29. X. Zhang, M. Dong, and J. Ou, "A distributed cooperative control strategy based on consensus algorithm in dc microgrid," in *2018 13th IEEE Conference on Industrial Electronics and Applications (ICIEA)*, May 2018, pp. 243–248.

30. S. Sahoo and S. Mishra, "An adaptive event-triggered communication-based distributed secondary control for dc microgrids," *IEEE Transactions on Smart Grid*, vol. 9, no. 6, pp. 6674–6683, Nov 2018.
31. D. Pullaguram, S. Mishra, and N. Senroy, "Event-triggered communication based distributed control scheme for dc microgrid," *IEEE Transactions on Power Systems*, vol. 33, no. 5, pp. 5583–5593, Sep. 2018.
32. V. Nasirian, A. Davoudi, F. L. Lewis, and J. M. Guerrero, "Distributed adaptive droop control for dc distribution systems," *IEEE Transactions on Energy Conversion*, vol. 29, no. 4, pp. 944–956, Dec. 2014.
33. T. V. Vu, D. Perkins, F. Diaz, D. Gonsoulin, C. S. Edrington, and T. El-Mezyani, "Robust adaptive droop control for dc microgrids," *Electric Power Systems Research*, vol. 146, pp. 95 - 106, 2017. Online]. Available: http://www.sciencedirect.com/science/article/pii/S0378779617300305
34. P. Prabhakaran, Y. Goyal, and V. Agarwal, "Novel nonlinear droop control techniques to overcome the load sharing and voltage regulation issues in dc microgrid," *IEEE Transactions on Power Electronics*, vol. 33, no. 5, pp. 4477–4487, May 2018.
35. A. Khorsandi, M. Ashourloo, H. Mokhtari, and R. Iravani, "Automatic droop control for a low voltage dc microgrid," *IET Generation, Transmission Distribution*, vol. 10, no. 1, pp. 41–47, 2016.
36. F. Chen, R. Burgos, D. Boroyevich, and W. Zhang, "A nonlinear droop method to improve voltage regulation and load sharing in dc systems," in *2015 IEEE First International Conference on DC Microgrids (ICDCM)*, June 2015, pp. 45–50.
37. S. Peyghami, H. Mokhtari, and F. Blaabjerg, "Decentralized load sharing in a low-voltage direct current microgrid with an adaptive droop approach based on a superimposed frequency," *IEEE Journal of Emerging and Selected Topics in Power Electronics*, vol. 5, no. 3, pp. 1205–1215, Sep. 2017.
38. H. Wang, M. Han, J. M. Guerrero, J. C. Vasquez, and B. G. Teshager, "Distributed secondary and tertiary controls for i–v droop-controlled-paralleled dc–dc converters," *IET Generation, Transmission Distribution*, vol. 12, no. 7, pp. 1538–1546, 2018.
39. F. Chen, R. Burgos, D. Boroyevich, E. Rodriguez-Diaz, L. Meng, J. C. Vasquez, and J. M. Guerrero, "Analysis and distributed control of power flow in dc microgrids to improve system efficiency," in *2016 4th International Symposium on Environmental Friendly Energies and Applications (EFEA)*, Sep. 2016, pp. 1–6.
40. M. A. Mumtaz, M. M. Khan, F. Xianghong, A. Karni, and M. T. Faiz, "An improved cooperative control method of dc microgrids based on finite gain controller," in *2018 20th European Conference on Power Electronics and Applications (EPE'18 ECCE Europe)*, Sep. 2018, pp. 1–9.
41. Y. Yang, M. M. Khan, and J. Yu, "An improved cooperative control method of dc microgrid based on nearest neighbors communication," in *IECON 2017 - 43rd Annual Conference of the IEEE Industrial Electronics Society*, Oct 2017, pp. 792–797.

42. D. Dam and H. Lee, "An adaptive power distributed control method to ensure proportional load power sharing in dc microgrid considering equivalent line impedances," in *2016 IEEE Energy Conversion Congress and Exposition (ECCE)*, Sep. 2016, pp. 1–6.
43. P. Ghalebani and M. Niasati, "A distributed control strategy based on droop control and low-bandwidth communication in dc microgrids with increased accuracy of load sharing," *Sustainable Cities and Society*, vol. 40, pp. 155–164, 2018. [Online]. Available: http://www.sciencedirect.com/science/article/pii/S221067071731627X
44. S. Augustine, M. K. Mishra, and N. Lakshminarasamma, "Adaptive droop control strategy for load sharing and circulating current minimization in low-voltage standalone dc microgrid," *IEEE Transactions on Sustainable Energy*, vol. 6, no. 1, pp. 132–141, Jan 2015.
45. F. Gao, S. Bozhko, G. Asher, P. Wheeler, and C. Patel, "An improved voltage compensation approach in a droop-controlled dc power system for the more electric aircraft," *IEEE Transactions on Power Electronics*, vol. 31, no. 10, pp. 7369–7383, Oct. 2016.
46. S. Moayedi and A. Davoudi, "Unifying distributed dynamic optimization and control of islanded dc microgrids," *IEEE Transactions on Power Electronics*, vol. 32, no. 3, pp. 2329–2346, March 2017.
47. J. Hu, J. Duan, H. Ma, and M. Chow, "Distributed adaptive droop control for optimal power dispatch in dc microgrid," *IEEE Transactions on Industrial Electronics*, vol. 65, no. 1, pp. 778–789, Jan. 2018.
48. S. Sahoo and S. Mishra, "A distributed finite-time secondary average voltage regulation and current sharing controller for dc microgrids," *IEEE Transactions on Smart Grid*, vol. 10, no. 1, pp. 282–292, Jan. 2019.
49. S. Sahoo, S. Mishra, S. M. Fezeli, F. Li, and T. Dragičević, "A distributed fixed-time secondary controller for dc microgrid clusters," *IEEE Transactions on Energy Conversion*, pp. 1–1, 2019.
50. S. Moayedi, V. Nasirian, F. L. Lewis, and A. Davoudi, "Team-oriented load sharing in parallel dc–dc converters," *IEEE Transactions on Industry Applications*, vol. 51, no. 1, pp. 479–490, Jan. 2015.
51. Q. Xu, J. Xiao, X. Hu, P. Wang, and M. Y. Lee, "A decentralized power management strategy for hybrid energy storage system with autonomous bus voltage restoration and state-of-charge recovery," *IEEE Transactions on Industrial Electronics*, vol. 64, no. 9, pp. 7098–7108, Sep. 2017.
52. M. Tucci, L. Meng, J. M. Guerrero, and G. Ferrari-Trecate, "Stable current sharing and voltage balancing in dc microgrids: A consensus-based secondary control layer," *Automatica*,vol. 95, pp. 1–13, 2018.
53. R. Olfati-Saber, J. A. Fax, and R. M. Murray, "Consensus and cooperation in networked multi-agent systems," *Proceedings of the IEEE*, vol. 95, no. 1, pp. 215–233, Jan 2007.
54. L. Meng, T. Dragicevic, J. M. Guerrero, and J. C. Vasquez, "Dynamic consensus algorithm based distributed global efficiency optimization of a droop

controlled dc microgrid," in *2014 IEEE International Energy Conference (ENERGYCON)*, May 2014, pp. 1276–1283.
55. B. Van Scoy, R. A. Freeman, and K. M. Lynch, "Exploiting memory in dynamic average consensus," in *2015 53rd Annual Allerton Conference on Communication, Control, and Computing (Allerton)*, Sep. 2015, pp. 258–265.
56. S. S. Kia, B. V. Scoy, J. Cortés, R. A. Freeman, K. M. Lynch, and S. Martínez, "Tutorial on dynamic average consensus: the problem, its applications, and the algorithms," *CoRR*,vol. abs/1803.04628, 2018.
57. R. Olfati-Saber and R. M. Murray, "Consensus problems in networks of agents with switching topology and time-delays," *IEEE Transactions on Automatic Control*, vol. 49, no. 9, pp. 1520–1533, Sep. 2004.
58. T. Morstyn, B. Hredzak, G. D. Demetriades, and V. G. Agelidis, "Unified distributed control for dc microgrid operating modes," *IEEE Transactions on Power Systems*, vol. 31, no. 1, pp. 802–812, Jan. 2016.
59. K. D. Hoang and H. Lee, "Accurate power sharing with balanced battery state of charge in distributed dc microgrid," *IEEE Transactions on Industrial Electronics*, vol. 66, no. 3, pp. 1883–1893, March 2019.
60. S. Corrigan, "Introduction to the controller area network (can)," Texas Instruments, TI, Tech. Rep., May 2016. [Online]. Available: https://www.ti.com/lit/an/sloa101b/sloa101b.pdf

8

Dynamic Analysis and Reduced-Order Modeling Techniques for Power Converters in DC Microgrid

Divya Navamani J.[1]*, Lavanya A.[1], Jagabar Sathik[2], M.S. Bhaskar[2] and Vijayakumar K.[1]

[1]Dept. of Electrical Engineering, SRM Institute of Science and Technology, Chennai, Tamilnadu, India
[2]Renewable Energy Lab., College of Engineering, Prince Sultan University, Riyadh, Saudi Arabia

Abstract
In the present-day electrical systems, the two main problems are depletion of available conventional resources and its cause to the environment. To overcome this situation, distributed energy resources offer support to confront the present situation. Power electronic interface is added to the dc microgrid to increase the efficiency of the system. To improve its performance, the dynamic analysis of the power converters needs to be studied and the issues to be addressed to increase the reliability of the system. Several modelling techniques reported in the literature are discussed and applied to the conventional boost converter. The main problem in the modelling technique is the derivation of the transfer function for the higher-order system. This can be overcome by incorporating reduced-order modelling techniques to the power converters. Furthermore, the solution of the resolvent matrix in the state-space averaging technique is eased by Leverrier's algorithm. Improved pole clustering method is proposed for dynamic analysis of quadratic boost converter with switched inductor cell exclusively to unravel the resolvent matrix and prove the methodology's easiness. Simulation software results validate the proposed method for power converters and point out that the higher-order converters can be dealt with using the proposed method and analysed. Additionally, a sight on various controllers applied for dc-dc converters is given for researchers to have this as a guide for their analytical study on mathematical modelling.

**Corresponding author*: divyateddy1@gmail.com; divyanaj@srmist.edu.in

Nikita Gupta, Mahajan Sagar Bhaskar, Sanjeevikumar Padmanaban and Dhafer Almakhles (eds.)
DC Microgrids: Advances, Challenges, and Applications, (241–272) © 2022 Scrivener Publishing LLC

***Keywords*:** Modeling, dynamic analysis, time-domain, frequency-domain, resolvent matrix, reduced-order, controller

8.1 Introduction

In present-day electrical systems, the two main problems are depletion of available conventional resources and the effect on the environment. To overcome this situation, distributed energy resources offer support. There are several merits for DC microgrid compared to AC systems:

- DC microgrid operates more reliably in both grid-connected and island mode.
- Many manufacturing factories like paper and aluminum production require DC supply for the smelting process where it is necessary to convert AC to DC.
- The blooming technology like charging stations for electric vehicles is finely fitted with DC power.
- Slowly, the loads in residence and commercial buildings are switching over to DC loads like LED lighting.

Recently, traditional centralized AC systems are being replaced with the promising DC microgrid fed with distributed energy resources as shown in Figure 8.1. In a DC microgrid, power conditioning units play a vital

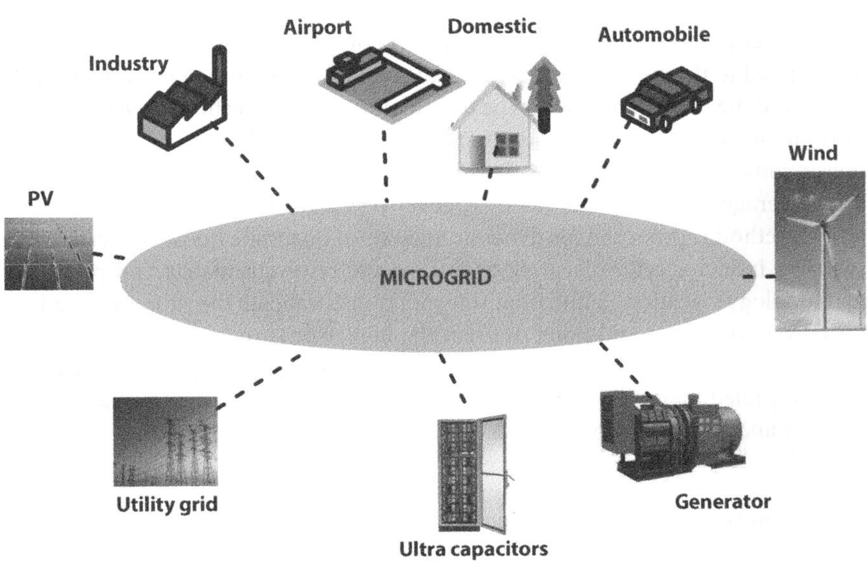

Figure 8.1 DC microgrid.

role where the dc-dc converter is placed as an intermediate module and its performance becomes significant in the overall system. Since the dc-dc converter is a key part of DC microgrid, it is essential to derive a topology with high gain and high efficiency.

8.2 Need of Dynamic Analysis for Power Converters

Most of the derived dc-dc converters for green energy conversion applications are analyzed in steady-state mode to validate their performance. Usually, steady-state analysis is performed with volt-sec and ampere-sec balance principles to study the converter's performance for any particular application theoretically. Still, this analysis will not be sufficient to study the converter's performance. Henceforth, it is required to analyse the converter in dynamic state. It is more essential to read the converter's performance when it is subjected to disturbances. To accomplish this dynamic analysis, the converter's mathematical model is crucial and also to design the feedback control for the derived topology. There are several modeling methods such as State-space averaging (SSA), Signal flow graph (SFG), circuit-averaging techniques, and so on. Many researchers are studying the dynamic performance of their suggested topology with the traditional State-space averaging technique because of its accuracy in designing the controller for the system.

The first problem in the State-space averaging method is finding the solution for $(sI\text{-}A)^{-1}$ resolvent matrix, Where I signify identity matrix, s denotes the self-adjoint nxn matrix and A represents state matrix. n symbolizes the order of the matrix, i.e., number of passive elements in the derived converter. In [1], modified cuk converter with reduced redundant power processing technique is proposed with dynamic analysis. This derived converter has eight passive elements which complicates the solution of the resolvent matrix with 8x8.

The power electronic engineers propose new controller design approaches for studying the dynamic performance of sophisticated dc-dc converters. The Tchebyshev polynomial approach is introduced for the design of feedback controller for the fifth boost converter [2]. To ensure the regulation condition and maximize the closed-loop bandwidth, the outer voltage and inner current loop are designed for the proposed topology [3]. The audio-susceptibility of the converter can be reduced by input-voltage feed forward control. In [4], research is carried out on high-order converter topologies with input-voltage feedforward control. Furthermore, the dynamic response is enhanced with a newly proposed PWM-based geometric control scheme for dc-dc converters [5]. From the literature as mentioned earlier, the significance of the dynamic study of the high gain dc-dc converter is highlighted and it is also observed that this is not yet completely explored in the literature.

To increase the voltage gain of the converter, the voltage multiplier cell and coupled inductor with the clamp circuit is added, which complicates the circuit and the modeling approach. A converter with voltage multiplier cell has the capacitors with similar energy-saving which can be eliminated in the modeling procedure and it simplifies the order of the system [6]. The fourth-order model is reduced to the second-order model and the analysis of voltage multiplier integrated boost converter is performed [7]. The state-space model with a reduced-order model is introduced for a multilevel dc-dc boost converter with voltage-mode control law for regulation [8] [9]. Reduced-order linear differential equation-based modeling method is proposed buck derived topology for optimizing the design of the converter [10]. In [6–10], it is emphasized that there is a need for reduced-order modeling study for a high gain dc-dc converter with an increased number of state variables.

It is indicated in Figure 8.2, that the papers published on dynamic analysis is observed to be less compared to the articles with topological derivation and steady-state analysis. The authors consider this chapter will provide insight into the modeling techniques and the significance of the reduced-order model. In the next section, we will give a glimpse of the various modeling techniques and the procedures involved in modeling the converter.

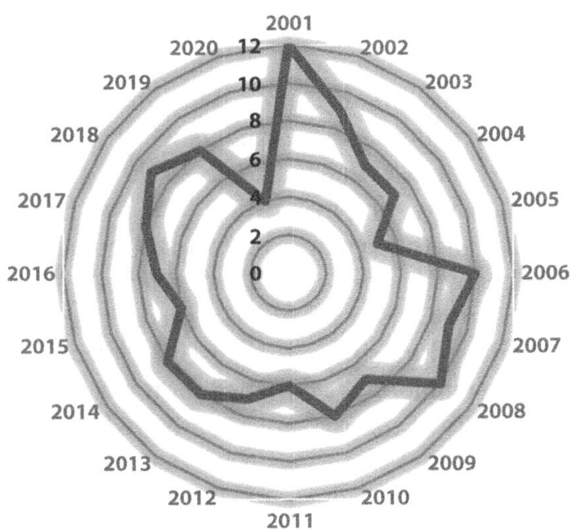

Figure 8.2 Annals of publication on the modeling of DC-DC converter.

This will be a guide to researchers who will then have an overview of several modeling techniques, and they can incorporate this for their study.

8.3 Various Modeling Techniques

In this section, various modeling techniques reported in the literature are identified and a glimpse of the technique is presented. Furthermore, the techniques are applied to conventional boost converter and the derived model is illustrated in Table 8.1. This table will give an overview of the modeling techniques used for the non-isolated and isolated dc-dc converter.

First and foremost, the method of modeling the dc-dc converter is the state-space averaging technique. The advantages of this modeling technique are various transfer functions can be derived easily and compact representation of the equations [11]. Compared to circuit averaging techniques [13], it can bring out several transfer functions such as line voltage to output voltage, control to output voltage, output current to output voltage, etc. Mason introduced a formula to analyze the system with a signal flow graph used as a switching flow graph to perform the converter's dynamic analysis [12].

In the unified topological modeling approach, the switching elements are assumed as a short circuit and open circuit in the ON and OFF states, respectively. In both linear circuit models of a converter, one can distinguish two types of branches [14]: Input voltage source is represented as G branch and a parallel connection of a load resistance output smoothing capacitor is represented as O branch. Current Injected Equivalent Circuit Approach (CIECA) depicts the small signal of switching dc-dc converters. The advantage of both the current injected type approach and the state-space averaging approach is combined in the CIECA approach. The equivalent circuit model obtained from this approach enables the designers to understand the converter operation and make better design [15, 16].

In the energy conservation principle modeling method, parasitic components, offset voltage of the switches is considered [17]. The equivalent circuit is simplified by reflection rules, i.e., shifting the circuit components for the reduction of the circuit. Similar to other modeling techniques, it can also be used in the circuit simulator for analysis. The main drawback of this method is it doesn't include the losses of switches and passive components. An improvisation is carried out by including the switching element losses by using the same energy conservation principle [18]. The PWM modeling method is found to be a much easier and simpler method for DC-DC converter in CCM [19]. In this method, semiconductor

Table 8.1 Modeling techniques with example.

Modeling technique	Procedure for modeling the converter	Example-boost converter
Input to output transfer function= $v_O(s)/v_g(s)$		
State-space averaging [11]	$\dot{X} = AX + BU$ $Y = CX + DU$ $\frac{v_O(s)}{v_g(s)} = C[SI - A]^{-1}B + D$ Where, $v_0(s)$ Output voltage $v_g(s)$ Input voltage U represents the input variables D represents the feed – forward matrix X represents the state variables Y represents the Output	$A = \begin{bmatrix} 0 & -(1-d)/L \\ (1-d)/C_O & -1/R_OC_O \end{bmatrix}; B = \begin{bmatrix} 1/L \\ 0 \end{bmatrix}$ Where, A represents the state matrix of the converter B represents the input matrix of the converter C represents the output matrix of the converter d represents the duty cycle R_0, C_0, L represents the output resistance ,capacitance and inductance in the converter respectively
Switching flow graph [12]	$\frac{v_O(s)}{v_g(s)} = \sum_K \frac{G_K \Delta_K}{\Delta}$ G_K=Forward path gain Δ_K = 1-(loop gain that does not touch the forward path) Δ = 1+ (Σ loop gain)- (Σ two non-touching loop)…… K= number of the forward path	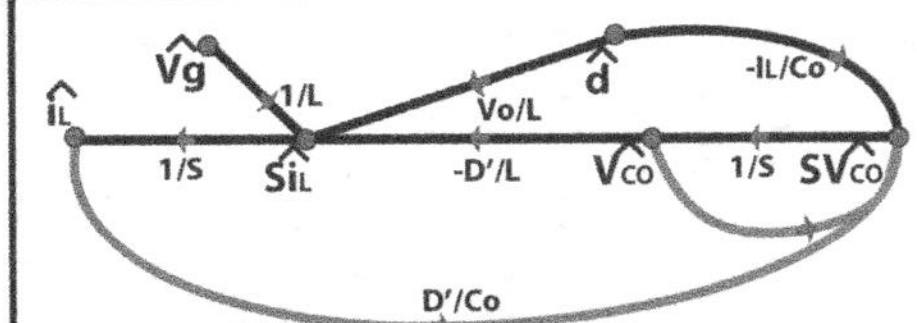

(*Continued*)

Table 8.1 Modeling techniques with example. (*Continued*)

Modeling technique	Procedure for modeling the converter	Example-boost converter
Circuit averaging [13]	Steps: 1. AC inductor loop equation is constructed. 2. AC capacitor node equation is constructed. 3. Combine the two circuits using an ideal AC transformer.	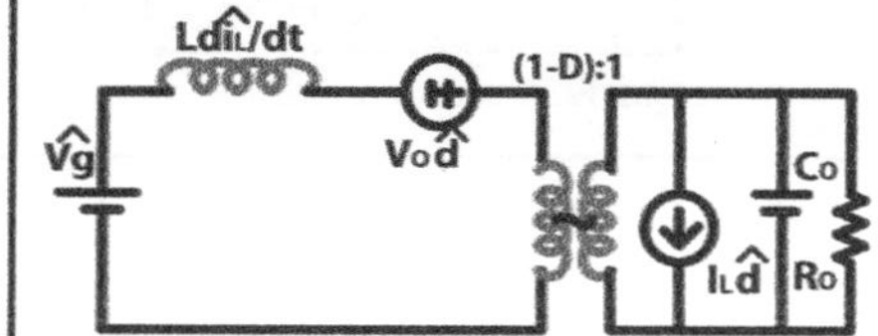
Unified topological modeling [14]	Property: DC-DC converter operating in CCM can be represented into a topological graph with the set of G, O and C branches and L branch forms a chord set. G branch: input voltage source O branch: load resistance and output capacitor. L branch: Energy transferring inductor	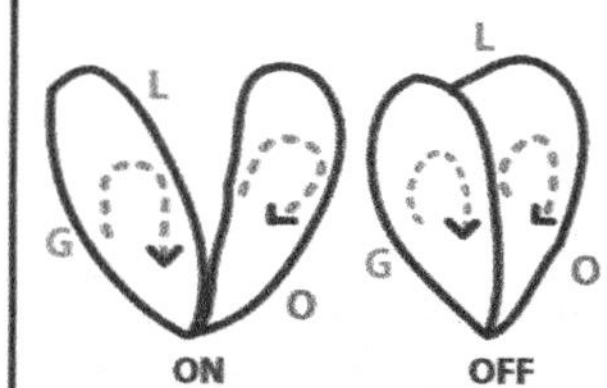 C branch: Energy transferring capacitor.
Current Injected Equivalent Circuit Approach (CIECA) [15, 16]	Property: 1. In a switching period, the average current in the first mode is injected into the second mode. 2. The relation between average injected inductor current and output voltage (v) is determined. $v = i_{ave}Z$ Where Z represents the impedance	The average injected inductor current (i_{ave}) injected into the output circuit during a switching period for duty cycle (D) is given by $i_{ave} = i_L\ (1\text{-}D)$ The derivative of the inductor current is specified by $L\,di_L/dt = Vg - V_O\ (1\text{-}D)$ $V_o = i_{ave}\dfrac{R_O}{1+SCR_O}$

(*Continued*)

Table 8.1 Modeling techniques with example. (*Continued*)

Modeling technique	Procedure for modeling the converter	Example-boost converter
Energy conservation principle [17]	Steps: 1. The equivalent averaged component is determined using the principle of energy conservation. 2. Large-signal averaged circuit model is obtained. 3. Reflection rules are established to simplify the circuit model.	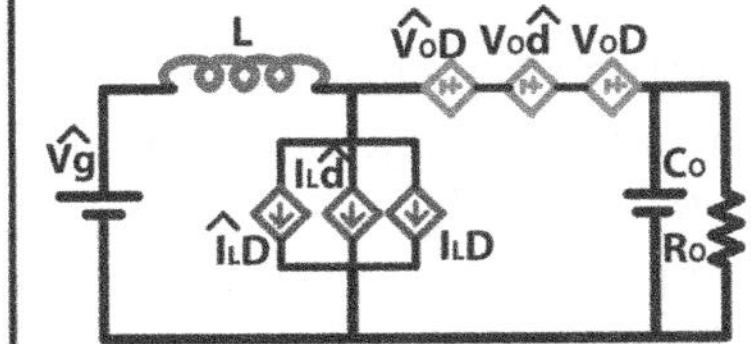
Averaged small-signal model using switching loss [18]	Similar to the principle of conservation of energy. Power losses are incorporated into the analysis.	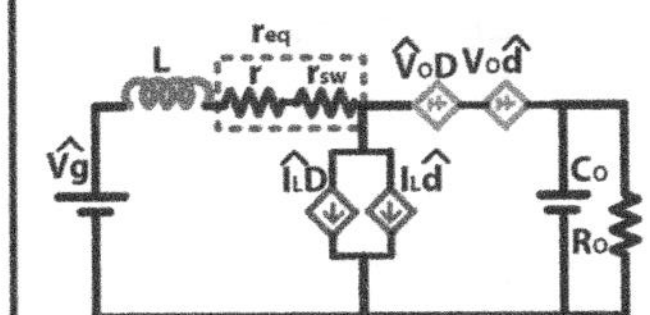
PWM switch model [19]	Property: 1. Averages the PWM switch of the converter. 2. Time averaged model is obtained by averaged terminal currents and voltages.	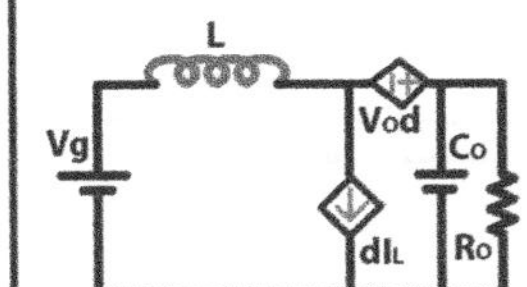
Alternate form of the PWM switch [20, 21]	Step 1: Higher-order converter even with coupled inductor can be deduced into PWM switch model. 2. PWM switch of the converter is averaged and simplified using reflection rule.	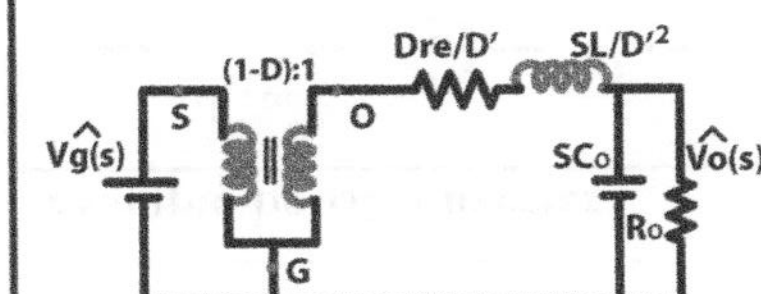

switches are averaged whereas the whole converter is averaged in the state-space-averaging method. By observing the converter circuit, the averaged model of the switches is obtained. The main advantage of this modeling method is that the model can be simulated in any circuit simulator. It has a drawback in that it can be applied only to the converter operating in CCM.

General PWM switch model is referred as p-PWM switch model where p is the common terminal. However, the alternate PWM switch model has two more models, such as a- PWM and c-PWM, where a and c is considered a common terminal. It can be applied to any type of converter with high order system. The main advantage of this technique is that it can be applied to DC-DC converter in DCM [20, 21].

8.3.1 Analysis from Modeling Method

The analysis for different modelling techniques are illustrated in Table 8.2. This table gives an overview of the modeling techniques, type of analysis carried out and the salient features.

It is noted from Figure 8.3 the plots derived to check the stability of the system are bode [25–27, 31, 32], root locus [22, 33], step response [23, 33–35] and pole-zero mapping [30, 33]. These plots are used to carry forward the analysis to design the controller and perform the closed-loop study. All the dc-dc power converters are incorporated with output LC filter which inherently makes the system unstable. Inference from bode plot analyses is used for ensuring a stable system. It is also used to design the controller with the desired performance at a particular operating point. For example, if the system has a phase margin of 45 degrees this infers that it provides a good response with little overshoot and essentially no ringing. The positive phase and gain margin represent a stable system. Hence, it is used to study the robustness of the control loop.

In the derived closed-loop transfer functions, the denominator is considered as a characteristic equation that can be equated to zero to determine the system's stability. The roots of the numerator and denominator of the transfer function are zeroes and poles of the system. The stability can be analyzed with the placement of poles and zeroes in the pole-zero map. Figure 8.4 presents the placement of roots and their inference for the stability study.

Time-domain response of dc-dc converter when it is subjected to step unit function, we can study various time-domain characteristics and it is tabulated in Table 8.3. The parameters listed in this table appraise the stiffness of the converter when subjected to the disturbances. The system's

Table 8.2 Type of analyses carried out in modeling methods.

Ref.	Modeling technique	Type of analysis	Description
[22]	State-space averaging technique	Root locus	The control to inductor current transfer function is derived to design the closed loop controller.
[23]	Energy factor mathematical modeling	Unit-step and impulse response	Step and impulse function responses are analyzed for various damping constant.
[24]	Large signal dynamic averaged state-space model	Eigenvalues	Open-loop stability analysis is performed.
[25]	AC equivalent circuit model-circuit averaging	Bode plot	Duty cycle to inductor current and output voltage are derived for the controller design
[26]	Multifrequency averaging	Time response and bode plot	Closed-loop controller is designed with duty cycle to output transfer function
[27]	State-space averaging technique	Transient response and bode plot	Overshoot and undershoot transient voltage is studied for change in load.
[28]	Averaged switch modeling	Frequency domain characterization	Large-signal time domain and small-signal frequency domain characteristics are analyzed.
[29]	Bi-switching status modeling	Transient response	The study is performed with load resistance disturbance
[30]	Sampler decomposition technique	Pole-zero mapping and pole trajectories	Stability maps are presented to study the stability and system dynamics.

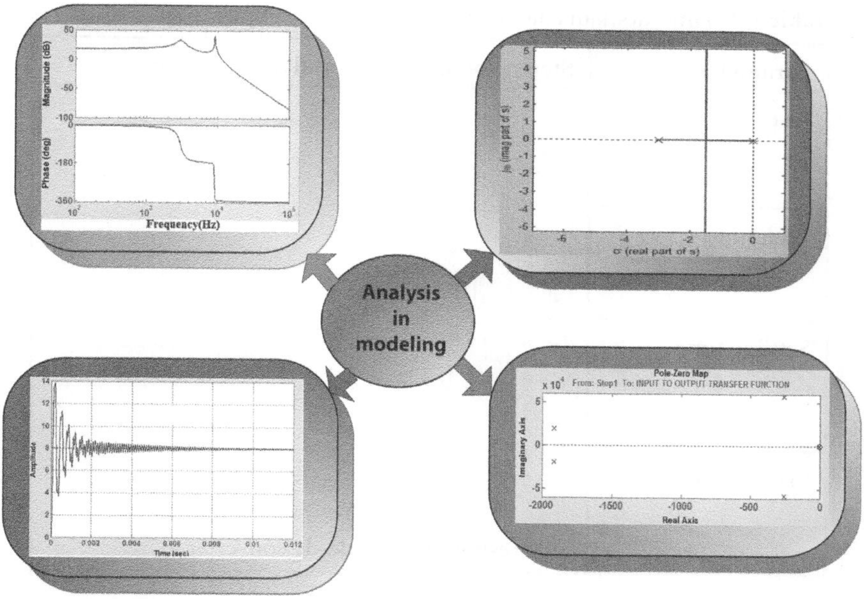

Figure 8.3 Different analysis derived from modeling techniques.

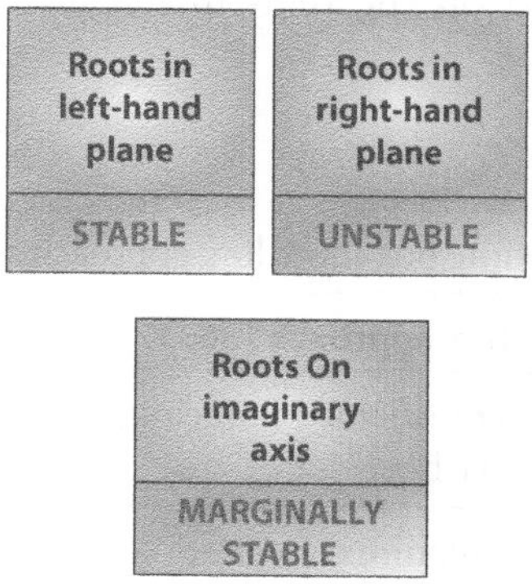

Figure 8.4 Positions of roots in the pole-zero map for stability analysis.

response is changed when the damping ratio is modified. It affects the settling time and overshoot of the system. For example, if the damping ratio is increased, it will reduce the oscillation. However, it increases the settling time. To design a good control system, it is necessary to how the damping ratio and natural resonant frequency affect the performance of the converter.

Table 8.3 Time-domain characteristics.

Parameters	Step response from simulation software
Rise time	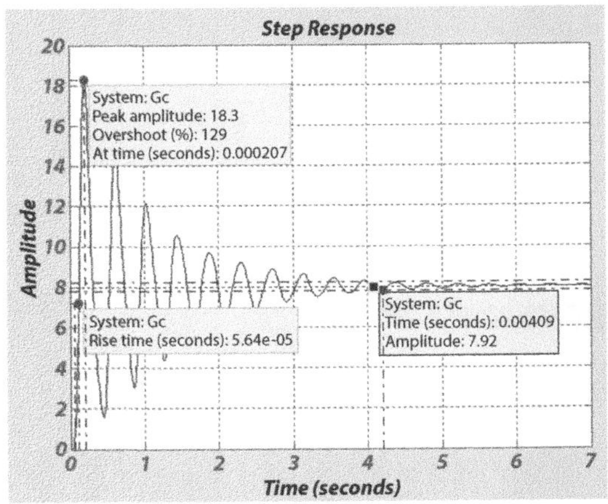 Step response of the dc-dc converter
Peak time	
Settling time	
% overshoot	

In addition to these theoretical analyses, hardware measurement is done with a network analyzer to validate the modeling techniques and also the derived analysis. One more comparison is done with the modeling techniques based on the real-time results and it is tabulated in Table 8.4.

Table 8.4 Black box vs. white box modeling approach.

Parameters	Black-box approach [36, 37]	White-box approach
Modeling approach	Hardware-oriented modeling approach	Software-oriented modeling approach
Required parameters	Doesn't require converter information.	Based on relations between internal state variables.
Derived model	G-parameter model is derived	Several models are derived based on the modeling technique.
Accuracy	Observed to be more accurate since it includes all parasites and disturbances.	Most of the ripples and parasites are neglected to simplify the analysis.

8.4 Reduce-Order Modeling

There are several methods available to reduce the order of the transfer function of the converter which will ease the derivation of the compensator of the system. Initially, in this section, the simplest algorithm to find the transfer function of the converter is presented and discussed without finding the resolvent matrix $[(SI\text{-}A)^{-1}]$ by finding the determinant and adjoints of the matrix manually. If we take a higher-order system (nxn), the mathematical calculation will be very tedious and cumbersome. Faddeev Leverrier has proposed an algorithm to find the resolvent matrix which can be applied to the modeling of the dc-dc converter with the state-space averaging (SSA) technique where finding the inverse of a matrix is very complex [38].

8.4.1 Faddeev Leverrier Algorithm

Recently, researchers have been concentrating on deriving high gain dc-dc converter with several passive components which makes the order of the matrix bigger in SSA technique. Conventional matrix inversion is observed to be tedious and it even leads to a miscalculation with high time consumption.

8.4.1.1 Procedure for Faddeev Leverrier Algorithm

One of the important tasks in the dynamic study is to compute the large-scale plant's resolvent matrix. The Leverrier algorithm suggests a simple procedure where the resolvent matrix of nth order can be easily computed. Consider a state system matrix A and the resolvent matrix is given by

$$(sI-A)^{-1}=\frac{Adj[sI-A]}{[sI-A]}=\frac{X_{n-1}s^{n-1}+X_{n-2}s^{n-2}+\ldots+X_1s+X_o}{s^n+Y_{n-1}s^{n-1}+Y_{n-2}s^{n-2}+\ldots+Y_1s+Y_o} \tag{8.1}$$

The coefficients of the numerator and denominator polynomial (X and Y) are derived with the following procedure.

Step 1: The coefficient of Y_{n-1} is calculated using the below formula by equating X_{n-1} to nth order identity matrix

$$Y_{n-1} = -\text{trace}\,(X_{n-1}\,A) \text{ where } X_{n-1} = I_n \tag{8.2}$$

Step 2: Calculate the next coefficient Y_{n-2} and the next matrix X_{n-2}.

$$Y_{n-2} = (-1/2)\text{trace}\,(X_{n-2}\,A) \text{ and } X_{n-2} = X_{n-1}\,A + Y_{n-1}\,I_n \tag{8.3}$$

Continue this procedure till the calculation ends up to Xo and Yo.

Step 3: In the end, calculate the coefficient Yo and the next matrix Xo.

$$Yo = (-1/n)\text{trace}\,(Xo\ A) \text{ and } Xo = X_1A + Y_1\,I_n \tag{8.4}$$

Using this algorithm, it is very simple to find the following transfer function.

The line to output transfer function or Audio-susceptibility

$$\frac{v_{o(s)}}{v_g(s)} = C(sI - A)^{-1}B \tag{8.5}$$

Control to output transfer function

$$\frac{v_{o(s)}}{d(s)} = C(sI - A)^{-1}B_d \tag{8.6}$$

8.4.1.2 *Illustrative Example with Switched-Inductor-Based Quadratic Boost Converter*

An example is taken and is discussed clearly in Table 8.5 with Leverrier's algorithm. Switched-inductor-based quadratic boost converter is considered for explaining the Leverrier algorithm and the circuit diagram is presented in Figure 8.5. In this example, the resolvent matrix is solved using this algorithm with the simple matlab coding given in Table 8.5. The operating principle of the converter is analysed and state system matrices for ON and OFF conditions are obtained. Similarly, the input and output are also determined for both shoot and non-shoot period. Finally, the A, B and C matrices are obtained by

$$A = A_{ON}D + A_{OFF}(1 - D) \tag{8.7}$$

$$B = B_{ON}D + B_{OFF}(1 - D) \tag{8.8}$$

Table 8.5 Example for Leverrier's algorithm.

<table>
<tr><th>State matrices/SSA</th><th>Leverrier's algorithm (Matlab coding)
Line to output transfer function Vo(s)/Vg(s)</th></tr>
<tr><td colspan="2">

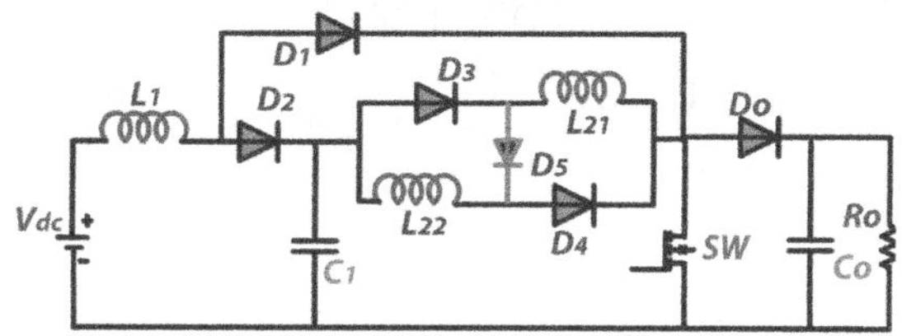

Figure 8.5 Switched-inductor based quadratic boost converter.

</td></tr>
<tr><td>

$$A=\begin{pmatrix} 0 & 0 & -D'/L_1 & 0 \\ 0 & 0 & 1+D/2L_{21} & -D'/2L_{21} \\ D'/C & -[1+D]/C & 0 & 0 \\ 0 & D'/C_O & 0 & -1/R_OC_O \end{pmatrix}$$

</td><td>

syms L1 L21 D C1 Ro Co S
A=[0 0 -(1-D)/L1 0;0 0 (1+D)/(2*L21) -(1-D)/(2*L21);(1-D)/C -(1+D)/C 0 0;0 (1-D)/Co 0 (-1/(Ro*Co))];
B=[1/L1 0 0];
C=[0 0 0 1];
Q3=eye(4);
a3=-trace(Q3*A);
Q2=(Q3*A)+(a3*Q3);
a2=-0.5*trace(Q2*A);
Q1=(Q2*A)+(a2*Q3);

</td></tr>
</table>

(Continued)

Table 8.5 Example for Leverrier's algorithm. (*Continued*)

State matrices/SSA	Leverrier's algorithm (Matlab coding) Line to output transfer function Vo(s)/Vg(s)

$B=\begin{pmatrix} 1/L_1 \\ 0 \\ 0 \\ 0 \end{pmatrix}; \; C=\begin{pmatrix} 0 & 0 & 0 & 1 \end{pmatrix}$	a1=-(1/3)*trace(Q1*A);
	Qo=(Q1*A)+(a1*Q3);
	ao=-(1/4)*trace(Qo*A);
	Denominator=S^4+(a3*S^3)+(a2*S^2)+(a1*S)+ao
	Numerator=B*[(Q3*S^3)+(Q2*S^2)+(Q1*S)+Qo]*C

$y(s)=C[sI-A]^{-1}Bu(s)$	
$\frac{\widetilde{v_O}(s)}{\widetilde{v_g}(s)}=\frac{1+D}{[1-D]^2}\left\{\frac{1}{\left[1+s\left\{\frac{G_V{}^2L_1}{R_L}+\frac{2L_{21}}{[1-D]^2R_O}\right\}+s^2\left\{G_V{}^2L_1C_O+\frac{L_1C}{[1-D]^2}+\frac{2L_{21}C_O}{[1-D]^2}\right\}+s^3\left\{\frac{2L_1L_{21}C}{[1-D]^4R_L}\right\}+s^4\left\{\frac{2L_1L_{21}CC_O}{[1-D]^4}\right\}\right]}\right\}$	

$$C = C_{ON}D + C_{OFF}(1 - D) \tag{8.9}$$

G_v represents the gain of the power converter

The A, B and C matrices for switched-inductor based quadratic boost converter is given in Table 8.5. With the Matlab coding, the numerator and denominator of the resolvent matrix are obtained. It is multiplied with the C and B to obtain input to output transfer function. In Table 8.5, the derived audio susceptibility is given in the last column.

8.4.2 Order Reduction of Transfer Function

Designing the controller and modeling the large-scale converter will be cumbersome and time-consuming. This can be overcome by reducing the order of the converter without disturbing features like static and transient response and the stability of the full-order system. By using reduction technique, the derived reduced-order model converter will be simpler and efficient for further computation.

Model order diminution is playing a significant role in the following fields of engineering:

- Chemical Engineering - Chemical vapour decomposition process
- Power Engineering - Wind conversion system in wind farm
- Space Engineering - Spacecraft network
- Electromagnetic field - Large-scale electromagnetic systems

As mentioned above, the fields are given as examples to illustrate the significance of model order reduction. The need for model order reduction is now more essential in the field of the power converter, especially in the derivation of novel topologies.

8.4.3 Techniques for Model Order Reduction

Recently, model order reduction has extended in various fields to ease the study on dynamic analysis, which is essential when designing a system for a real-time application where the system is more prone to a dynamic situation. There are several techniques in frequency and time domain conditions, as tabulated in Table 8.6.

Table 8.6 shows that the most widely used method for large-signal linear time-invariant systems is the pole clustering method.

Table 8.6 Techniques for model order reduction.

Frequency-domain	Time-domain
Pole clustering	Aggregation method
Padé approximation	Singular perturbation
Routh approximation	Balanced realization
Routhhurwitz	Krylov subspace
Stability equation	Arnoldi
Time moment matching	Hankel norm approximation

8.4.4 Pole Clustering Method

This method is well suited for a multivariable system. In the reduced model, this method confirms the steady-state and dynamic response of the full order model in the reduced-order model by preserving the stability of the model [39]. The main drawback of this method is that it requires an additional mathematical calculation to match the reduced-order model to the full order model by determining the tuning and gain adjustment factor. To overcome this issue, an improved pole clustering technique is proposed to simplify the mathematical computation.

8.4.5 Procedure for Improved Pole Clustering Technique

This technique is used to reduce the order of the denominator polynomial. However, the numerator polynomial is determined with the simple mathematical calculation [40]. The main objective of this technique is to simplify the analysis in reducing the dimension of the model. Consider a statement of the problem with higher-order transfer function for single input and single output linear time-invariant system and it is presented in the full order form of

$$G_{FO}(s)=\frac{x_o+x_1s+x_2s^2+\ldots x_{n-1}s^{n-1}+x_ns^n}{y_o+y_1s+y_2s^2+\ldots y_{m-1}s^{m-1}+x_ms^m} \tag{8.10}$$

Always $m \geq n$

Similarly, the matching lower-dimensional system is in the form of

$$G_{RO}(s)=\frac{u_o+u_1s+u_2s^2+\ldots u_{r-1}s^{r-1}+u_rs^r}{v_o+v_1s+v_2s^2+\ldots v_{q-1}s^{q-1}+v_qs^q} \quad (8.11)$$

Where q≥r.

Equation (8.11) closely matches (8.10) and maintains the characteristics of the original large-scale dynamical system. Furthermore, the response is approximated as close as possible. The poles of the derived transfer function govern the time and frequency-domain behaviour of the model. For most reduction methods, the poles lying to the imaginary axis are considered since they affect the time response. In pole clustering techniques, the following rules are considered for clustering the poles.

- Separate clusters are formed with poles lying in the left and right half of the s-plane.
- Real and complex poles are clustered separately.
- Poles lying at the origin and purely imaginary poles are taken directly in reduced-order model computation.

Model order reduction of transfer function consists of two steps:

- Denominator polynomial are obtained using improved pole clustering technique.
- Numerator polynomial are determined by means of simple mathematical computation.

8.4.5.1 *Computation of Denominator Polynomial of Lower-Dimensional Model*

An improved pole clustering method reported in [40] is considered for determining the poles of the reduced-order model and the procedure is elaborated as follows

Let m number of poles (p) are available in the derived transfer function as per equation (8.10).

Step 1: Arrange the poles in the ascending order of the magnitude.

$$\lfloor p_1 \rfloor < \lfloor p_2 \rfloor < \lfloor p_3 sss \rfloor \ldots\ldots \lfloor p_m \rfloor \quad (8.12)$$

Step 2: Group the poles as clusters and set N=1. Pole clusters are formed by placing the 1st pole in the 1st cluster, 2nd pole in the 2nd cluster and the same procedure is traced till the end of the last pole. After grouping the poles as clusters, the cluster centers of the clusters are determined as mentioned in the next step.

Step 3: Determine the pole cluster center Pc

$$p_N = \left[\left(\frac{\frac{-1}{p_1} + \sum_{i=2}^{k} \frac{-1}{p_i - p_1}}{k} \right) \right]^{-1} \tag{8.13}$$

Where N= Number of clusters

Step 4: Need to check for N=k. If so, the process is completed with the final cluster center Pc= $P_{N.}$

Step 5: Set N= N+1.

Step 6: The next cluster center will be

$$p_N = \left[\left(\frac{\frac{-1}{p_1} + \frac{-1}{p_N}}{2} \right) \right]^{-1} \tag{8.14}$$

Step 7: Again, check for N=k. If so, then the final cluster center Pc= $P_{N.}$ If not, jump to step 5.

In a case where the poles of the system are complex poles, then the computation of complex poles is carried out by considering the real and imaginary part of the poles separately. Real and complex pole clusters are fixed based on the order of the reduced-order system. The denominator polynomial of the lower dimensional system is obtained as follows after computing the cluster centers.

$$D_r(s) = (S-(a_1 \pm jb_1))\,(S-(a_2 \pm jb_2))\,(S-(a_3 \pm jb_3))\ldots\ldots\ldots\ldots\,(S-(a_r \pm jb_r)) \tag{8.15}$$

8.4.5.2 *Computation of Numerator Polynomial of Lower-Dimensional Model*

Using simple mathematical computation [41], the numerator of the reduced-order model is determined. It is obtained by comparing the transfer function of the full order and reduced-order model as mentioned below.

$$\frac{x_o + x_1 s + x_2 s^2 + \ldots x_{n-1} s^{n-1} + x_n s^n}{y_o + y_1 s + y_2 s^2 + \ldots y_{m-1} s^{m-1} + x_m s^m} = \frac{u_o + u_1 s + u_2 s^2 + \ldots u_{r-1} s^{r-1} + u_r s^r}{v_o + v_1 s + v_2 s^2 + \ldots v_{q-1} s^{q-1} + v_q s^q}$$

Equate the S'power from S^0 to S^r on both sides and get 'r' number of equations.

$$x_o v_o = y_o u_o$$

$$x_o v_1 + x_1 v_o = y_o u_1 + y_1 u_o$$

$$x_o v_2 + x_1 v_1 + x_2 v_0 = y_o u_2 + y_1 u_1 + y_2 u_0 \tag{8.17}$$

By solving the equations (8.17), the coefficients of the numerator polynomial are obtained. Since the coefficients of the denominator polynomial of the lower-dimensional model are known, it is simple to calculate the numerator polynomial coefficients. Furthermore, this method also retains the characteristics of the pole clustering method.

8.4.5.3 *Design of Controller*

A suitable mathematical model is required for designing and analysing the controller for a dynamic system. In a higher-order system, the design of the controller is a more tedious and cumbersome task. The cost of the controller design increases with the increase in order of the large-scale plant. It also increases the complexity of the system design in analysing its dynamic and transient behaviour. This can be overcome by reducing the system's order with good matching and retaining all the fundamental features of the original full-order system. After the model reduction, the compensator or controller design can be carried out with a good approximated reduced-order model.

8.5 Illustrative Example with the Power Converter

Consider a high gain converter to explain the operation of the improved pole clustering method. For this example, diode-capacitor-inductor (DCL) based quadratic boost converter is taken for analysis and the circuit diagram of the converter is depicted in Figure 8.6.

For analysis, the following ratings are considered for the converter:

- Input voltage = Vg= 12V
- Output voltage= Vo= 96 V
- Switching frequency= fs= 60 kHz
- Duty cycle=D= 0.5
- Load resistance= R_L= 230 Ω
- Inductor L_1= 15 μH, L_{21}= L_{22}= 60 μH

Applying volt-sec and amp-sec balance law, the voltage acroos the inductors and current through the capacitors are obtained. From the acquired equations, A, B and C matrix are determined.

$$A = \begin{pmatrix} 0 & 0 & -(1-D)/L_1 & 0 \\ 0 & 0 & 1+2D/3L_{21} & -(1-D)/3L_{21} \\ (1-D)/C_1 & -[1+D]/C_1 & 0 & 0 \\ 0 & (1-D)/C_3 & 0 & -1/R_L C_3 \end{pmatrix};$$

$$B = \begin{pmatrix} 1/L_1 \\ 0 \\ 0 \\ 0 \end{pmatrix}; C = [0\,0\,0\,1]$$

Using the above specifications and solving the resolvent matrix by Leverrier's algorithm, the line to output transfer function of DCL-QB converter is obtained.

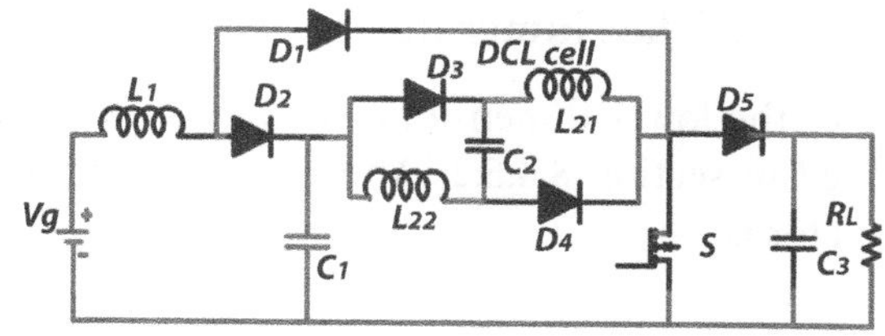

Figure 8.6 Circuit diagram of DCL-QB converter.

The full order transfer function of DCL-QB converter is

$$\frac{v_o(s)}{v_g(s)} = \frac{115}{6.21e^{-17}s^4 + 2.7e^{-13}s^3 + 7.7e^{-8}s^2 + 1.5e^{-4}s + 14.37} \tag{8.18}$$

Apply pole clustering techniques mentioned in section 8.4.5 and acquire the reduced model's denominator and numerator.

8.5.1 Derivation of the Denominator

Initially, obtain the poles of the full order transfer function given in equation 8.18.

- Pole 1: −1585 + 31647i
- Pole 2: −1585 − 31647i
- Pole 3: −839 + 15161i
- Pole 4: −839 − 15161i

Arrange the real and imaginary poles in the ascending order separately. After that, the poles are clustered and the procedure described in the section 8.4.5.1 is followed step by step to find the real and imaginary poles of the reduced-order transfer function. The procedure is applied distinctly for real poles and imaginary poles.

The calculated reduced-order poles (μ_1 and μ_2) are

- μ_1= −813.6 + 15471.8i
- μ_2= −813.6 − 15471.8i

The derived two poles are solved to obtain the denominator of the reduced-order transfer function

$$\text{Den (s)} = s^2+1.63e^3s+2.4e^9 \tag{8.19}$$

8.5.2 Derivation of the Numerator

The numerator of the lower dimensional system is obtained by the procedure mentioned in section 8.4.5.2. After simplifying the equation, the obtained numerator is

$$\mathrm{Num}\,(s) = 1.92e^9 \tag{8.20}$$

Combining the equations (8.19) and (8.20), the transfer function of the reduced-order model is

$$\frac{v_O(s)}{v_g(s)} = \frac{1.92e^9}{s^2 + 1.63e^3 s + 2.4e^9} \tag{8.21}$$

To validate the derived manipulation for improved pole clustering method, step responses of DCL-QB converter are obtained for both full order and reduced-order line to output transfer function. From Figure 8.7, it is clearly noted that both the responses are matching, which in turn proves the application of pole clustering techniques to power converters.

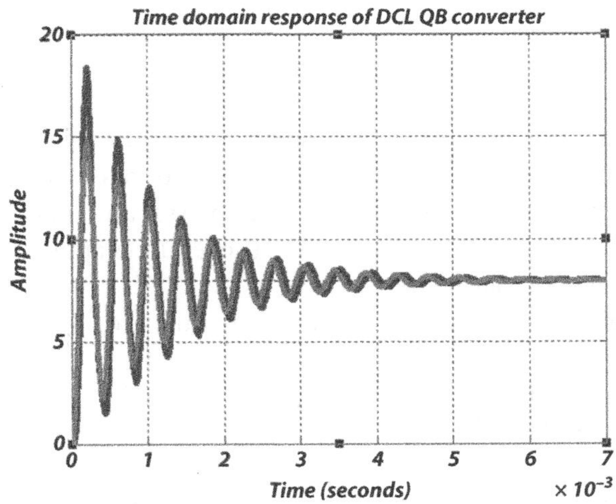

Figure 8.7 Step response of DCL-QB converter.

8.6 Controllers for Power Converter

In most of the industries, the need for digitally controlled power converters is increasing. Design of controller for this non-linear power converter is a great challenge for power electronics engineers and also it is essential to meet the requirements of the consumer. The proper functioning of power converters is mainly affected by load and input variations. Hence, the design of controllers is more crucial for the power converters for all the applications.

8.6.1 Need of Controller

DC-DC converter is a very good example for a non-linear system that puts forth a big challenge in the controller design. There are two types of controller: Linear and non-linear controller. Linear controller design relay on a small-signal model and the controller design is confined to a specific operating point. However, in the case of large parameter variations, its performance is not satisfactory. Hence, to overcome this issue, non-linear controllers are introduced with the fast dynamic response to large variations and it is also more robust in control. Closed-loop control in power converters is to ensure a good regulation of input and output voltage variations. Figure 8.8 represents the types of non-linear controllers. A few non-linear controllers reported in the literature are presented in the next section.

8.6.2 Types of Controller

Mostly, voltage-mode and current-mode controller are used for dc-dc converters to regulate the output voltage. For the regulation of input and

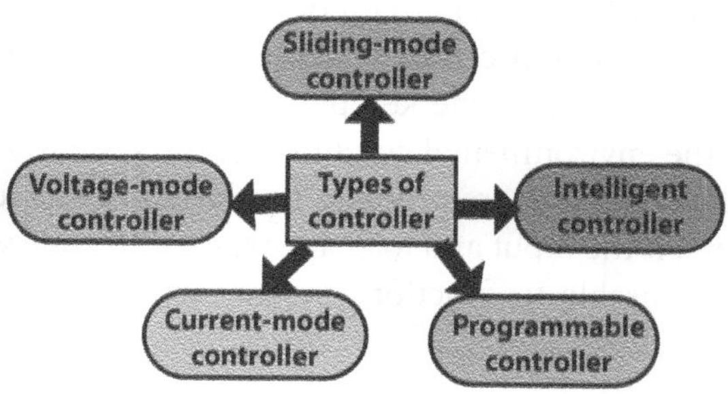

Figure 8.8 Types of controller used in power converters.

output voltage, a voltage-mode controller is proposed for a multilevel dc-dc converter [42]. Digital average current control technique is introduced by reducing the error between the reference and average inductor current to vary the appropriate duty cycle for voltage regulation [43]. In addition to these conventional controllers, several programable ICs are implemented to increase the power density of the entire system. A programmable controller IC is designed for dc-dc converter to regulate input current and voltage variation and it is built with 0.25μm CMOS [44]. Similarly, a silicon chip digital controller implemented for good voltage regulation [45]. An auto tunable digital PID controller is based on online frequency measurement to maximize the closed-loop bandwidth of the controller [46]. Furthermore, intelligent controllers are introduced based on trial and error and expert knowledge.

The performance of the fuzzy controller depends on membership functions, number of rules and linguistic variables. Fuzzy logic controller is proposed for dc-dc converter and also shows how its tracking is well fitted to the load and line variation [47]. Additionally, genetic algorithm and neural networks reduce the design procedure of fuzzy controller. Fuzzy logic is integrated with the neural network to control dc-dc converter and evaluate its performance by conducting two tests such as load and line regulation [48]. An adaptive PI controller is proposed with inner current control and outer voltage control loop to provide good steady-state and fast dynamic response [49]. To achieve desired dynamic response, the controlled variables track the reference path in the sliding mode controller. It is designed for Super-lift Luo converter to provide performance under certain parameter variations [50–52].

There are two sliding-mode controller methods: infinite switching frequency (f_{sw}) and fixed frequency sliding-mode controller. The main drawback of direct or infinite f_{sw} sliding-mode controller is power loss due to high switching frequency. This issue can be overcome by introducing fixed frequency with a constant timer circuit or PWM-based sliding-mode controller. From this section, it can be observed that the controller type can be selected depending upon the key parameters to be controlled, and it also hinges on the environmental condition of the system, since the operation of the converters is more prone to the variation due to temperature etc. Furthermore, if the input and load are exposed to wide variation then, it is more indispensable to function the converter with the closed-loop operation.

8.7 Conclusion

Modeling methods of power converters are discussed with the procedure for handling those methods. The significance of the dynamic analysis of power converters is emphasized at the beginning of the chapter. All the modeling techniques are elaborated with the boost converter as an explanatory example. The results derived from the modeling techniques are explained with the interpretation of the result and its implications. The issues dealt with the computation of resolvent matrix are studied for a higher-order system. Leverrier's algorithm is proposed for dynamic analysis of power converter exclusively to unravel the resolvent matrix. Moreover, the need of a reduced-order model for the design of controller is presented. An improved pole clustering method is proposed for reducing the order of the line to output transfer function of the power converter. With these two algorithms, the following simplifications can be carried out in dynamic analysis:

- Computation of resolvent matrix
- Controller design can be carried out with the lower-order transfer function.

A descriptive example is given to elucidate the application of the above-mentioned algorithm to dc-dc converter and its feasibility. From these illustrative examples, it is observed that both the techniques are more suitable for power converters in mathematical modeling. It is prominent that the Leverrier's algorithm and pole clustering method will definitely ease the mathematical modeling of power converters for dynamic study and to design the controller for closed-loop operation. Additionally, a sight on various controllers applied for dc-dc converters is given for researchers to have this as a guide for their analytical study on mathematical modeling.

References

1. T. Miguel Klein Faistel, C. Henrique Illa Font, A. M. S. S. Andrade and M. Lúcio da Silva Martins, "Modelling and control of DC–DC Ćuk converter with R^2P^2 and voltage multiplier cells for PV applications," in *IET Power Electronics*, vol. 12, no. 9, pp. 2214–2223, 7–8, 2019.

2. M. Veerachary, "Digital Controller Design for Low Source Current Ripple Fifth-Order Boost Converter," in *IEEE Transactions on Industrial Electronics*, vol. 61, no. 1, pp. 270–280, Jan. 2014.
3. J. A. Morales-Saldaña, R. Loera-Palomo, E. Palacios-Hernández and J. L. González-Martínez, "Modelling and control of a DC-DC quadratic boost converter with R2P2," in *IET Power Electronics*, vol. 7, no. 1, pp. 11–22, January 2014.
4. Y. Li, P. Jia and T. Q. Zheng, "Research on input-voltage feedforward control for high-order converter topologies," in *IET Power Electronics*, vol. 7, no. 11, pp. 2778–2790, November 2014.
5. I. Galiano Zurbriggen, M. Ordonez and M. Anun, "PWM-Geometric Modeling and Centric Control of Basic DC–DC Topologies for Sleek and Reliable Large-Signal Response," in *IEEE Transactions on Industrial Electronics*, vol. 62, no. 4, pp. 2297–2308, April 2015.
6. S. Hasanpour, A. Baghramian and H. Mojallali, "Reduced-order small signal modelling of high-order high step-up converters with clamp circuit and voltage multiplier cell," in *IET Power Electronics*, vol. 12, no. 13, pp. 3539–3554, July 2019.
7. M. Palaveashem and K. Anbukumar, "Reduced order linear quadratic regulator controller for voltage multiplier cells integrated boost converter," in *IET Circuits, Devices & Systems*, vol. 10, no. 6, pp. 536–548, November 2016.
8. W. Jiang, S. H. Chincholkar and C. Chan, "Investigation of a Voltage-Mode Controller for a dc-dc Multilevel Boost Converter," in *IEEE Transactions on Circuits and Systems II: Express Briefs*, vol. 65, no. 7, pp. 908–912, July 2018.
9. A. M. Lopez, D. E. Quevedo, R. P. Aguilera, T. Geyer and N. Oikonomou, "Limitations and Accuracy of a Continuous Reduced-Order Model for Modular Multilevel Converters," in *IEEE Transactions on Power Electronics*, vol. 33, no. 7, pp. 6292–6303, July 2018.
10. L. Jia, S. Lakshmikanthan, X. Li and Y. Liu, "New Modeling Method and Design Optimization for a Soft-Switched DC–DC Converter," in *IEEE Transactions on Power Electronics*, vol. 33, no. 7, pp. 5754–5772, July 2018.
11. R.D. Middlebrook, Slobodancuk, A general unified approach to modeling switching converter power stages, *IEEE power electronics specialist conference*, 1976.
12. L. K. Wong and T. K. Man, "Small signal modelling of open-loop SEPIC converters," in IET Power Electronics, vol. 3, no. 6, pp. 858–868, Nov. 2010, doi: 10.1049/iet-pel.2009.0300.
13. Wester, G.W.; Middlebrook, R.D., Low-Frequency Characterization of Switched DC-DC Converters, *IEEE Transactions on Aerospace and Electronic Systems*, pp. 376–385, 1973.
14. A. Pietkiewicz and D. Tollik, "Unified Topological Modeling Method of Switching DC-DC Converters in Duty-Ratio Programmed Mode," in *IEEE Transactions on Power Electronics*, vol. PE-2, no. 3, pp. 218–226, July 1987, doi: 10.1109/TPEL.1987.4766362.

15. P.R.K. Chetty, Current Injected Equivalent Circuit Approach to Modeling Switching DC-DC Converters, *IEEE Transactions on Aerospace and Electronic Systems* Vol. AES-17, No. 6, 1981.
16. P.R.K. Chetty, Current Injected Equivalent Circuit Approach to Modeling Switching DC-DC Converters in discontinuous inductor conduction mode, *IEEE Transactions on Industrial Electronics*, Vol. IE-29, no. 3, 1982.
17. Reatti.a, Kazimiercuk. M.K; Small-signal model of PWM converters for discontinuous conduction node and its application for boost converter, *IEEE Transactions on Circuits and Systems I:Fundamental Theory and Applications*, pp. 65–73, 2003.
18. A. Ayachit and M. K. Kazimierczuk, "Averaged Small-Signal Model of PWM DC-DC Converters in CCM Including Switching Power Loss," in *IEEE Transactions on Circuits and Systems II: Express Briefs*, vol. 66, no. 2, pp. 262–266, Feb. 2019, doi: 10.1109/TCSII.2018.2848623.
19. Edwin van Dijk, Herman J. N. Spruijt, Dermot M. O'Sullivan, and J. Ben Klaassens, PWM-Switch modeling of DC-DC Converters, *IEEE Transactions on Power electronics*, vol. 10, no. 6, Nov. 1995.
20. Ngo.K.,D.T, Alternate form of the PWM switch model.,*IEEE Transactions on Aerospace and Electronic system*; pp.1283–1292, 1999.
21. Chen, J.; K.D.T. Alternate forms of the PWM switch model in discontinuous conduction mode, *IEEE Transactions on Aerospace and Electronic Systems*, pp. 754–758, 2001.
22. A. Elserougi, I. Abdelsalam, A. Massoud and S. Ahmed, "A Non-Isolated Hybrid-Modular DC-DC Converter for DC Grids: Small-Signal Modeling and Control," in *IEEE Access*, vol. 7, pp. 132459–132471, 2019, doi: 10.1109/ACCESS.2019.2941249.
23. F. L. Luo and H. Ye, "Small Signal Analysis of Energy Factor and Mathematical Modeling for Power DC–DC Converters," in *IEEE Transactions on Power Electronics*, vol. 22, no. 1, pp. 69–79, Jan. 2007, doi: 10.1109/TPEL.2006.886652.
24. G. J. Kish and P. W. Lehn, "Modeling Techniques for Dynamic and Steady-State Analysis of Modular Multilevel DC/DC Converters," in *IEEE Transactions on Power Delivery*, vol. 31, no. 6, pp. 2502–2510, Dec. 2016, doi: 10.1109/TPWRD.2015.2508445.
25. O. Hegazy, J. V. Mierlo and P. Lataire, "Analysis, Modeling, and Implementation of a Multidevice Interleaved DC/DC Converter for Fuel Cell Hybrid Electric Vehicles," in *IEEE Transactions on Power Electronics*, vol. 27, no. 11, pp. 4445–4458, Nov. 2012, doi: 10.1109/TPEL.2012.2183148.
26. V.A. Caliskan ; O.C. Verghese ; A.M. Stankovic "Multifrequency averaging of DC/DC converters," in *IEEE Transactions on Power Electronics*, vol. 14, no. 1, pp. 124–133, Jan. 1999.
27. M. Das and V. Agarwal, "Generalized Small Signal Modeling of Coupled-Inductor-Based High-Gain High-Efficiency DC–DC Converters," in *IEEE*

Transactions on Industry Applications, vol. 53, no. 3, pp. 2257–2270, May-June 2017, doi: 10.1109/TIA.2017.2657488.
28. A. Davoudi, J. Jatskevich, P. L. Chapman and A. Khaligh, "Averaged-Switch Modeling of Fourth-Order PWM DC–DC Converters Considering Conduction Losses in Discontinuous Mode," in *IEEE Transactions on Power Electronics*, vol. 22, no. 6, pp. 2410–2415, Nov. 2007, doi: 10.1109/TPEL.2007.909270.
29. J. Han, B. Zhang and D. Qiu, "Bi-switching Status Modeling Method for DC–DC Converters in CCM and DCM Operations," in *IEEE Transactions on Power Electronics*, vol. 32, no. 3, pp. 2464–2472, March 2017, doi: 10.1109/TPEL.2016.2574894.
30. A. Villarruel-Parra and A. J. Forsyth, "Enhanced Average-Value Modeling of Interleaved DC–DC Converters Using Sampler Decomposition," in *IEEE Transactions on Power Electronics*, vol. 32, no. 3, pp. 2290–2299, March 2017, doi: 10.1109/TPEL.2016.2559449.
31. Byungcho Choi, Wonseok Lim and Seungwon Choi, "Control design and closed-loop analysis of a switched-capacitor DC-to-DC converter," in *IEEE Transactions on Aerospace and Electronic Systems*, vol. 37, no. 3, pp. 1099–1107, July 2001, doi: 10.1109/7.953261.
32. N. Kondrath and M. K. Kazimierczuk, "Control-to-output transfer function of peak currentmode controlled pulse-width modulated dc–dc buck converter in continuous conduction mode," in *IET Power Electronics*, vol. 5, no. 5, pp. 582–590, May 2012, doi: 10.1049/iet-pel.2011.0282.
33. Divya Navamani, Vijayakumar Krishnaswamy, Jegatheesan Ramiah," Stability Analysis of a Novel Switched Inductor Based Quadratic Boost DC-DC Converter", *Advances in Electrical and Electronics Eng.*, vol 15, no 5, pp. 788–798, Dec 2017, doi: 10.15598/aeee.v15i5.2436.
34. Byungcho Choi, "Step load response of a current-mode-controlled DC-to-DC converter," in *IEEE Transactions on Aerospace and Electronic Systems*, vol. 33, no. 4, pp. 1115–1121, Oct. 1997, doi: 10.1109/7.624347.
35. Y. Chi, X. Lai and H. Du, "Fast transient response high-accuracy current-sensing technique for step-up DC–DC converter," in *Electronics Letters*, vol. 51, no. 7, pp. 577–579, 2 4 2015, doi: 10.1049/el.2014.4102.
36. L. Amedo, R. Burgos, F. Wang and D. Boroyevich, "Black-Box Terminal Characterization Modeling of DC-to-DC Converters," *APEC 07 - Twenty-Second Annual IEEE Applied Power Electronics Conference and Exposition, Anaheim, CA, USA, 2007*, pp. 457–463, doi: 10.1109/APEX.2007.357553.
37. Xiancheng Zheng, H. Ali, Xiaohua Wu, O. Kuseso, S. Khan and H. Zaman, "Frequency response measurements based reduced order identification for dc-dc converter," *2016 International Conference on Intelligent Systems Engineering (ICISE), Islamabad, 2016*, pp. 78–82, doi: 10.1109/INTELSE.2016.7475165.
38. M. M. Garg and Y. V. Hote, "Leverrier algorithm based reduced order modeling of dc-dc converters," *2014 IEEE 6th India International Conference*

on Power Electronics (IICPE), Kurukshetra, 2014, pp. 1–6, doi: 10.1109/IICPE.2014.7115838.

39. Singh, C.B.Vishwakarma, K.Chattterjee, "Biased reduction method by combining improved modified pole clustering and improved Padé approximations", *Appl. Math. Model.* 40, 1418–1426, 2016.
40. Ramesh Komarasamy, Nirmalkumar Albhonso and Gurusamy Gurusamy, "Order reduction of linear systems with an improved pole clustering", *Journal of Vibration and Control* 18(12) 1876–1885, 2011.
41. A.K. Prajapati, R. Prasad, Order reduction of linear dynamic systems by improved Routh approximation method. *IETE J. Res* (2018). https://doi.org/10.1080/03772063.2018.1452645.
42. W. Jiang, S. H. Chincholkar and C. Chan, "Investigation of a Voltage-Mode Controller for a dc-dc Multilevel Boost Converter," in *IEEE Transactions on Circuits and Systems II: Express Briefs*, vol. 65, no. 7, pp. 908–912, July 2018, doi: 10.1109/TCSII.2017.2723660.
43. Y. Qiu, H. Liu and X. Chen, "Digital Average Current-Mode Control of PWM DC–DC Converters without Current Sensors," in *IEEE Transactions on Industrial Electronics*, vol. 57, no. 5, pp. 1670–1677, May 2010, doi: 10.1109/TIE.2009.2032130.
44. W. Liou *et al.*, "A Programmable Controller IC for DC/DC Converter and Power Factor Correction Applications," in *IEEE Transactions on Industrial Informatics*, vol. 9, no. 4, pp. 2105–2113, Nov. 2013, doi: 10.1109/TII.2012.2235845.
45. J. Roh, "Analogue-to-digital interface technique for digital controllers in DC-DC converters," in *Electronics Letters*, vol. 41, no. 18, pp. 1034–1036, 1 Sept. 2005, doi: 10.1049/el:20051452.
46. M. Shirazi, R. Zane and D. Maksimovic, "An Autotuning Digital Controller for DC–DC Power Converters Based on Online Frequency-Response Measurement," in *IEEE Transactions on Power Electronics*, vol. 24, no. 11, pp. 2578–2588, Nov. 2009, doi: 10.1109/TPEL.2009.2029691.
47. A. R. Ofoli and A. Rubaai, "Real-Time Implementation of a Fuzzy Logic Controller for Switch-Mode Power-Stage DC–DC Converters," in *IEEE Transactions on Industry Applications*, vol. 42, no. 6, pp. 1367–1374, Nov.-dec. 2006, doi: 10.1109/TIA.2006.882669.
48. A. Rubaai, A. R. Ofoli, L. Burge and M. Garuba, "Hardware implementation of an adaptive network-based fuzzy controller for DC-DC converters," in *IEEE Transactions on Industry Applications*, vol. 41, no. 6, pp. 1557–1565, Nov.-Dec. 2005, doi: 10.1109/TIA.2005.857461.
49. H. Wang, M. Han, R. Han, J. M. Guerrero and J. C. Vasquez, "A Decentralized Current-Sharing Controller Endows Fast Transient Response to Parallel DC–DC Converters," in *IEEE Transactions on Power Electronics*, vol. 33, no. 5, pp. 4362–4372, May 2018, doi: 10.1109/TPEL.2017.2714342.
50. M. Mahdavi, M. Shahriari-kahkeshi and N. R. Abjadi, "An Adaptive Estimator-Based Sliding Mode Control Scheme for Uncertain POESLL

Converter," in *IEEE Transactions on Aerospace and Electronic Systems*, vol. 55, no. 6, pp. 3551–3560, Dec. 2019, doi: 10.1109/TAES.2019.2908272.
51. Y. Gu, N. Bottrell and T. C. Green, "Reduced-Order Models for Representing Converters in Power System Studies," in *IEEE Transactions on Power Electronics*, vol. 33, no. 4, pp. 3644–3654, April 2018, doi: 10.1109/TPEL.2017.2711267.
52. L. Guo, P. Li, X. Li, F. Gao, D. Huang and C. Wang, "Reduced-order modeling and dynamic stability analysis of MTDC systems in DC voltage control timescale," in *CSEE Journal of Power and Energy Systems*, vol. 6, no. 3, pp. 591–600, Sept. 2020, doi: 10.17775/CSEEJPES.2019.02350.

9

Matrix Converter and Its Probable Applications

Khaliqur Rahman

Department of Electrical Engineering, Aligarh Muslim University, Aligarh, India

Abstract

An electrical drive system consists of a converter to control the voltage and frequency that is necessary to change the speed of the AC drive to meet the load condition. In the existing drive system, back-to-back converters based on voltage source are generally used. These back-to-back converters consist of bulky energy-storing capacitors for the conversion of power. These capacitors are likely to have poor performance and be less reliable. Therefore, the use of a capacitor degrades the reliability of the converter system for drives. These problems associated with drives can be mitigated by a matrix converter (MC) at the place of the back-to-back converter. There is no energy-storing element in the MC between the rectification and inversion stage of the converter that can control power without the restriction of the output and input frequency. A matrix converter offers some advantages over inverter-based drives without the limitation of operation frequency. The major advantages offered by MCs are: i) no need of capacitors for power exchange, ii) four-quadrant operation is possible, iii) unity power factor operation at the input side of the MC. In this chapter, a detailed study of MC has been prepared. The control of multiphase MC for multiphase drive applications is discussed to make the system more reliable. The control aims at the selection of space vectors from the available pool of space vectors to get sinusoidal output and to lessen total harmonic distortion. A brief description of each part of the drive system fed from the matrix converter is presented. Moreover, several issues associated with the matrix converter as modulation, control and mitigating effects of non-idealities are discussed. The different possible applications, especially in more electric aircraft applications, wind power conversion, and applications in electrical drives are also discussed.

Keywords: Five-phase induction motor, matrix converter, AC-AC converter, multiphase drive, space-vector PWM

Email: er.khaliqurrahman@gmail.com

Nikita Gupta, Mahajan Sagar Bhaskar, Sanjeevikumar Padmanaban and Dhafer Almakhles (eds.) DC Microgrids: Advances, Challenges, and Applications, (273–298) © 2022 Scrivener Publishing LLC

9.1 Introduction

Venturini and Alesina offered the first study of the AC-AC conversion without the energy storing element in 1980 [1]. They presented the basic circuit configuration as an organization of switches that connect any of the output phases to the input. That is named as matrix converter and they offered the first control concept based on the scalar approach. For better performance, a filter at the input of the MC is essential to reduce the high-frequency current components and decrease the effect of perturbation of the grid. The control theory was named Venturini modulation. It describes the low-frequency behavior of the matrix converter currents and voltages. In this technique, the voltage transfer to load was limited to half and there was a constraint in the control of the input power factor. Venturini and Alesina further offered a modified modulation with the increased voltage transfer limit [2]. In the modified modulation the voltage gain improvement becomes possible by injection of third-harmonics in the reference output voltage and input currents. This control method is named the improved Venturini or optimum Venturini method. In both modulation methods of Venturini, the output voltage is expressed as,

$$V_o(t) = M(t) v_i(t) \tag{9.1}$$

Where M (t) is a transferred matrix consisting of switches duty ratios. The main disadvantage of the MC is the lower transfer ratio than unity. Different control approaches, whether based on space-vector or scalar, need an interpretation of maximum attainable voltage at the output. The classification of different modulation schemes for the MC is presented in Figure 9.1. The problem of the lower voltage transfer to the load has

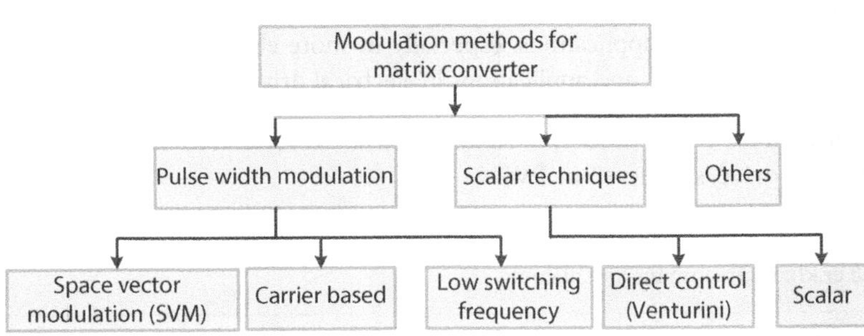

Figure 9.1 Classification of modulation schemes for MC.

motivated strenuous exploration which leads to the development of new control techniques to improve the voltage transfer ratio.

Rodriguez proposed a control technique in 1983 that was named the indirect control method. This proposed control method was based on a "fictitious dc link" [3]. In this control, the voltage is rectified then inverted with the required frequency. Here, the switching pattern is such that each output line is connected to the most positive or most negative input lines. The developed control technique is discussed in detail in [4].

The voltage transfer limit is increased up to 1.053 with much sacrifice of low-frequency distortion in input current and output voltage. A "scalar" control was presented in 1987 by Roy and April [5]. In it, the switching signals and their operating time are calculated directly from the measurement of input phase voltages. Still, the output voltage was limited to 0.5 to the input. It can be increased to 0.866 by the sacrifice of low-frequency distortion in load voltages and source currents with modification in the modulation [5]. Moreover, the input power factor control was restricted.

Complete control of voltage transfer to the load and input power factor of the scalar method is presented in [5]. In it, the voltage transfer is limited to 0.866 with the controlled input power factor followed by space vector PWM. First, space vector PWM was used by Huber *et al.* in [6], in which an MC is controlled based on the principles of space vector. In it, instantaneous representation of output voltages and input currents are considered. With the help of the space vector control technique has full control on the output voltage and input power factor. This control technique has been employed successfully in direct and indirect MC. The research contribution in drive application [7], to the reduction of switching commutations [8] control of the input power factor [9] and non-sinusoidal or unbalanced supply voltage conditions [10, 11]. Other techniques are carrier-based control, sliding mode control, and recently predictive control [12, 13]. In [14] the control techniques which give output voltage higher than input voltage commonly suffer from low-frequency distortion in output as well as input parameters.

9.2 Classification of Matrix Converter

Depending upon the conversion stages, MC can be broadly classified into two categories. These are:

1. Direct matrix converter
2. Indirect matrix converter

Direct and indirect can be distinguished by power conversion stages from source to load. In direct MC, any input-phase can be connected directly with any of the output phases [15–17]. That is also named as single-stage conversion. In this, the load is connected with the source, with the help of bidirectional switches, Figure 9.2(a). It offers variable frequency and amplitude control dependent on the switching operation.

Indirect MC topology consists of two-stage conversion. It consists of a controlled rectification (stage-I) and an inversion stage (stage-II), as shown in Figure 9.2(b), [18, 19]. There are '2m' number of bidirectional switches in stage-I and 'n' number of bidirectional switches in stage-II. In these MC, performance indices like voltage gain, THD of load voltage and input power factors are identical. Here, rectifier and inverter stages are controlled independently.

Further classification of MC converters can be done based on phases in input and outputs. This can be either three-to-three phase, three-to-n phase and m-to-n phase. Here, m and n can be any real number. Indirect MC can be further classified on the basis of active components used. The classification can be done as classical, sparse, very sparse and ultra-sparse MC.

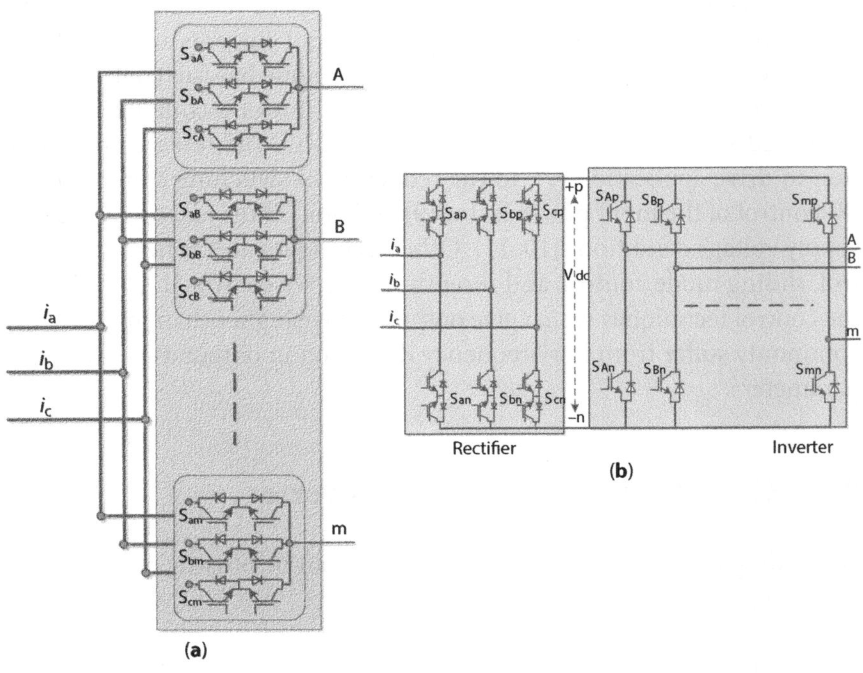

Figure 9.2 Circuit configuration of three-to-n phase (a) direct MC (b) indirect MC.

9.2.1 Classical Matrix Converter

Classical three-to-three phase MC comprises 9 bi-directional switches. The rectification stage has 6 bi-directional and the inversion stage has 6 unidirectional switches. Since two different stages have to be controlled independently, the control of the full converter becomes easy. Zero current commutation switching technique at dc-link is employed. For phase transition on the input side, at first stage-II is switched into a freewheeling state and then at stage-I, the rectifier will commutate [20].

9.2.2 Sparse Matrix Converter

In stage-I of the indirect MC, both the polarities of DC link voltage can be controlled, hence four quadrants operation is possible [21]. Sparse MC is displayed in Figure 9.3. In this topology of indirect MC, there is a reduced number of active components in each leg. Three-to-three phase sparse MC consists of 15 transistors and 18 diodes. Sparse MC is similar to the classical indirect MC with a lower component count.

9.2.3 Very Sparse Matrix Converter

The active switches of indirect MC could be further reduced to 12 resulting in a new arrangement and its control, termed a very sparse matrix converter [22], Figure 9.4. It has lower switching losses due to the usages of less active switches, though the conduction loss becomes higher as more diodes are conducting.

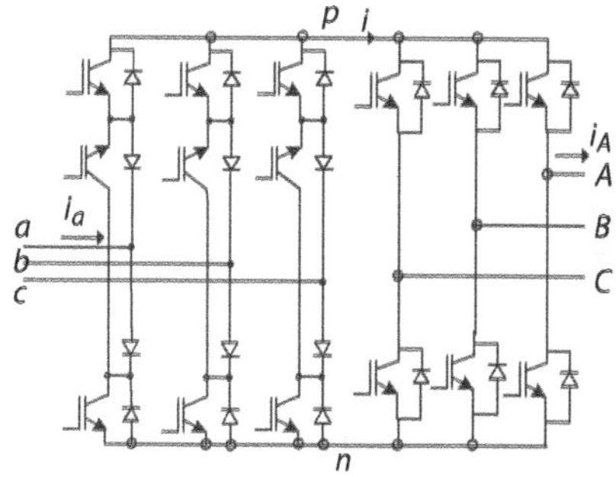

Figure 9.3 Switch configuration of sparse MC.

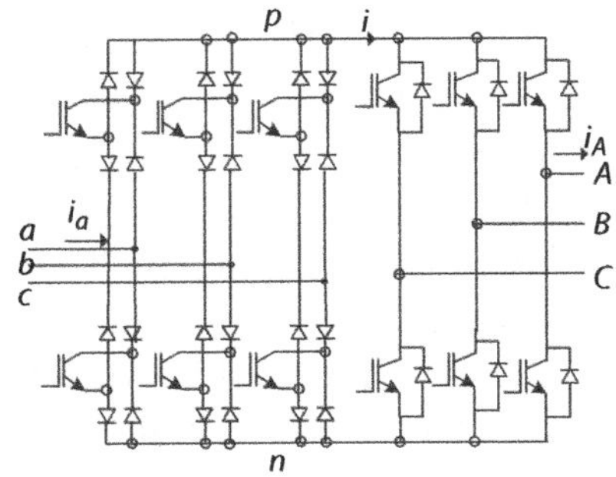

Figure 9.4 Switch configuration of very sparse MC.

9.2.4 Ultra-Sparse Matrix Converter

Ultra-parse MC can be applied only with unidirectional power flow for example in aerospace [23]. Related to other indirect MCs, ultra-sparse comprises 9 unidirectional active switches only and a lower quantity of diodes, displayed in Figure 9.5. However, overvoltage overshoots can be there due to possible reverse current flow. Hence, it becomes important to use a clamp circuit to overcome the voltage spikes.

Most of the matrix converter modulation strategies have a voltage output less than input. However, the topologies with voltage transfer more than one suffer from the low-frequency distortion of the output and input waveforms. Finally, a summary of the comparison of the voltage gain of different MCs is presented in Table 9.1.

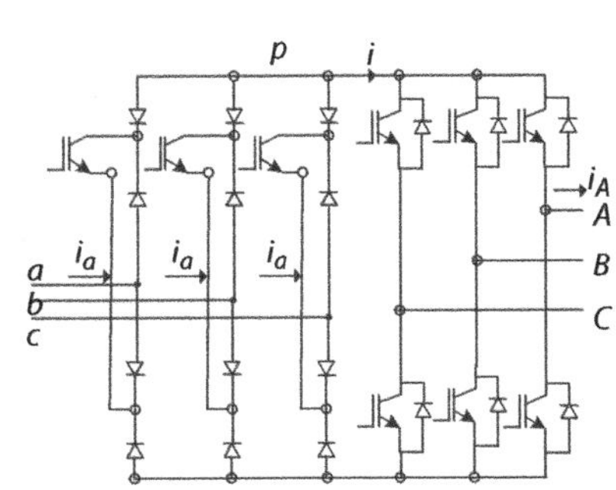

Figure 9.5 Switch arrangement of the ultra-sparse MC.

Table 9.1 Voltage gain in different types of three-to-three phase matrix converter topologies [24].

Type of converter	Control strategy	Voltage gain	Transistor	Diodes	Isolated power supply
Voltage source MC	Classical Venturini	0.5	18	18	9
	Improved Venturini	0.866	18	18	9
	Scalar	0.866	18	18	9
	SVM	0.866	18	18	9
Indirect IM	SVM	0.866	18	18	9
Sparse MC	SVM	0.866	15	18	9
Very sparse MC	SVM	0.866	12	30	6
Ultra sparse MC	SVM	0.866	9	18	6

9.3 Problems Associated with the MC and the Drives

The proposed topic is worth investigating since several issues remained unanswered, such as modulation, control and mitigating effects of non-idealities in MCs.

9.3.1 Commutation Issues

The switching commutation of bidirectional switches is one of the main problems in the MC. The problem is due to the deficiency of a freewheeling path in the matrix converter. Hence it becomes challenging to commutate safely from one switch to another [24, 25–27]. A commutation technique is proposed to reduce the losses produced by the reverse leakage current. The proposed control technique is based on certain voltage polarities of bidirectional switches and output current direction [28].

9.3.2 Modulation Issues

After Venturini's direct transfer function approach, there are two types of modulation of matrix converter, which are scalar and vector control. The instant specific input phase voltages ratio is being used to generate the switching signals of the MC's switches. The sine PWM technique is the simplest approach of the MC control. However, due to some deficiencies, it becomes important to work for advanced control. A prevailing solution that is space vector modulation is applied to the control of a matrix converter [29]. Though, one of the major challenges with multiphase MC followed on space vector PWM is to select the suitable set of space vectors for improved modulation and control.

9.3.3 Common-Mode Voltage and Common-Mode Current Issues

In high-frequency pulse width modulated converter fed motor drives, a very high-frequency voltage component is generated between winding to winding and between winding to iron (due to the parasitic winding capacitance). These voltage components cause leakage currents of high-frequency which may flow through the motor bearing. Thus, the converter operation for the control of induction motor (IM) may create different forms of bearing currents, i.e., circulating bearing current, rotor ground current, electric discharge machining current, and

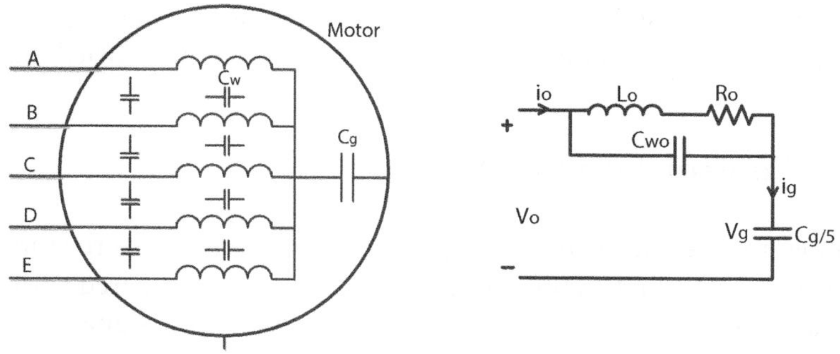

Figure 9.6 Shortened high-frequency model of (a) a five-phase IM (b) phase 'A' of the five-phase IM.

capacitive bearing current [30–32]. Among these, the capacitive bearing current is very small, almost 5-10 mA, which can be ignored. But the remaining currents are noticeable. These currents depend on common mode voltage or its rate of change. Usually, these currents flow from the stator winding's neutral of the converter (negative DC rail). No actual connection between motor neutral and converter neutral of the DC bus exists. Therefore, the measurement of common mode current is not possible in this system. Although, an approximate relation with the zero-sequence current can be derived. The model of high-frequency for the five-phase IM is shown in Figure 9.6. *Ro* is the per phase stator winding resistance, *Lo* is self-inductance, and C_w is the capacitance between the stator winding neutral (*n*) and a stator phase winding. C_g indicates the stray capacitance between the frame and the stator winding's neutral part.

9.3.4 Protection Issues

The semiconductor switches of the MC must be protected against overcurrent and overvoltage. A form of protection circuit consisting of diode bridges at the output and input side of the converter was proposed in [33–35]. The diode bridge output is connected to a capacitor which is then connected with resistance for the discharging of the capacitor. This protection circuit further increases the active parts of the system. Alternatively, protection based on varistor and a suppressor diode was suggested by Mahlein *et al.* in [35], which was used only for a low-capacity system. A protection circuit is proposed by Itoh *et al.*, consisting of a small capacitor and resistance [36]. A small clamping circuit is essential in avoiding a sudden change in overvoltage either due to switch commutation or switch failure.

9.4 Control Techniques

Two main pulse width control implementation techniques are employed in the MC. In the first technique, a reference signal is compared with a triangular signal and their intersection defines the switching signals, whereas in another technique that is based on SVPWM, the time of operation of the different states is pre-calculated for each carrier cycle. The different states are obtained by using the space vector theory. The SVPWM technique is computationally intensive, which offers superior performance in drives. This technique has the advantage of taking account of interaction among the phases when load neutral is isolated from the supply. SVPWM can be used to minimize harmonics of the isolated neutral load because in the application time of zero voltage vectors T_{31} and T_0 are identical. Moreover, the easy digital implementation of SVPWM makes it one of the most competitive solutions with the rest of the modulation schemes for MC.

Space vector PWM for three-three phase MC can be extendable to the multiphase topology. However, it is extremely multifaceted because of the available large number of space vectors [15]. For example, in a three-five phase MC, the total possible switching states are 2^{15}=32768. However, applying the different switching constraints of the MC, only 3^5 =243 vectors corresponding to 243 switchings are useful. Out of these 243 only 93 are there those can be controlled and sinusoidal operation can be possible.

However, real-time processing still poses challenges in terms of voltage gain, space vector selection, sequence and duration of operation of the selected vector, and number of switching transitions in the case of space vector modulation. The output voltage is limited to 78.86% of the input magnitude for linear modulation range in a three-five phase MC. The voltage transfer limits further decrease with increased phase number. The magnitude of output voltage increases for *n-to-m* MC ($n>m$) and it reaches 100% for a specific case of six-three phase MC. This can be increased further by taking the advantage of over-modulation. No work has been reported yet in this area of over-modulation of the multiphase MC. However, the sinusoidal output from space vector PWM for multiphase MC can be achieved by optimum selection of dwell time to generate the required voltage reference.

9.5 Basic Components of the Matrix Converter Fed Drive System

The basic assembly of a three-five phase MC is given in Figure 9.7, which shows different parts of the MC fed drive system. In this section of the chapter, a brief description of every part of the system is given.

(a) Three-five phase MC
In the direct MC, every input phase can be connected with any of the output phases. The connection can be established by switching states. There is a total of fifteen bidirectional switches for a three-to-five phase MC. As every switch can have two states ON or OFF, hence total switching combinations can be 2^{15}.

Applying the constraints for safe switching operation that are:

- First switching constraint is 'Saj + Sbj + Scj = 1'. At an instant, only any one phase from 'a' 'b' 'c', should be connected with the 'j' output phase, to avoid the source short-circuiting.
- Output phases should not be opened during the operation as the load is inductive and interruption of inductive current at any value might cause high di/dt.

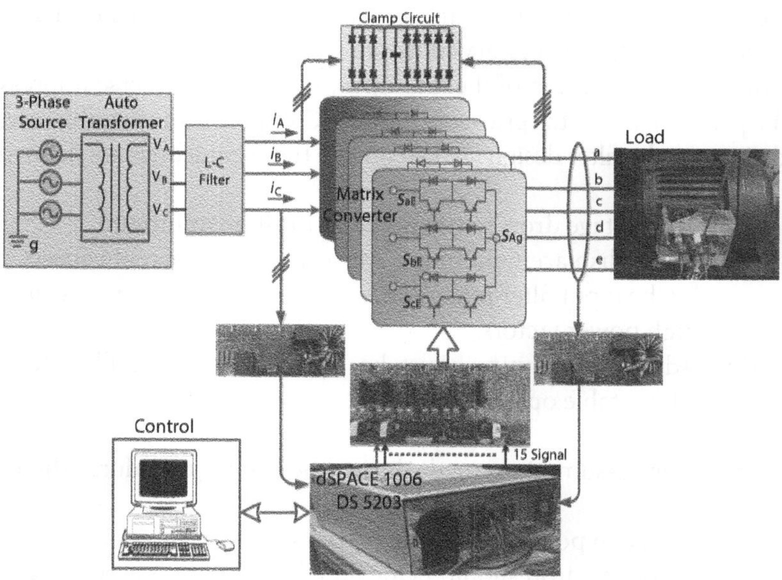

Figure 9.7 Drive system fed from Three-five phase MC.

There are 3^5 i.e., 243 different possible safe switching combinations. These combinations of output phase with input phases are shown in Figure 9.8. The remaining 150 switching combinations are there which have connections of output phases with all of the three input phases simultaneously.

(a) Load
Full system of the MC was tested with R-L load as well as an induction motor.

(b) The input filter
Current drawn by the MC usually consists of large harmonic components in addition to fundamental components due to switching action. When the MCs are interconnected to the utility grid, it injects harmonics in the grid. To meet IEEE regulations code for input current, input filter design becomes essential. The input filter interfaces between the matrix converter and the AC mains. Its essential feature is to avoid significant changes in the converter input voltage. There are many papers that discuss the problem of the input filter design for the MC [37–39]. The design has been proposed in different configurations. Input filter configuration used in the MC of the three-phase prototype is displayed in Figure 9.9. The converter's input terminals are joined with the grid that is a voltage source, and generally an output is inductive. Therefore, an inductive filter in the current fed system and the capacitor filter in the voltage fed system are required. AC/AC converter, especially MC lacks in the usage of a large capacitor. Hence it becomes intrinsically necessary to use an input filter.

However, the inclusion of the filter must not have an adverse impression on the performance of the grid-connected MC system. To ensure this, certain guidelines for the filter design are laid down. These guidelines are defined as:

(i) The voltage drop across the filter must be minimum.
(ii) Phase displacement in the converter input current caused by the input filter must be minimized to allow operation at high power factors.
(iii) Adequate damping must be introduced in the filter to allow stable operation.

There are two assumptions made during the design procedure. These are:

1. There is no power loss in the matrix converter,
2. The voltage drop should be negligible that occurs across the filter elements.

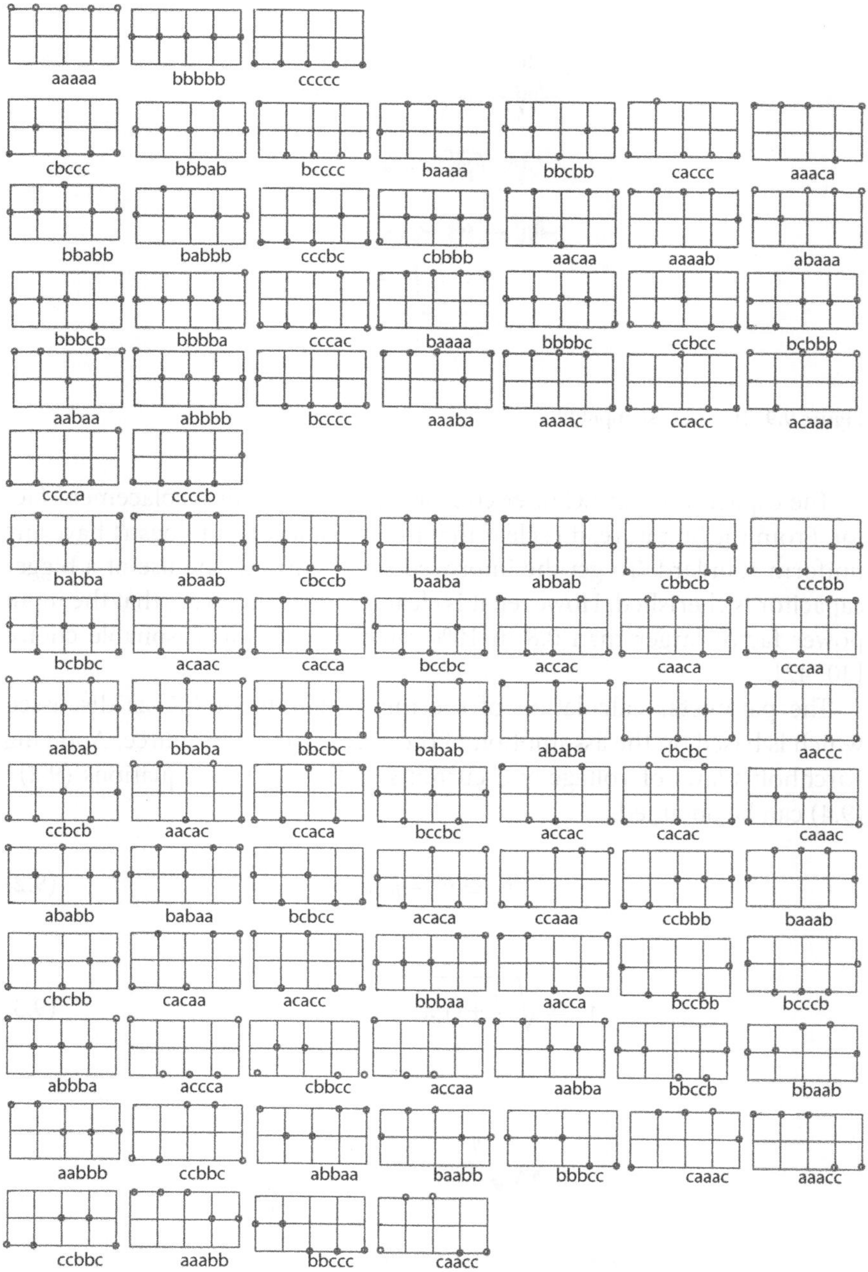

Figure 9.8 Switching combinations of three-to-five phase MC.

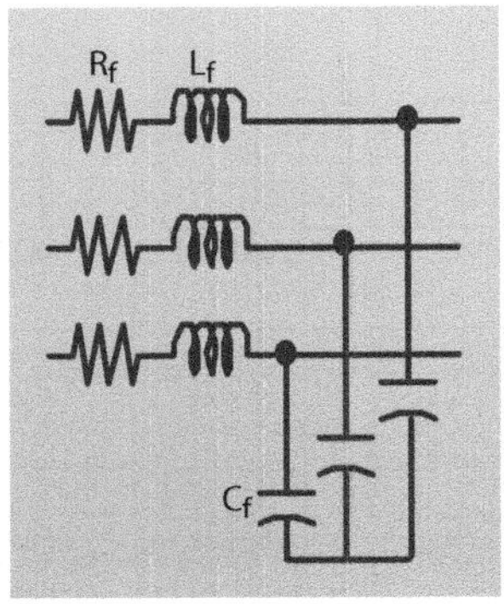

Figure 9.9 Three-phase input filter.

The capacitor of an LC filter contributes to the input displacement factor. From the literature, it is clear that input displacement doesn't have any uniform standard. To get the improved input current, the use of a bigger capacitor is cherished. However, it is clear from the literature that the input power factor larger than 0.8 in 10% rated power is a reasonable choice [40–42].

The parameter calculation of the input filter for a MC is discussed which is based on the assumption of the MC as a current source. Applying Kirchhoff's laws of voltage and currents in Figure 9.10, equations (9.2)–(9.4) can be obtained.

$$v_c = v_i - j\omega l i_i \tag{9.2}$$

$$v_c = \sqrt{v_i^2 + (\omega l i_i)^2}\, e^{-j\,arctg\frac{\omega l i_i}{v_i}} \tag{9.3}$$

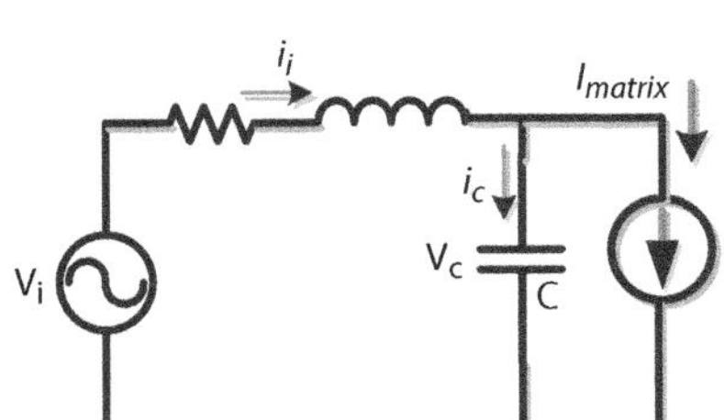

Figure 9.10 Equivalent circuit of a MC.

Also, the input current can be interrelated with the MC current

$$i_{matrix} = \sqrt{\left[i_i\left(1-\omega^2 lc\right)\right]^2 + \left(\omega c v_i\right)^2 e^{-j arctg \frac{\omega c v_i}{i_i\left(1-\omega^2 lc\right)}}} \tag{9.4}$$

The displacement factor of the MC can be attained as:

$$-j arctg\frac{\omega l i_i}{v_i} + -j arctg\frac{\omega c v_i}{i_i\left(1-\omega^2 lc\right)} > 0.8 \tag{9.5}$$

$$c < \frac{tg\left(0.8 + arctg\frac{\omega l i_i}{v_i}\right)}{\omega v_i}\left(1-\frac{\omega^2}{\omega_c^2}\right)i_i \tag{9.6}$$

Assuming; $arctg\frac{\omega l i_i}{v_i} << 0.8$ and $\frac{\omega^2}{\omega_c^2} \approx 0$

By neglecting the displacement factor due to filter inductance and making an allowance for the cut-off frequency around 1700 Hz (34th harmonics), the maximum value of filter capacitance depends on desired output power and the input current amplitude.

$$c \leq \frac{tg(0.8)\times\sqrt{2}}{3\omega v_{i_{ef}}^2}P_{out} \leq 21\times10^{-9}P_{out} \tag{9.7}$$

For a 2 kW MC, the capacitance can be calculated and should be less than 42×10^{-6}. The capacitor voltage ripple can be obtained by (9.8);

$$\frac{dv_c}{dt} = \frac{i_i - i_{matrix}}{c} \tag{9.8}$$

Considering that the MC output current is sinusoidal, so neglect power loss in the converter;

$$P_{in} = P_{out}$$

$$3v_{i_{ef}} i_{i_{ef}} = 5v_{o_{ef}} i_{o_{ef}} \cos\varphi_o \tag{9.9}$$

The current drawn by MC is irregular and can be either time-varying or zero. Consequently, the maximum value of ripple in capacitor voltage can be attained when the matrix converter current i_{matrix} and input current i_i has the opposite signs, also the ripple increases for lower modulation index of the converter and lower values of $\cos\phi_0$. The limit in inductance value can be established by considering the filter cut-off frequency in (9.10),

$$f_c = \frac{1}{2\pi\sqrt{lc}} \tag{9.10}$$

To minimize the resonance effect at a cut-off frequency (f_c), the f_c should be sufficiently high and lower than the switching frequency. The series connected resistance with filter inductance for damping purpose can improve the stability of the MC, however, its value is very small in milliohms [40].

(c) Clamp Circuit

The traditional DC link converter has a static freewheeling path for the output motor whereas in the case of MC there is no automatic static freewheeling path, so there is a possibility of commutation-dependent risk. AC supply has voltage surge which may lead to the generation of overvoltage externally, switch commutation error or timing inaccuracies if MC is used for AC drive application. However, MC switches require a freewheeling path during dead time operation, during switching transitions or an emergency shutdown of the motor where destructive overvoltage has appeared. The switches can be protected from overvoltage stress with the help of a clamp circuit. A simple clamp circuit is presented in Figure 9.11 consisting of a diode bridge rectifier along with capacitor and resistance combination.

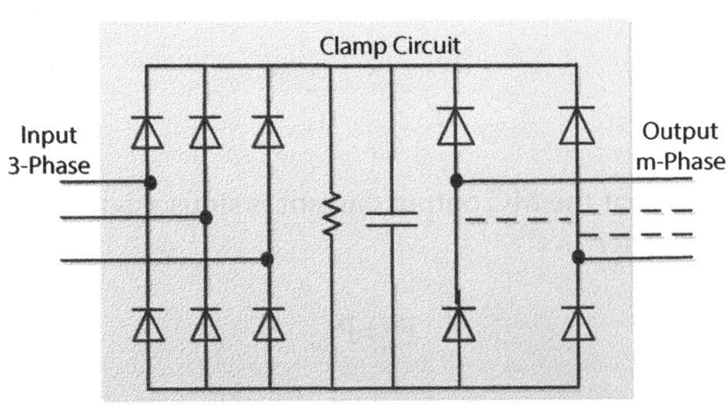

Figure 9.11 Clamp circuit to protect the MC switches.

In case of overcurrent at a prescribed value of current, a relay is energized and the circuit breaker disconnects the circuit. Disconnection of the circuit can also be achieved by using semiconductors if the motor current can get a freewheeling path. Hence, the overvoltage protection scheme somehow also includes an overcurrent protection scheme. The switches need protection from the input as well as from output side surges which are produced when the converter shuts down in an emergency condition. A clamp circuit is employed for protection as proposed in [43, 44]. The clamp circuit has capacitors and diodes bridges connected to the output and input side junctions of the MC.

If a capacitor is used in the clamp circuit, then no critical overvoltage is generated since the energy that was stored in the inductive load is transported to the capacitor. This energy in the load is due to the interruption of inductive currents in the motor. The clamp circuit also protects from voltage spikes that are generated due to the commutation of the inductance current. During switching operation capacitor voltage unceasingly increases, so there is a need for an element to discharge the capacitor. Resistance is connected in shunt to discharge the capacitor. However, there are some drawbacks to this scheme:

- The requirement of the number of semiconductor devices is increased to 12 fast-recovery diodes in the MC of three-to-three phase;
- It increases the number of reactive components needed;
- It is very important to know the circuit parameters for the optimal design of a clamp circuit.

9.6 Industrial Applications of Matrix Converter

- A low voltage transfer limit is the biggest drawback of a matrix converter in the industrial drive. Though the control in over modulation can increase the voltage transfer limit, but unavoidably, with the sacrifice of the quality of power. In those applications where specified load motors are required and properly designated, the voltage transfer ratio doesn't remain a problem. In many drive applications where a motor is incorporated with the converter, the advantages of lower size and weight of the MC can be preferred over challenging voltage source inverter technologies.

- The reduced weight, size, and absence of dc-link capacitor give applicability in aircraft applications. Different projects' description is present in the literature for the development of aircraft actuator. An alliance with Smiths Aerospace runs the development of a 7 kW MC to drive a 10000 rpm permanent magnet synchronous motor combined with an electro-hydrostatic actuator.
- An indirect MC-based drive system was established for an EMA usage designated in collaboration with MOOG.
- A type of MC was used to regulate a wind turbine generator and a single-phase transformer which is coupled with a dc transmission through a rectifier. An industrial interest is outlined. The features of the MC from a manufacturer's viewpoint in terms of size, competitiveness, and cost is provided.

Matrix converter products in the market:

- So far, the drive producers Fuji Electric Systems and Yaskawa offer MC products. Fuji Electric Systems has manufactured the FRENIC-MC series of MCs which has specifications of input voltage 230 V in 15 kW and 30 kW power ratings. Moreover, rated voltage 480 V has 15 kW, 30 kW, and 45 kW power ratings. A product of Yaskawa, AC7 offers all the characteristics with additional benefits that a programmable industrial vector drive can have. Due to a completely regenerative drive, it can be applied in conveyors, hoists, escalators and lifts.

Yaskawa also proposes drives fed from MC at a medium voltage range. The FSDrive-MX1S is intended on 3 kV and 6 kV. Energy conservation and saving due to regenerative drive is a major selling point. The MX1S power range for a 3 kV system is from 0.2 MW to 3 MW and for a 6 kV system from 0.4 MW to 6 MW.

(a) Aircraft applications:
To power the loads in aircraft such as lighting, in-flight entertainment and avionics, an electrical system is used. The total power level is usually over 1 MW for all these systems and for a comparatively small platform it is a large power. A design of distinct power systems for these four power requirements on an aircraft has evolved. A dedicated load is suitable for each system. If a single power source is considered for the whole aircraft,

the reduced weight and improved efficiency will be there. The most suitable single source can be an electrical power system that offers several advantages over other systems in terms of the range of applications and flexibility of control. Thus arises the concept of More Electric Aircraft (MEA). The use of an MEA system is a historical change in technology to lessen fuel consumption, operating cost and environmental impact. Many more essential features can help to improve aircraft's ease of use and reduce the need for unscheduled maintenance. The electrical power system capacity of the present aircraft for instance Boeing 737 and Boeing 787 is almost 100 kW and more than 1 MW, respectively. A system with higher voltage can be considered so that the current level and hence the cable weight can be reduced significantly. Some of the evolving standards are 540 V dc and 230 V ac at the variable frequency between 320–800 Hz. A combination of AC and DC supplies is being used in various large aircraft. Generally, 28 V DC supply is used for the critical loads, for instance avionics, whereas the higher loads are supplied from the DC or AC or the combination of both, shown in Figure 9.12.

The increasing demands of electrical energy provide an opportunity for the redesign of the generation systems in more electric aircraft. To make an MEA system more competitive with currently used systems in terms

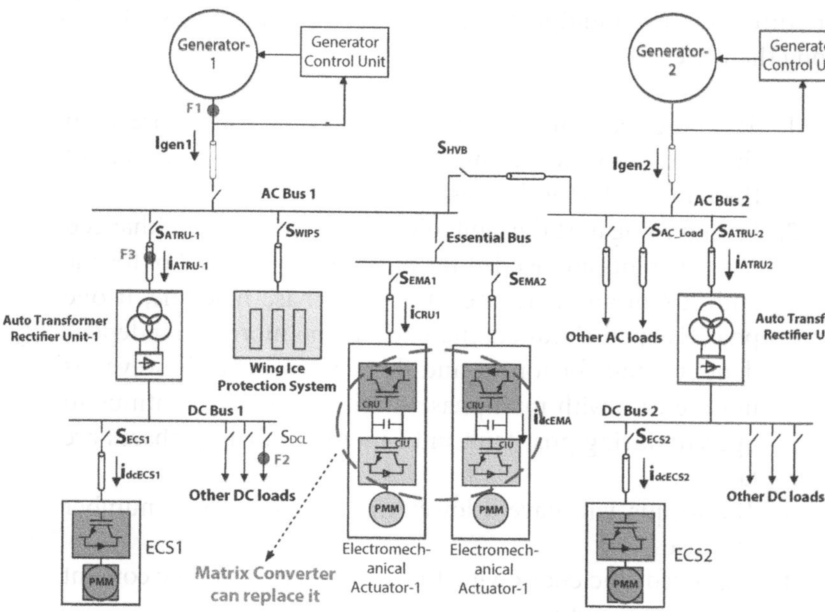

Figure 9.12 Schematic of electrical power requirements in the aircraft.

of reliability and weight, the industry is looking for several alternatives in the generation of electrical energy at optimum voltage and frequency. Currently, a complex and very expensive mechanical gearbox is being used in most civilian aircraft to generate a constant speed. To get the constant frequency, a shaft drives the generator at a constant speed. The voltage is controlled using an exciter with a feedback control loop for this purpose. In this way, a controlled voltage with a fixed frequency is obtained.

The frequency will change during the phases of flight of the aircraft as a response to changing demands for the gas turbine speed. Then this variable frequency can be managed by an appropriately designed AC/AC power converter to yield a fixed voltage and frequency supply for the electrical systems of the aircraft. The AC-AC power converter can be back-to-back inverters, cycloconverters or matrix converters, although this system has a drawback of high power rated converter because it mandatory routes all the produced power. Thus the converter should be of full power rating. Regrettably, the converter switching components limit the rating of the power rating of the converter.

(b) Drives:

The power electronic converters provide an interface for decoupling machines and three-phase mains. The number of phases for variable speed drives should not be limited to three only because machines having more than three-phases (multiphase) possess many advantages [45]. A few among them are:

1. The required power ratings can be split into more than three-phases which ultimately reduces the current ratings of the semiconductor devices.
2. Immunity against the fault tolerance of the drive is enhanced (two currents are needed for independent control of torque and flux in any machine). In three-phase machine, if one phase is open circuited, there will be no more two independent currents for torque and flux control. This, however, is not the case with multiphase machines as they continue to operate during post-fault until more than (n-3) phases are faulted.
3. The torque will have a lower ripple as the phase number increases.
4. Improved efficiency owing to the reduced harmonic content of magneto-motive force.

Among several benefits of the matrix converter technology, the most important is the larger power density of these converter because of the nonexistence of the DC link capacitor. Therefore it has the advantage in the optimization of the filter size, by considering appropriately high switching frequency of matrix converter. At present, the three-phase drives are fed from three-phase direct AC-AC matrix converters. The main advantages of this topology compared to the VSI include the single-stage conversion, which enables minimization (theoretically, omission) of the bulky capacitor in the DC link of the VSI. Matrix converters have been under extensive development for the last 35 years for three-phase drives. Recently, it has been realized that it should be possible to achieve multiphase supply equivalent, similar to the operation obtainable with multiphase VSIs, by using matrix converters. The first application of a three-n phase MC is presented in [46]. Matrix converters with multiphase outputs have gained deep attention during the last few years, which is quite visible with many papers in this period dealing with the topic.

(c) Wind energy applications:
A wind energy system consists of any one of the squirrel cage induction generator, doubly-fed induction generator, permanent magnet synchronous generator and synchronous generator to generate variable frequency and voltage. The grid is generally a three-phase system consisting of a generator that needs some system for phase conversion, as shown in Figure 9.13.

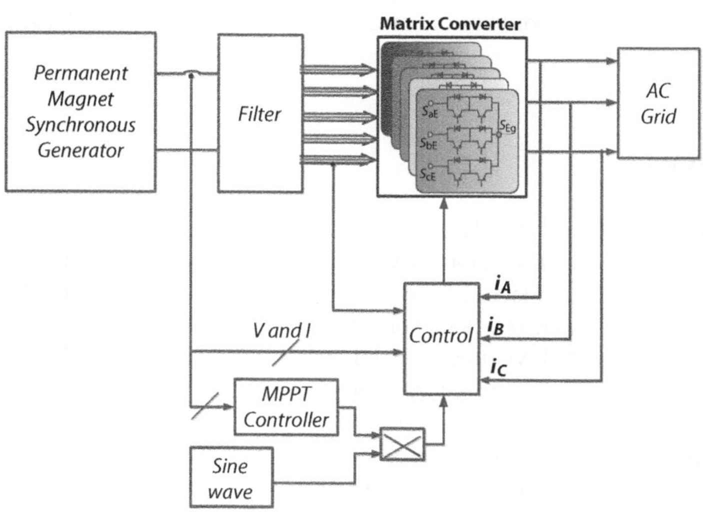

Figure 9.13 Application of matrix converters in the wind energy systems.

Therefore, an MC is used to interface with the grid through a permanent magnet synchronous generator, which may have multiphase to get the inherited benefit of multiphase system such as reduced phase's losses, higher output power, reduced component size and low-rated components compared to three-phase counterpart.

A matrix converter not only has modular nature to be applicable for any number of phases to control; however, it is a powerful topology for AC-AC power conversions. MC offers bidirectional power flow that is very imperative in nearly unity power factor operation, regeneration process. A permanent magnet synchronous generator is used in the wind energy conversion system to get power in electrical form from the wind turbine. A matrix convert helps to control the frequency and voltage amplitude which will be fed into the grid.

9.7 Summary

In this chapter, an introduction of the MC is presented with a literature review regarding the invention and development in the modulation of the matrix converter. A different perspective of classification of the matrix converter is presented. The basic topology of a three-five phase MC fed drive system is given. A brief description of each part of the system is presented. Moreover, several issues associated with the matrix converter such as modulation, control and mitigating effects of non-idealities are discussed. The different possible applications especially in more electric aircraft applications, wind power conversion and applications in electrical drives have been discussed.

References

1. M. Venturini and A. Alesina, "The generalised transformer: A new bidirectional, sinusoidal waveform frequency converter with continuously adjustable input power factor," in *PESC Record - IEEE Annual Power Electronics Specialists Conference*, 1980, vol. 1980–January, pp. 242–252.
2. A. Alesina and M. Venturini, "Solid-state power conversion: A Fourier analysis approach to generalized transformer synthesis," *IEEE Trans. Circuits Syst.*, vol. 28, no. 4, pp. 319–330, Apr. 1981.
3. J. Rodriguez, "References." Proceedings of control in power electronics and electrical drives conference, pp. 203–208, 1983.

4. P. D. Ziogas, S. I. Khan, and M. H. Rashid, "Analysis and Design of Forced Commutated Cycloconverter Structures with Improved Transfer Characteristics," *IEEE Trans. Ind. Electron.*, vol. IE-33, no. 3, pp. 271–280, 1986.
5. G. Roy and G. E. April, "Direct Frequency Changer Operation Under a New Scalar Control Algorithm," *IEEE Trans. Power Electron.*, vol. 6, no. 1, pp. 100–107, 1991.
6. L. Huber and D. Borojevic, "Space vector modulated three-phase to three-phase matrix converter with input power factor correction," *IEEE Trans. Ind. Appl.*, vol. 31, no. 6, pp. 1234–1246, 1995.
7. C. Klumpner, P. Nielsen, I. Boldea, and F. Blaabjerg, "A new matrix converter motor (MCM) for industry applications," *IEEE Trans. Ind. Electron.*, vol. 49, no. 2, pp. 325–335, 2002.
8. B.-H. Kwon, B.-D. Min, and J.-H. Kim, "Novel commutation technique of AC–AC converters," *IEE Proc. - Electr. Power Appl.*, vol. 145, no. 4, p. 295, 1998.
9. M. Milanovic and B. Dobaj, "Unity input displacement factor correction principle for direct AC to AC matrix converters based on modulation strategy," *IEEE Trans. Circuits Syst. I Fundam. Theory Appl.*, vol. 47, no. 2, pp. 221–230, 2000.
10. J. K. Kang, H. Hara, A. M. Hava, E. Yamamoto, E. Watanabe, and T. Kume, "The matrix converter drive performance under abnormal input voltage conditions," *IEEE Trans. Power Electron.*, vol. 17, no. 5, pp. 721–730, 2002.
11. P. Szczepankowski, "Direct matrix converter control based on the triangular shape function of finite element under input voltage source unbalance condition," *Prz. Elektrotechniczny*, vol. 90, no. 6, pp. 160–165, 2014.
12. H. Mohamed Basri and S. Mekhilef, "Experimental evaluation of model predictive current control for a modified three-level four-leg indirect matrix converter," *IET Electr. Power Appl.*, vol. 12, no. 1, pp. 114–123, Jan. 2018.
13. P. Correa, J. Rodríguez, M. Rivera, J. R. Espinozav, and J. W. Kolar, "Predictive control of an indirect matrix converter," *IEEE Trans. Ind. Electron.*, vol. 56, no. 6, pp. 1847–1853, 2009.
14. M. Chai, D. Xiao, R. Dutta, and J. E. Fletcher, "Space vector PWM techniques for three-to-five phase indirect matrix converter in the overmodulation region," *IEEE Trans. Ind. Electron.*, vol. 63, no. 1, pp. 550–561, 2016.
15. A. Iqbal, S. M. Ahmed, and H. Abu-Rub, "Space Vector PWM Technique for a Three-to-Five Phase Matrix Converter," *IEEE Trans. Ind. Electron.*, vol. 48, no. 2, pp. 697–707, 2012.
16. P. Szcześniak, "Review of AC–AC Frequency Converters." In: *Three-phase AC-AC Power Converters Based on Matrix Converter Topology. Power Systems.* Springer, London, 2013, pp. 17–86.
17. J. Rodriguez, M. Rivera, J. W. Kolar, and P. W. Wheeler, "A Review of Control and Modulation Methods for Matrix Converters," *IEEE Trans. Ind. Electron.*, vol. 59, no. 1, pp. 58–70, Jan. 2012.

18. T. D. Nguyen and H.-H. Lee, "Development of a Three-to-Five Phase Indirect Matrix Converter With Carrier-Based PWM Based on Space-Vector Modulation Analysis," *IEEE Trans. Ind. Electron.*, vol. 63, no. 1, pp. 13–24, Jan. 2016.
19. H. Wang *et al.*, "Topology and Modulation Scheme of a Three-Level Third-Harmonic Injection Indirect Matrix Converter," *IEEE Trans. Ind. Electron.*, vol. 64, no. 10, pp. 7612–7622, Oct. 2017.
20. Q.-H. Tran and H.-H. Lee, "A Three-Vector Modulation Strategy for Indirect Matrix Converter Fed Open-End Load to Reduce Common-Mode Voltage With Improved Output Performance," *IEEE Trans. Power Electron.*, vol. 32, no. 10, pp. 7904–7915, Oct. 2017.
21. J. W. Kolar, F. Schafmeister, S. D. Round, and H. Ertl, "Novel Three-Phase AC–AC Sparse Matrix Converters," *IEEE Trans. Power Electron.*, vol. 22, no. 5, pp. 1649–1661, Sep. 2007.
22. F. Schafmeister, C. Rytz, and J. W. Kolar, "Analytical calculation of the conduction and switching losses of the conventional matrix converter and the (very) sparse matrix converter," in *Twentieth Annual IEEE Applied Power Electronics Conference and Exposition, 2005. APEC 2005.*, vol. 2, pp. 875–881.
23. A. M. Bozorgi, A. Hakemi, M. Farasat, and M. Monfared, "Modulation Techniques for Common-Mode Voltage Reduction in the Z-Source Ultra-Sparse Matrix Converters," *IEEE Trans. Power Electron.*, vol. 34, issue 1, pp. 958-970, 2018.
24. P. Szcześniak, "Review of AC–AC Frequency Converters." In: *Three-phase AC-AC Power Converters Based on Matrix Converter Topology. Power Systems.* Springer, London, 2013, pp. 17–86.
25. K. Sun, D. Zhou, L. Huang, K. Matsuse, and K. Sasagawa, "A Novel Commutation Method of Matrix Converter Fed Induction Motor Drive Using RB-IGBT," *IEEE Trans. Ind. Appl.*, vol. 43, no. 3, pp. 777–786, 2007.
26. B.-H. Kwon, B.-D. Min, and J.-H. Kim, "Novel commutation technique of AC–AC converters," *IEE Proc. - Electr. Power Appl.*, vol. 145, no. 4, p. 295, 1998.
27. X.-F. Zhang, B. He, H. Lin, and W.-X. Wei, "Safe commutation strategies for matrix converter," *Zhongguo Dianji Gongcheng Xuebao/Proceedings Chinese Soc. Electr. Eng.*, vol. 28, no. 18, 2008.
28. L. Zhang, C. Watthanasarn, and W. Shepherd, "Analysis and comparison of control techniques for AC–AC matrix converters," *IEE Proc. - Electr. Power Appl.*, vol. 145, no. 4, p. 284, 1998.
29. D. Casadei, G. Grandi, G. Serra, and a Tani, "Space vector control of matrix converters with unity input power factor and sinusoidal input/output waveforms," *Power Electron. Appl. 1993., Fifth Eur. Conf.*, pp. 170–175 vol.7, 1993.
30. D. Busse, J. Erdman, R. Kerkman, D. Schlegel, and G. Skibinski, "Characteristics of shaft voltage and bearing currents," *IEEE Ind. Appl. Mag.*, vol. 3, no. 6, pp. 21–32, 1997.

31. A. Muetze and A. Binder, "Practical Rules for Assessment of Inverter-Induced Bearing Currents in Inverter-Fed AC Motors up to 500 kW," *IEEE Trans. Ind. Electron.*, vol. 54, no. 3, pp. 1614–1622, Jun. 2007.
32. J. Jung *et al.*, "Monitoring Journal-Bearing Faults: Making Use of Motor Current Signature Analysis for Induction Motors," *IEEE Ind. Appl. Mag.*, vol. 23, no. 4, pp. 12–21 Jul. 2017.
33. P. Nielsen, F. Blaabjerg, and J. K. Pedersen, "New protection issues of a matrix converter: design considerations for adjustable-speed drives," *IEEE Trans. Ind. Appl.*, vol. 35, no. 5, pp. 1150–1161, 1999.
34. Z. Malekjamshidi, M. Jafari, J. Zhang, and J. Zhu, "Design and analysis of protection circuits for safe operation of direct matrix converters," in *2017 20th International Conference on Electrical Machines and Systems (ICEMS), 2017*, pp. 1–4.
35. J. Mahlein, M. Bruckmann, and M. Braun, "Passive protection strategy for a drive system with a matrix converter and an induction machine," *IEEE Trans. Ind. Electron.*, vol. 49, no. 2, pp. 297–303, Apr. 2002.
36. J. Itoh, I. Sato, A. Odaka, H. Ohguchi, H. Kodachi, and N. Eguchi, "A Novel Approach to Practical Matrix Converter Motor Drive System With Reverse Blocking IGBT," *IEEE Trans. Power Electron.*, vol. 20, no. 6, pp. 1356–1363, Nov. 2005.
37. A. K. Sahoo, K. Basu, and N. Mohan, "Systematic Input Filter Design of Matrix Converter by Analytical Estimation of RMS Current Ripple," *IEEE Trans. Ind. Electron.*, vol. 62, no. 1, pp. 132–143, Jan. 2015.
38. P. Wheeler and D. Grant, "Optimised input filter design and low-loss switching techniques for a practical matrix converter," *IEE Proc. - Electr. Power Appl.*, vol. 144, no. 1, p. 53, 1997.
39. M. Hamouda, F. Fnaiech, and K. Al-Haddad, "Input filter design for SVM dual-bridge matrix converters," in *IEEE International Symposium on Industrial Electronics*, 2006, vol. 2, pp. 797–802.
40. D. Casadei, G. Serra, A. Tani, A. Trentin, and L. Zarri, "Theoretical and Experimental Investigation on the Stability of Matrix Converters," *IEEE Transactions on Industrial Electronics*, vol. 52, no. 5, pp. 1409-1419, October 2005.
41. C. Klumpner, P. Nielsen, I. Boldea, and F. Blaabjerg, "New Matrix Converter Motor (MCM) for Industry Applications," *IEEE Transactions on Industrial Electronics*, vol. 49, no. 2, pp. 325-335, April 2002.
42. D. Casadei, G. Serra, A. Tani, A. Trentin, and L. Zarri, "Theoretical and Experimental Investigation on the Stability of Matrix Converters," *IEEE Transactions on Industrial Electronics*, vol. 52, no. 5, pp. 1409-1419, October 2005.
43. A. L. Julian and G. Oriti, "A Novel Clamp Circuit for a Regenerative Rectifier Using AC/AC Matrix Converter Theory," *IEEE Trans. Ind. Appl.*, vol. 41, no. 1, pp. 68–74, Jan. 2005.

44. I. Sarkar, S. K. Show, and P. Syam, "Stepless current commutation in Matrix Converter," in *PEDES 2012 - IEEE International Conference on Power Electronics, Drives and Energy Systems*, 2012.
45. E. Levi, "Multiphase electric machines for variable-speed applications," *IEEE Trans. Ind. Electron.*, vol. 55, no. 5, pp. 1893–1909, 2008.
46. R. F. Chu and J. J. Burns, "Impact of Cycloconverter Harmonics," *IEEE Trans. Ind. Appl.*, vol. 25, no. 3, pp. 427–435, 1989.

10

Multilevel Converters and Applications

P. Prem[1], Jagabar Sathik[2] and K.T. Maheswari[3]*

[1]Switchgear Electromechanical-India, Chennai, India
[2]Renewable Energy Lab, College of Engineering, Prince Sultan University, Riyadh, Saudi Arabia
[3]Department of Electrical and Electronics Engineering, Bannari Amman Institute of Technology, Sathyamangalam, India

Abstract

The output voltage or current levels of voltage source inverters (VSI) with two levels are 0 and $\pm V_{dc}$. To reduce the ripple present at the output, the voltage source inverters are operated at high frequency and improved PWM schemes. However, the two-level inverters are not applicable for operating in high voltage and medium power uses, mainly because of switching loss and device constrictions. In recent years, Multilevel inverters (MLIS) have become the most essential power electronic converters for medium voltage high power energy control due to their capability of producing high power at the output with low device rating, increased number of voltage level with reduced total harmonic distortion, decreased common mode voltage (CMV), decreased dv/dt stress and less electromagnetic interference. In this chapter, a complete and widespread analysis of different types of MLIs is presented. There exist three conventional types of MLI topologies: diode-clamped multilevel inverter (DCMLI), (Neutral Point Clamped Multilevel Inverter), Capacitor-Clamped (Flying Capacitor) Multilevel Inverter (CCMLI) and Cascaded H-bridge Multilevel Inverter (CHBMLI). The diode-clamped multilevel inverter (DCMLI) is the most promising topology used in industrial applications. As the switching devices are operated at frequency of fundamentals, the efficiency of DCMLI is greater. However, Capacitor voltage imbalance and unequal switching device rating are the disadvantages of DCMLI which may cause device failure under temperature overload conditions. In FCMLI, the voltage balancing at different output voltage levels are accomplished by switching redundancy and the voltage difference between the cell and capacitor is sustained within the safe limit. But the factors that

**Corresponding author*: maheswarikt@gmail.com

Nikita Gupta, Mahajan Sagar Bhaskar, Sanjeevikumar Padmanaban and Dhafer Almakhles (eds.)
DC Microgrids: Advances, Challenges, and Applications, (299–324) © 2022 Scrivener Publishing LLC

limit the use of FCMLI are high switching frequency and switching loss, higher number of capacitors is needed and complex control. Even though the CHBMLI needs the least number of components when compared to the other two topologies of multilevel converters to obtain the same level of voltages, the main drawback that limits its application is that it needs distinct voltage sources for individual cell. Nowadays, the research is focusing on Active Neutral-Point clamped multilevel inverter (ANPC) that overcomes the shortcomings of existing MLI topologies. This chapter proposed an ANPC multilevel inverter that has the advantages of both flying capacitor and Neutral point-clamped multilevel inverter such as flexibility of switching redundancy and robustness to produce multilevel voltages. The operation of the proposed ANPC Multilevel inverter is explained in detail.

***Keywords*:** Active neutral point clamped multilevel inverter, capacitor voltage, multilevel inverters, ripple, switching frequency, switching loss, total harmonic distortion

10.1 Introduction

Currently, the solar PV system has gained unique significance to generate electricity free from pollution. The solar PV system converts the incident solar light energy into electric current through the help of an array of solar cells [1, 2]. However, the electricity generated by the solar PV system is direct current. To connect the output voltage from the solar PV system to the grid a two-level inverter is needed. Nevertheless, it is inexperienced in producing a waveform similar to sinusoidal and thus pays for high harmonic content in current and voltages, rendering the inverter enormous in size. It is not compatible for conversion purposes in the high-power market, so two-level inverters have been replaced by multilevel inverters suitable for high-power conversion. Recently, the attention of researchers over multilevel inverters has been increased due to its advantages [3]. Generally, multilevel inverters are the power converters that generate the desired output waveform from various values of inputs.

The distortion of the output waveform is minimized by enhancing the amount of levels in the output voltage and it is rendered to be almost sinusoidal. The multilevel inverter (MLI) has therefore received tremendous consideration in the high-voltage trade in the high-power sector [4]. In general, as the level of existing MLI increases, the quantity of switches and component size increases. These power converters are also applicable to single-phase and three-phase systems. However, this chapter proposed an Active Neutral-Point (ANPC) inverter which consists of a reduced number of switches and diodes when compared to existing multilevel inverters. Only single-phase circuits are used for many applications such as solar PV systems [5].

10.2 Multilevel Inverters

In the early 1980s, thyristor-based semiconductors were first developed, and the CSIs become popular. In industrial applications, CSIs are used as load commutated converters which comprises the switches like gate turn-off thyristors (GTO) and Insulated Gate Commutated thyristors (IGCT) [6]. Normally, the CSIs could be operated in short circuit conditions using current controlled power switches and hence, minimum switching frequency is required. However, design of the gate driver circuit is complex.

In the 1990s, the Insulated-gate bipolar transistors (IGBTs) were introduced to be operated in Voltage source inverters. They offer minimum switching loss, high reliability, and high efficiency. Also, the current harmonic distortion of voltage source inverter is less as it is operating at high switching frequency [7]. Due to its number of advantages, the VSIs are preferred to be used in industrial applications. Recently, the Multilevel inverters have appeared as a precise significant substitute for obsolete technologies in the industry sector. Thus, MLIs are the major concern on research and will be further discussed in this chapter.

10.2.1 Multilevel Inverters vs. Two-Level Inverters

The foremost aim of a basic inverter is to produce an output voltage which is alternating in nature from a dc power supply. These types of outputs are mostly required in industrial applications like adjustable speed drives, UPS, etc. The conventional two-level VSIs are employed to get the controllable output power [8]. Here, the dc voltage at the input is transformed into ac voltage of desired parameters. The basic inverters consist of voltage sources and power switches which are used to control the output parameters as shown in Figure 10.1. Based on type of inverter, either four or six power switches can be used [9].

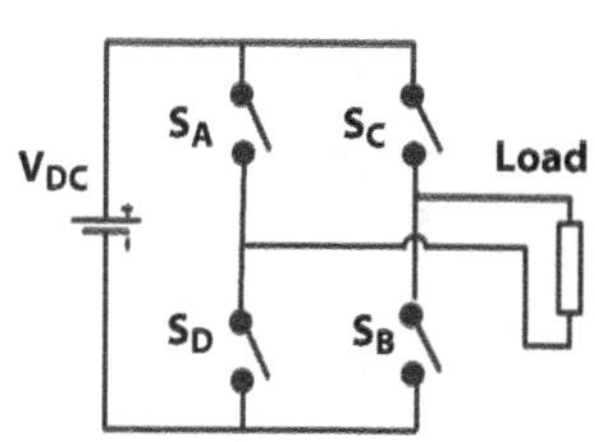

Figure 10.1 Basic inverter Topology.

The conventional inverters could not be operated below high voltage presentations, due to its restrictions that it could not be operated at high switching frequency. The disadvantages of conventional inverters are high switching losses and requirement of high rating of power devices. To attain high power handling capability the series and parallel connection of power switches has to be used. However, the use of more switches results in high power loss and reduced efficiency.

Hence, the multilevel inverters have become a good choice in the field of industrial applications [10, 11]. The essential for using MLIs is to offer high output power from a medium voltage source. Thus, high voltage can be obtained without using high power rated devices. Also, multilevel inverters can reduce voltage stress across power electronic switches that connect other switches in series. Hence, the main advantage of a multilevel inverter includes, reducing dv/dt, Voltage THD and Electromagnetic interference. Figure 10.2 shows the configuration multilevel inverters where two inverters are associated in sequence to get the output voltage with multilevel. By increasing the number of steps, the output waveform of good power quality can be obtained. Also, the switching frequency can be minimized for the PWM operation of switches [12]. In addition to the above-said advantages, some other advantages of MLIs can be grouped as follows:

1. Advantages of MLIs based on waveform,
2. Advantages of MLIs based on Topology.

While comparing the two advantages, one may think that these are complementary. Obviously, a MLI topology will produce a Multilevel

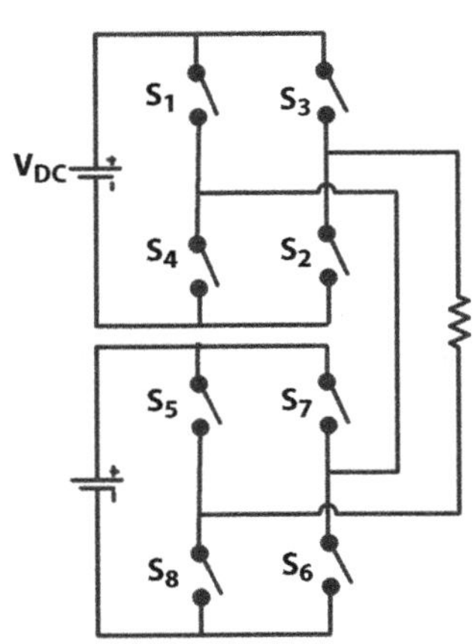

Figure 10.2 Two inverters connected in series.

Waveform. However, the emerging new topology of MLI must be investigated critically. Any topology may be capable of generating multilevel waveform, but it is not compulsory that it should offer all the advantages of MLI. A clear explanation on the benefits of MLIs based on waveform as well as topology is given in detail in the next section.

10.2.2 Advantages of Multilevel Converters Based on Waveforms

The following are the merits of MLIs based on its stepped waveform as shown in Figure 10.3 [13, 14]:

1. The stepped waveform provides good quality and a good harmonic profile. Thus, the need for filter has been reduced considerably. The modulation index and the carrier frequency are set as 0.85 and 1 Hz, respectively. From the results of MLI, it can be observed that as the number of steps in the output waveform increases, the THD and also the lower order harmonics decrease.
2. Reduced Electromagnetic compatibility problem. When the levels in the stepped waveform are raised, the dv/dt load stress is also reduced, thereby raising the load lifetime.
3. The control on the level of output waveform is increased based on the number of voltage, step size, and firing angles.

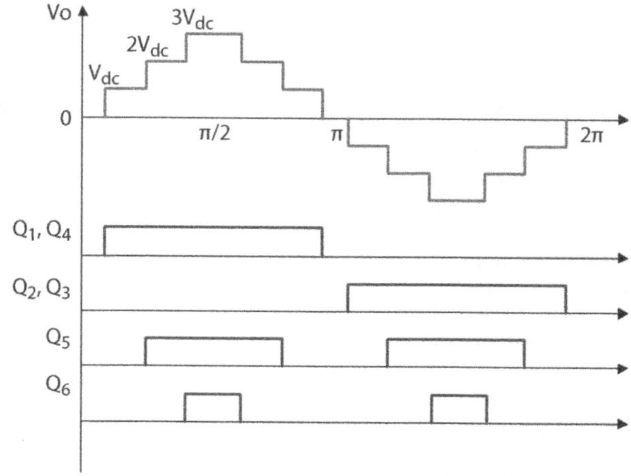

Figure 10.3 Stepped waveform of multilevel inverter.

4. By reducing the Common mode voltage, the bearing stresses of multilevel motor drive can be minimized.
5. The modulation of MLIs can be carried out with the fundamental as well as the high switching frequency.

10.2.3 Advantages of Multilevel Converters Based on Topology

The following are the advantages of MLIs based on topologies [15]:

1. The stresses on the switches are very much reduced when compared to overall operating voltage. Hence, the series parallel combination of switches is not required.
2. The redundancy in switching states is allowed in many topologies of MLIs. Such redundancies can be utilized to program under fault conditions. Using these redundant switching states, the power switches can be able to switch at optimal states of switching. Also, in three-phase topologies the redundant switching states are used to reduce CMV.
3. In MLIs, the distortion of input current is reduced.
4. MLIs can be easily interfaced to the non-conventional energy resources and can provide smooth control for load sharing. The advantages of MLI topologies are summarized as shown in Figure 10.4.

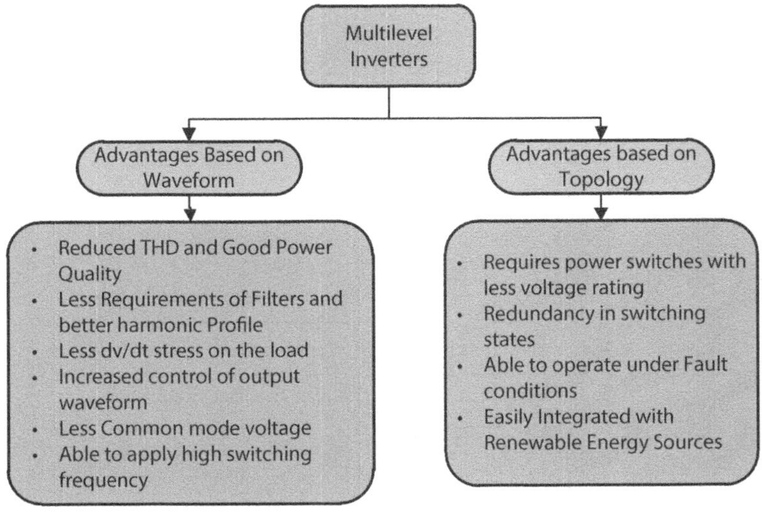

Figure 10.4 Advantages of multilevel inverters.

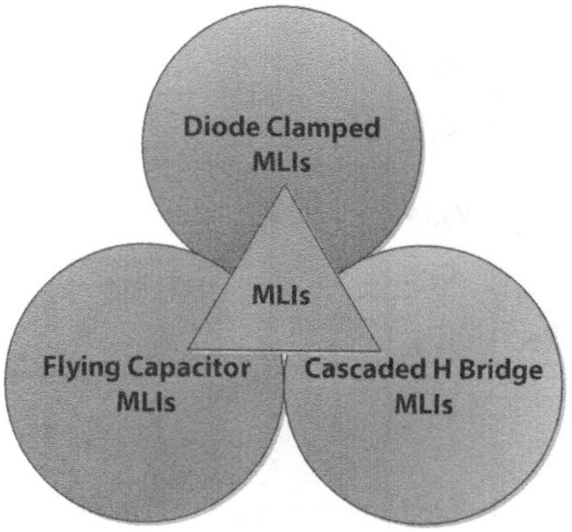

Figure 10.5 Basic classification of multilevel inverters (MLIs).

10.3 Traditional Multilevel Inverter Topologies

With the advent of three-level inverters, the MLI topologies became familiar in the early 20th century. As the use of MLIs increased, for the given number of levels the power rating of the device has also been improved, but the number of components required is reduced [16]. Recently, the utilization of MLIs in industrial sectors has been increased for high power rating applications. A major factor that paves the way for the continuous growth of MLIs is that this type of converter can be easily interfaced with adjustable speed drives, energy storage systems and capacitors. The three different topologies of multilevel inverters are shown in Figure 10.5 and are as follows:

1. Diode Clamped Multilevel Inverter (DCMLI)
2. Flying Capacitor Multilevel Inverter (FCMLI)
3. Cascaded H-Bridge Multilevel Inverter (CHMLI)

10.3.1 Diode Clamped Multilevel Inverter

The Neutral Point-Clamped Multilevel Inverter (NPCMLI) is otherwise known as Diode Clamped topology. In the 1980s, Nabae *et al.* started implementation of NPC topology [17]. A DCMLI consists of (n-1) number of capacitors and is able to generate n level output voltage between the phases. The main objective in NPCMLI is to employ diodes to limit the voltage stress on power devices. Figure 10.6 shows the one leg bridge of 5-level DCMLI. The switches present in the leg can be numbered as S_{11},

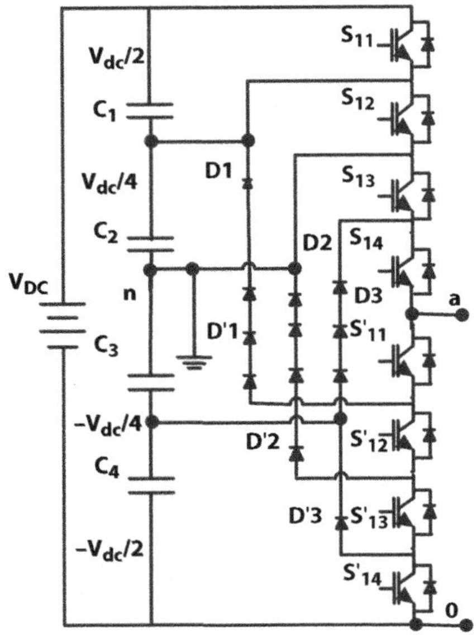

Figure 10.6 One leg of a bridge of diode clamped MLI.

S_{12}, S_{13}, S_{14}, S'_{11}, S'_{12}, S'_{13} and S'_{14}. There are four capacitors in the dc bus such as C_1, C_2, C_3, C_4 and the capacitor voltage is $V_{dc}/4$ at each capacitor, for the input voltage of V_{dc}. Consequently, the voltage stress on each device is limited to $V_{dc}/4$ through clamping diodes [18, 19]. On the whole a s level inverter needs (s-1) capacitors, 2(s-1) switches and (s-1) (s-2) number of clamping diodes.

Consider only one leg of an inverter that contributes to a multilevel inverter. Let the output voltage reference point be 0. Table 10.1 shows the

Table 10.1 Switching states of DCMLI.

Switching states								Output voltage Vo
S_{11}	S_{12}	S_{13}	S_{14}	S'_{11}	S'_{12}	S'_{13}	S'_{14}	
1	1	1	1	0	0	0	0	V5
0	1	1	1	1	0	0	0	V4
0	0	1	1	1	1	0	0	V3
0	0	0	1	1	1	1	0	V2
0	0	0	0	0	0	0	0	V1

switching table of DCMLI and its equivalent output levels of voltages. If the state of the switch is 1, that denotes that the power switch is turned ON and if the switching state is 0 the power switch is switched OFF. There are four pairs of complementary switches in the leg and each power switch is turned ON once per cycle. In one leg of the inverter the four complementary pairs of switches are (S_{11}, S'_{11}), (S_{12}, S'_{12}), (S_{13}, S'_{13}), and (S_{14}, S'_{14}). Thus, in the principle of operation of DCMLI, the four switches are switched together at the same time and the other switch in the same pair should be switched off if one of the switches in the complementary switch pair is switched ON. Figure 10.7 shows the output voltage waveform of Diode clamped multilevel inverter. The Total Harmonic Distortion (THD) of output voltage is shown in Figure 10.8. The measured value of THD is 16.84%.

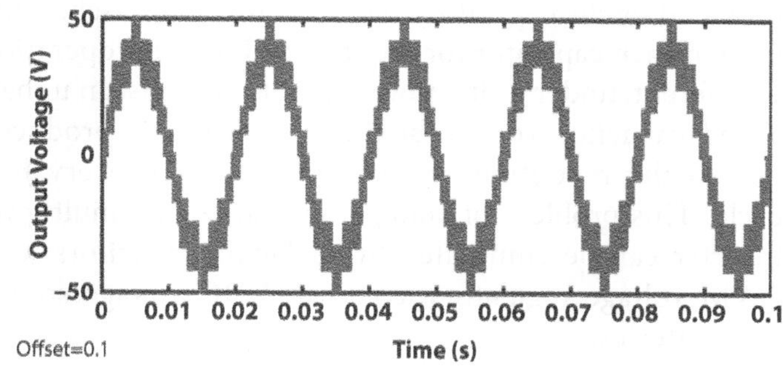

Figure 10.7 Output voltage waveform of diode clamped MLI.

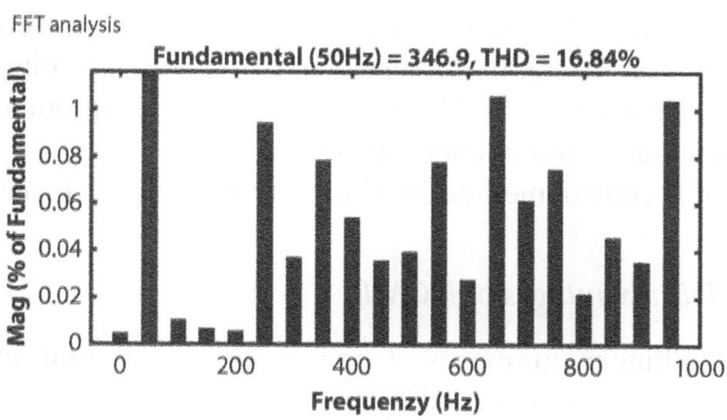

Figure 10.8 THD analysis of diode clamped MLI.

10.3.1.1 Features of DCMLI

1. Blocking Diodes with High voltage rating: Each switching device is desired to stop the level of voltage $V_{dc}/(n-1)$. However, the diodes clamped in the circuit possess several types of reverse voltage blocking ratings [20].
2. Switching device rating is unequal. It is observed from Table 10.1, the conducting period of the Switch S_{11} is only once when compared to that of switch S_{14} which is turned on fully except the time when Vo=0. This unequal conducting time of the switches results in the requirement of power switches with various current ratings. Hence, if in an inverter the duty cycle average is used to find the rating of the device, the switches at the top may be of bigger size than the bottom switches.
3. Capacitor Voltage unbalances. As the output levels at the terminals of the capacitors vary, the different current is provided by the capacitors. The charging and discharging time of each capacitor for rectifier and inverter operation is different, under unity power operation. Thus, an unbalanced capacitor voltage at different levels is produced due to the repetition of charging profile at every half cycle. This problem of voltage unbalance in a multilevel inverter can be eliminated by replacing capacitors with PWM voltage regulators, controlled DC voltage source and Batteries.

10.3.1.2 Advantages of DCMLI

1. The amount of harmonic is reduced when the levels at the output voltages are increased. Thus, the need for filters is avoided by accumulating the quantity of levels of voltage.
2. The efficiency of the MLI is increased as all components are operated at fundamental frequencies.
3. Simple control method for this type of multilevel inverter.

10.3.1.3 Disadvantages of DCMLI

1. For a higher number of stepped waveform, the number of clamping diodes needed is high.
2. Clamping Diodes with different voltage ratings are essential.

3. Due to the capacitor voltage imbalance, real power flow is complicated.
4. To block the reverse voltages, high voltage blocking diodes are required.
5. If the quantity of steps in the output waveform raises, the integer of switches, capacitors, and diodes needed in the network increases.

10.3.1.4 Applications of DCMLI

1. It is employed in Static Var Compensation.
2. It can be used as an adjustable speed drive with high power motors.
3. It can also be used in interconnection of high voltage systems.
4. It can be utilized at the interfaces of the high-power DC and AC Transmission lines.

10.3.2 Flying Capacitor Multilevel Inverter

The flying capacitor multilevel inverter, otherwise known as imprecated cell inverter, uses a capacitor clamped switching cells connected in series. The flying capacitor multilevel inverter is so named because the capacitor floats with respect to the earth's potential [22]. Though clamping diodes are required for its operation, operating states of FCMLI are almost analogous to that of DCMLI. Even though theoretically it can produce infinite levels, because of practical restrictions it only produces up to six voltage levels.

Figure 10.9 shows the one leg of single-phase fling capacitor multilevel inverter. Like DCMLI, there are eight power switches in the single leg such as S_{11}, S_{12}, S_{13}, S_{14}, S'_{11}, S'_{12}, S'_{13} and S'_{14}. Assuming that the voltage rating of the capacitors is equal and these are connected in series. The three inner loop balancing capacitors are C_a, C_b and C_c belongs to Phase a leg and are independent from other leg b. However, the energy storage Capacitors C_1, C_2, C_3 and C_4 are shared by both Phase leg a and Phase leg b for an inverter in single phase. Hence, for a s level inverter, the phase voltage of an s level inverter has s levels, whereas the line voltage has (2s-1) levels. In this topology, the voltage across each capacitor must be maintained at 0.5 V_{dc}, and hence the maximum voltage stress on switching device will be restricted to one capacitor voltage. Figure 10.10 shows the output voltage waveform of Capacitor clamped multilevel inverter. The Total Harmonic Distortion (THD) of output voltage is shown in Figure 10.11; the measured value of THD is 17.42%.

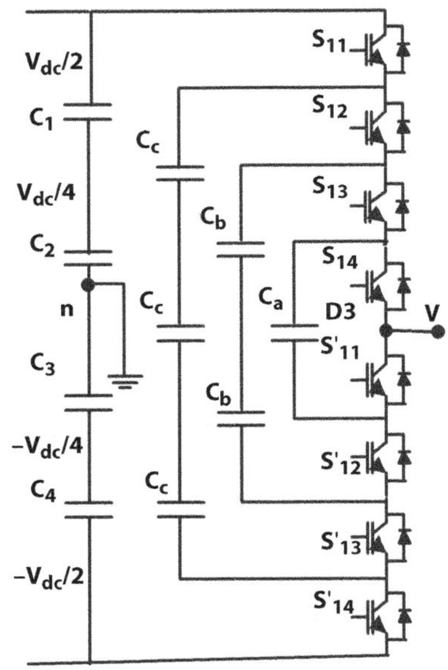

Figure 10.9 Flying capacitor multilevel inverter (one leg of a bridge).

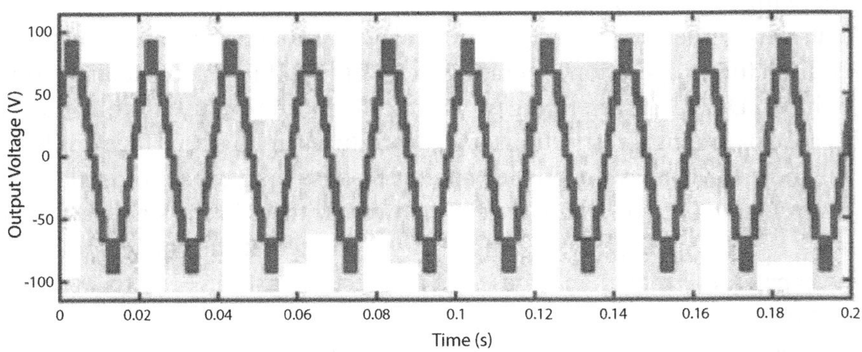

Figure 10.10 Output voltage waveform of flying capacitor MLI.

If the rating of capacitor is equal, for n-level converter the dc bus requires (n-1) capacitors $C_{1 \text{ to }} C_4$. The five-level output voltage can be produced with many possible combinations of operating states. Table 10.2 shows the switching states to obtain possible combinations of levels of voltage at the output. It could be detected from Table 10.2 that the switches are turned

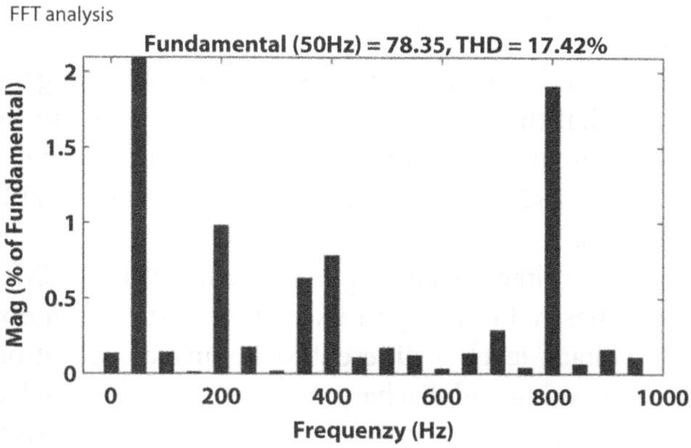

Figure 10.11 THD analysis of flying capacitor MLI.

Table 10.2 Switching states of FCMLI.

Switching states								Output voltage Vo
S_{11}	S_{12}	S_{13}	S_{14}	S'_{11}	S'_{12}	S'_{13}	S'_{14}	
1	1	1	1	0	0	0	0	V5
1	1	1	0	1	0	0	0	V4
1	1	0	0	1	1	0	0	V3
1	0	0	0	1	1	1	0	V2
0	0	0	0	1	1	1	1	V1

on at an unequal time period. The number of capacitors needed for each phase is given by

$$Mc = \sum_{j=1}^{n} (n-j)$$

In this inverter topology each level comprises switching devices which are mostly transistors and the cells in every limb are coupled in inner nested series. Every cell of this inverter is made of a single capacitor and two power electronic switches.

10.3.2.1 Features of FCMLI

1. A large number of capacitors is required for storage. Let us assume that the capacitor voltage rating is equal to the rating of switching components, then s-level converter needs ((s-1)* (s-2))/2 auxiliary capacitors per leg and (s-1) dc bus capacitors.
2. Capacitor voltage balancing is needed. Unlike DCMLI, the FCMLI has redundancy in its switching states. To obtain a same voltage level, a different switching combination that allows charging and discharging of capacitors is needed. The operation of the capacitor is balanced in one or more output cycles by employing two or more switching arrangements to obtain medium voltage level. Thus, the FCMLI may be employed in real power conversions by appropriate choice of switching arrangements.

10.3.2.2 Advantages of FCMLI

1. Able to work during power outages with the help of a large number of capacitors.
2. Redundancy in switching combinations is possible to balance voltage levels.
3. The harmonic content is less so the requirement of filter is reduced.
4. Real and reactive power control is possible.
5. Doesn't require clamping diode.
6. No need of Transformer to achieve the desired number of voltage levels.
7. Balancing capacitors share single DC source advantages.

10.3.2.3 Disadvantages of FCMLI

1. Output voltage will be partial of the input voltage.
2. Due to high switching frequency and switching losses will take place.
3. If it is employed for greater number of levels, the size of capacitors is bulky and will be more expensive.
4. Complex control of circuit operation.

10.3.2.4 *Applications of FCMLI*

1. Direct Torque control in induction motor drives.
2. Can be used for power conversion applications.
3. Rectifiers with Sinusoidal current.
4. Static Var Compensators.

10.3.3 Cascaded H Bridge Multilevel Inverter

There are three types of multilevel inverter. CHB Multilevel Inverter is mostly preferred because of its use of fewer components and many high-power industrial applications. CHBMLI was first introduced in 1975. The traditional CHBMLI consists of eight switches. The inverter is made up of IGBT power switches. The output levels are increased by increasing the number of H Bridges and thereby the number of switches also increases.

Each H bridge is coupled in sequence with individual dc source. The output voltage levels are 0, $+V_{DC}$,$-V_{DC}$. There are two types of DC sources, symmetric and asymmetric dc source. The CHBMLI can be operated in both configurations. Symmetric dc source means it has equal magnitude of voltage, whereas Asymmetric dc source has unequal magnitude of voltage. Asymmetric dc source is preferred most instead of symmetric because it has increased number of voltage levels. However, the stresses on the switches are not equal. Unequal loss sharing among the switches will cause varying temperatures for switches and also it requires switches with high voltage ratings.

The multilevel output voltage can be generated by linking the multiple H bridge inverters in series. It was Baker and Bannister who first filed the patent on MLI topology for generating multilevel voltage. Figure 10.12 shows the m-level CHB inverter where the output ac terminals of various inverters are merged in series. Here, each H bridge is commonly called as a "cell". Hence, each cell is made up of separate DC Source and each DC level are isolated from each other. Dissimilar to DCMLI and FCMLI, the cascaded H bridge MLI does not require more components. The H bridge MLI has two legs and each leg consists of two power switches. An n-level inverter consists of two diodes with a blocking voltage of,

$$V_D = \frac{n-1-k}{n-1} . V_{dc}$$

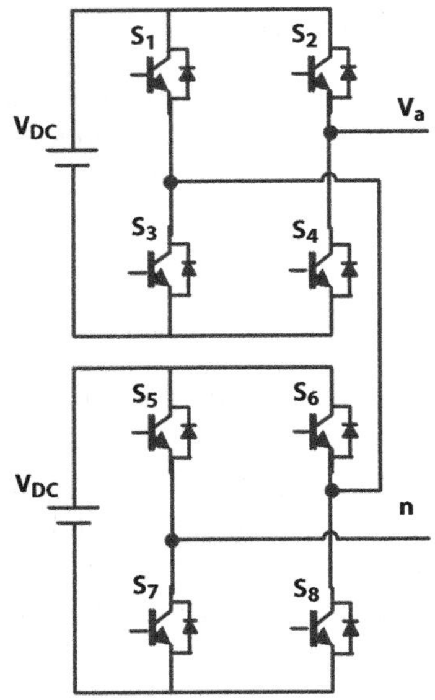

Figure 10.12 Cascaded H bridge multilevel inverter.

Where, *n* is the number of levels, *k* differs from 1 to (*n*-2), V_{dc} stands for total dc link voltage. The number of diodes needed for each phase is $N_D = (n-1)(n-2)$, if the diode blocking voltage is equal to switching devices. Figure 10.13 shows the output voltage waveform of Cascaded H bridge multilevel inverter. The Total Harmonic Distortion (THD) of output voltage is shown in Figure 10.14. And the measured value of THD is 41.74%. It is observed that as the number of levels in the output voltage is increased the THD of the output voltage has been reduced.

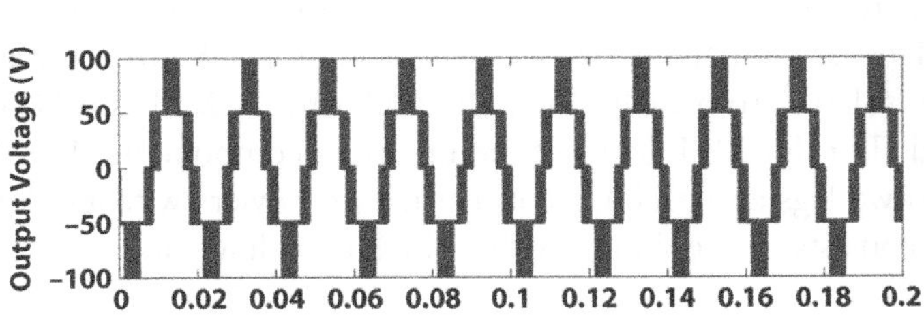

Figure 10.13 Output voltage waveform of cascaded H bridge multilevel inverter.

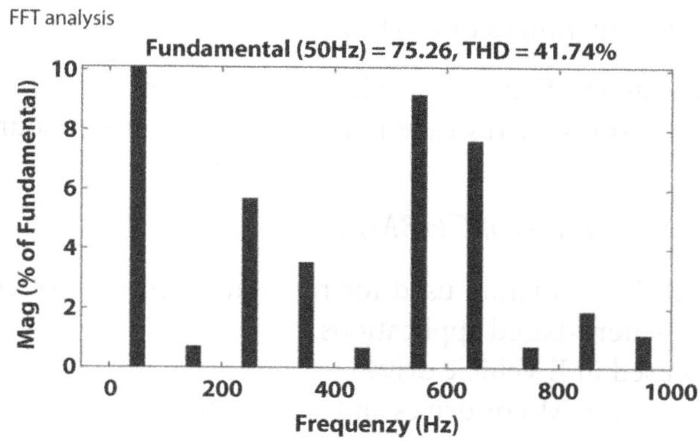

Figure 10.14 THD analysis of cascaded H bridge multilevel inverter.

10.3.3.1 Features of CHBMLI

1. The CHMLI needs discrete dc voltage source for power conversions.
2. The assembly of distinct dc sources is best suited for numerous non-conventional energy sources.
3. Two converters are not able to connect with individual dc source in a back-to-back fashion because when two back-to-back inverters are not connected and operated properly it will result in short circuit.

10.3.3.2 Advantages of CHBMLI

1. DC bus regulation is simple.
2. Modularity of control is obtained. The cascaded structure of the full bridge inverter is modulated individually.
3. To get the same amount of voltage levels necessitates the minimum quantity of devices.
4. Soft-switching is employed in this construction to prevent bulky and lossy resistor capacitor-diode snubbers.
5. The Output voltage levels are multiplied by the quantity of sources.
6. Manufacturing can be done effectively and rapidly.
7. The Packing and Layout is modularized. It can be easily controlled by a transformer.
8. It is cheap and it does not require any capacitors or diodes for clamping.

10.3.3.3 Disadvantages of CHBMLI

1. Separate DC Source is needed by each H Bridge.
2. Applications are restricted due to large number of sources.

10.3.3.4 Applications of CHBMLI

1. CHBMLI is mainly used for renewable resource of energy and battery-based applications.
2. It is used in E Vehicle drives.
3. It is used in Motor drives and Active filters.
4. It is used to utilize DC sources.
5. It is used for Power factor compensator.

10.4 Advent of Active Neutral Point Clamped Converter

In the conventional Neutral point clamped inverter clamping points will be there in DC link [21, 22]. The flying capacitor inverter is another conventional inverter topology. In this topology increased capacitors result in enlarged size of switches, and complicated control techniques are required to equalize the voltages of both intermediate capacitors and flying capacitors.

Recently, hybrid multilevel level inverters are of great interest in industrial applications as they combine the characteristics of various MLI topologies [23, 24]. One of the hybrid widespread MLIs that overcomes the drawbacks of conventional multilevel inverters is Active Neutral Point Clamped inverter (ANPC). ANPC combines the features of the conventional flying capacitor (FC) type and conventional Neutral Point Clamped type (NPC) inverter. It could be observed from the existing multilevel inverter topologies that the number of components essential to get the five-level output voltage waveform is more, which makes the system more expensive. However, for industrial applications, the main parameter considered is cost and it is most important to reduce the cost. The main intention of this chapter is to reduce the quantity of components. Centered on this requirement the innovative five-level ANPC inverter with seven switches is proposed in this chapter [25].

The reduced number of switches can be obtained by ignoring the inner switches S_5, S_6 and S_7 as shown in Figure 10.15 (a). The output voltage level is +2 or -2 when the power switches S_1, S_2 or S_3, S_4 are turned ON.

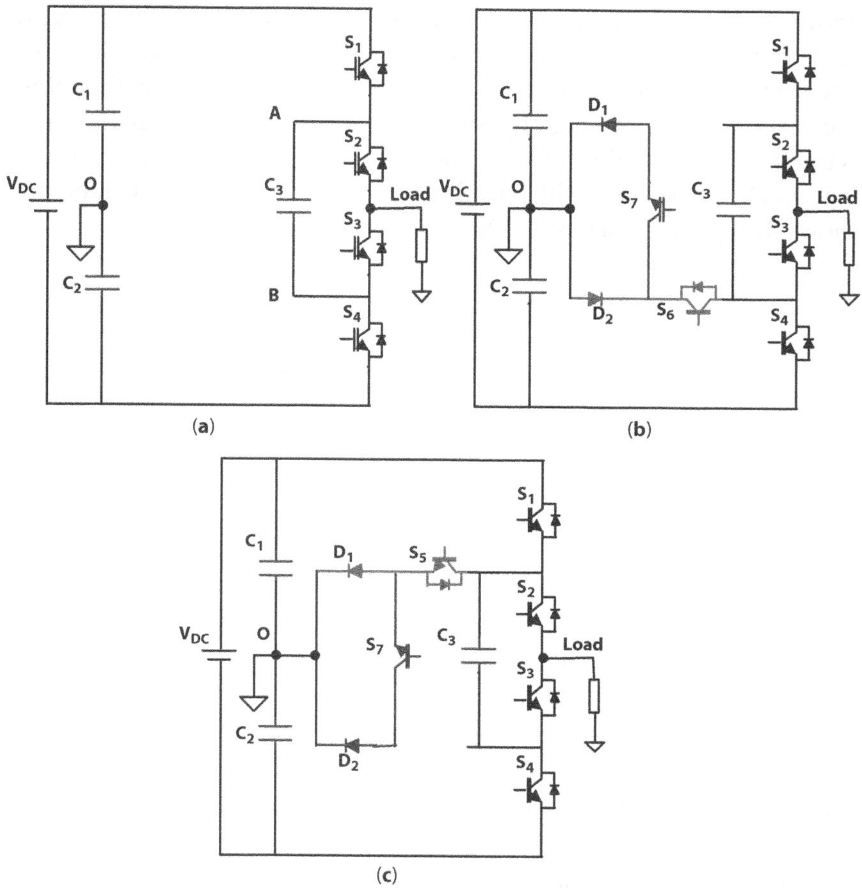

Figure 10.15 Formation of seven-switch five-level ANPC topology (a) Circuit I (b) Circuit II and (c) Circuit III.

Likewise, the output voltage levels of +1 or -1 can be obtained by turning on the switching combination of S_1, S_3 or S_2, S_4. Under the operation of these switching states the flying capacitor (FC) is charged with unity power factor. To maintain the FC voltage balancing, the redundant +1 switching state needs to be incorporated. This can be achieved by adding the switches S_6 and D_2 as shown in Figure 10.15(b). The diode D_2 is added to prevent addition of FC to C_1 when S_1 is turned ON. Also, to provide a reactive current path and to obtain the bidirectional current flow, D_1 and S_7 are added.

Thus the reactive current will flow through S_6, S_7 and D_1. Similarly, as shown in Figure 10.15 (c) redundant -1 and 0 level switching states are obtained by adding two switches S_5, S_7 and two diodes D_1, D_2. From the above circuit, by combining the switches, the final simplified circuit is presented in Figure 10.16 proposed topology. The main function of S_7 is to offer bidirectional current flow. In addition to the proposed methodology, the current flowing through the switch S_7 happens only when power factor is unity. Thus, when the power factor is sufficiently high, lower current rating switch can be used, which leads to the reduced system cost. Table 10.3 shows the operating states of ANPC inverter.

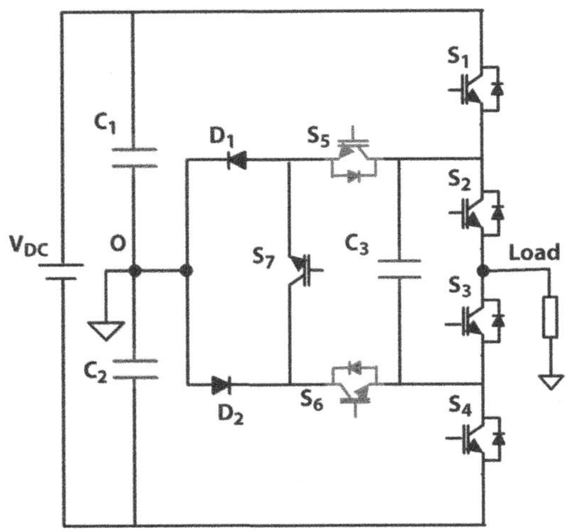

Figure 10.16 Proposed seven-switch five-level topology.

Table 10.3 Switching states of seven-switch five-level ANPC inverter.

Switching state	Condition of active switches							Voltage levels at the output
	S_1	S_2	S_3	S_4	S_5	S_6	S_7	
1	1	1	0	0	0	1	0	+2
2	1	0	1	0	0	1	0	+1
3	0	1	0	0	0	1	1	+1
4	0	0	1	0	0	1	1	+0
5	0	1	0	0	1	0	1	−0
6	0	0	1	0	1	0	1	−1
7	0	1	0	1	1	0	0	−1
8	0	0	1	1	1	0	0	−2

10.4.1 Comparison with Traditional Topologies

ANPC inverter topology increases the output levels by employing Flying Capacitors, which eliminates the use of clamping diodes. And apart from that, the ANPC inverter topology separates the intermediate link between capacitors so it is simpler to achieve voltage between these two capacitors. Another notable thing about ANPC is the redundant switching states can be utilized to balance Flying Capacitor voltage. Figure 10.17 shows the output voltage waveform of ANPC multilevel inverter. The Total Harmonic Distortion (THD) of output voltage is shown in Figure 10.18 and the value of THD is 12.05%. The THD of ANPC multilevel inverter is much less when compared to other topologies of MLI.

The switches in ANPC inverter will experience the same voltage and current stress and thus render equal power loss. Figure 10.19 represents the comparison of various parameters such as number of components

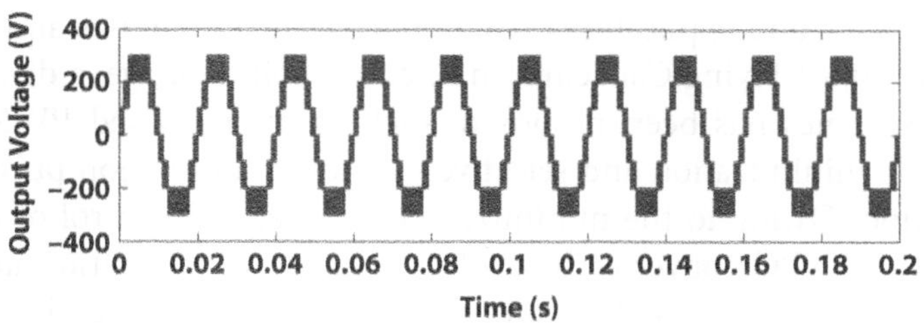

Figure 10.17 Output voltage waveform of proposed ANPC MLI.

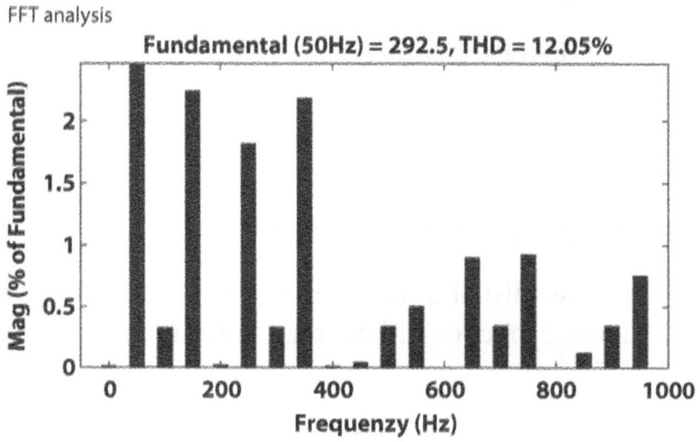

Figure 10.18 THD analysis of proposed ANPC MLI.

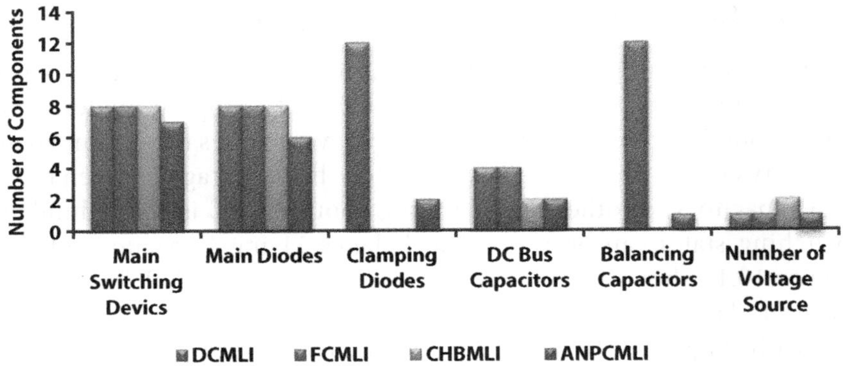

Figure 10.19 Comparison of various parameters in different types of MLIs.

used, and voltage sources for different multilevel inverter topologies of five level. Table 10.4 compares the advantages and disadvantages of ANPCMLI topology with conventional MLI topologies based on their advantages, disadvantages, and applications. To produce switching pulses and to concurrently adjust Flying Capacitor and neutral point voltage, a diversity of control schemes has been proposed, such as carrier-based PWM, real-time THD minimization and selective harmonic elimination pulse width modulation. Owing to the minimized costs, size, and control of complication, the ANPC converter has got cumulative consideration and is at present commercially used for medium- and high-power-level industrial application.

10.4.2 Advantages of ANPC MLI

1. Capable of operating under reactive power condition.
2. Low cost.
3. Small in size.

10.4.3 Disadvantages of ANPC MLI

1. Unequal loss distribution among semiconductor devices and DC link capacitor link voltage unbalancing.

Table 10.4 Advantages and disadvantages: a comparison of ANPC MLI with conventional MLI topologies.

Types of MLI / Parameters	Diode Clamped Multilevel Inverter	Flying Capacitor Multilevel Inverter	Cascaded H Bridge Multilevel Inverter	Active Neutral Point Clamped Converter
Advantages	• Reduced harmonics • Simple control methods • High efficiency	• Requirement of filter is reduced. • Real and reactive power control • Doesn't require clamping diode	• Simple DC bus regulation • Least number of components • Very cheap	• Operation under reactive power condition • Low cost. • Small in size
Disadvantages	• More clamping diodes are required. • Need high voltage rating diodes	• Complex control of circuit operation • Switching losses	• Each H Bridge requires a separate dc source • Limited applications	• Unequal power loss distribution
Applications	• Static var compensation. • Variable speed motor drives • HV DC and AC Transmission lines	• Sinusoidal current rectifiers • AC-DC and DC-AC conversion • Static var generation.	• Renewable resource of energy and battery-based applications. • E-Vehicle drives. • Power factor compensator.	• Medium power level industrial application

10.5 Conclusion

Multilevel inverters continue to gain more importance in medium and high-power industrial applications. Multilevel inverters are used in industrial motor drives and to interface utility systems. Also, multilevel inverters substitute the existing multipulse converter topologies with transformer. Many researchers on MLI have proposed various topologies of MLI for specific applications. The multilevel converters provide low THD, high efficiency and power factor. In this chapter, the operation and characteristics of three types of MLIs such as diode clamped MLI, Flying capacitor MLI and cascaded H bridge MLI have been explained in detail with their advantages, disadvantages and applications. However, the MLI balancing of series connected capacitor voltage is a major concern. In this chapter, a novel seven-switch, five-level ANPC topology has been proposed. When compared to the conventional topologies, for single-phase operation it requires seven power switches and it must be possible to select low current stating switch under high power factor operation. The operation and switching states of ANPC MLI are also presented. The detailed comparison of conventional multilevel inverter topologies with the proposed ANPC MLI has also been made. It has been concluded from the analysis that the ANPC MLIs are best suitable for industrial applications rather than the other topologies in terms of number of switches, capacitors, clamping components, THD, switching loss, control strategies, operating at unity power factor, etc.

References

1. Prem P, P Sivaraman, DJ Almakhles, S Padmanaban, Z Leonowicz, "A New Multilevel Inverter Topology with Reduced Power Components for Domestic Solar PV Applications", *IEEE Access* 8, 187483–187497, 2020.
2. P Sivaraman, P Prem, "PR controller design and stability analysis of single stage T-source inverter based solar PV system", *Journal of the Chinese Institute of Engineers* 40 (3), 235–245, 2017.
3. Jose Rodriguez; Leopoldo G. Franquelo; Samir Kouro; Jose I. Leon; Ramon C. Portillo; Ma Angeles Martin Prats; Marcelo A. Perez, "Multilevel Converters: An Enabling Technology for High-Power Applications", *Proceedings of IEEE*, Vol. 97, no. 11, pp. 1786–1817, Nov. 2009.
4. M. Steurer; F. Bogdan; M. Bosworth; O. Faruque; J. Hauer; K. Schoder; M. Sloderbeck; D. Soto; K. Sun; M. Winkelnkem, "Multifunctional megawatt scale medium voltage DC test bed based on modular multilevel converter (MMC) technology", International Conference on Electrical Systems for Aircraft, Railway, Ship Propulsion and Road Vehicles (ESARS), Mar. 2015.

5. Jeyraj Selvaraj; Nasrudin A. Rahim, "Multilevel Inverter For Grid-Connected PV System Employing Digital PI Controller", *IEEE Transactions on Industrial Electronics*, vol. 56, no. 1, pp. 149–158, July 2008.
6. Bin Wu; F.A. DeWinter, "Voltage stress on induction motors in medium-voltage (2300-6900-V) PWM GTO CSI drives", *IEEE Transactions on Power Electronics* , vol. 12, no. 2, pp. 213–220, Mar. 1997.
7. O. Ojo; I.E. Davidson, "PWM-VSI inverter-assisted stand-alone dual stator winding induction generator ", *IEEE Transactions on Industry Applications*, vol. 36, no. 6, pp. 1604–1611, Nov. 2000.
8. Zicheng Liu; Zedong Zheng; Scott D. Sudhoff; Chunyang Gu; Yongdong Li, "Reduction of Common-Mode Voltage in Multiphase Two-Level Inverters Using SPWM with Phase-Shifted Carriers", *IEEE Transactions on Power Electronics*, vol. 31, no. 9, pp. 6631–6645, Sep. 2016.
9. F. Blaabjerg, M. Liserre and K. Ma. "Power Electronics Converters for Wind Turbine Systems". *IEEE Transactions on Industry Applications*, vol. 48, no 2, pp. 708–719, March/April 2012.
10. P Prem, M. Jagabar Sathik, P.sivaraman, A. Matheswaran, Aleem, Abdel, "A new asymmetric dual source multilevel inverter topology with reduced power switches", *Journal of the Chinese Institute of Engineers* 42 (5), 460–472, 2019.
11. P Prem, V Sugavanam, AI Abubakar, JSM Ali, B C Sengodan, "A novel cross-connected multilevel inverter topology for higher number of voltage levels with reduced switch count", *International Transactions on Electrical Energy Systems*, 2020.
12. Venu Sonti; Sachin Jain; Subhashish Bhattacharya, "Analysis of the Modulation Strategy for the Minimization of the Leakage Current in the PV Grid-Connected Cascaded Multilevel Inverter", *IEEE Transactions on Power Electronics*, vol. 32, no.2, pp. 1156–1169, Feb. 2017.
13. P Ponnusamy, S Velliangiri, JSM Ali, "A Hybrid Switched Capacitor Multi-Level Inverter with High Voltage Gain and Self-Voltage Balancing Ability", *Electric Power Components and Systems* 48 (8), 755–768, 2020.
14. https://www.google.co.in/books/edition/Multilevel_Inverters/iZCZDgAAQBAJ?hl=en&gbpv=1&dq=Advantages+of+Multilevel+Converters+based+on+Waveforms&pg=PA26&printsec=frontcover
15. K T Maheswari, Analysis of New Three Phase Multilevel inverter with shared Power Switches, *International Journal of Research in Electrical Engineering* (IJREE), 4 2 67–70 Apr. 2017.
16. Maheswari K.T. Bharani Kumar R. Novel Three Phase Quasi Z Source Multilevel Inverter Topology for Wind Turbine Driven Permanent Magnet Generator, *Journal of Electrical Engineering* 18 3 1–10 Oct. 2018.
17. Ui-Min Choi; June-Seok Lee; Kyo-Beum Lee, "New Modulation Strategy to Balance the Neutral-Point Voltage for Three-Level Neutral-Clamped Inverter Systems", *IEEE Transactions on Energy Conversion*, vol. 29, no.1, pp. 91–100, Mar. 2014.

18. Xiaoqiang Guo; Marcelo C. Cavalcanti; Alexandre M. Farias; Josep M. Guerrero, "Single-Carrier Modulation for Neutral-Point-Clamped Inverters in Three-Phase Transformerless Photovoltaic Systems", *IEEE Transactions on Power Electronics*, vol. 28, no. 6, pp. 2635–2637, Oct. 2012.
19. X. Yuan and I. Barbi, "Fundamentals of a new diode clamping multilevel inverter," *IEEE Trans. Power Electron.*, vol. 15, no. 4, pp. 711–718, July 2000.
20. H. Wang, Y. Liu, and P. C. Sen, "A neutral point clamped multilevel topology flow graph and space NPC multilevel topology," in *Proc. IEEE ECCE*, 2015, pp. 3615–3621.
21. Jose Rodríguez, S. Bernet, P. Steimer and I. Lizama. "A survey on Neutral-Point-Clamped Inverters". *IEEE Transactions on Industrial Electronics*, vol. 57, no 7, pp. 2219–2230, July 2010.
22. K. Wang, Y. Li, and Z. Zheng, "A neutral-point potential balancing algorithm for five-level ANPC converters," in *Proc. ICEMS*, Aug. 2011, pp. 1–5.
23. T. Brückner, S. Bernet and H. Güldner. "The Active NPC Converter and Its Loss-Balancing Control". *IEEE Transactions on Industrial Electronics*, vol 52, no. 3, pp. 855–868, June 2005.
24. P. Barbosa, P. Steimer, J. Steinke, M. Winkelnkemper, and N. Celanovic, "Active-neutral-point-clamped (ANPC) multilevel converter technology," in *Proc. Eur. Conf. Power Electron. Appl.*, 2005, vol. 10, pp. 1–10.
25. Hongliang Wang, Lei Kou, Yan-Fei Liu, and Paresh C. Sen, "A Seven-Switch Five-Level Active-Neutral-Point-Clamped Converter and Its Optimal Modulation Strategy", *IEEE Transactions on Power Electronics*, DOI 10.1109/TPEL.2016.

11

A Quasi Z-Source (QZS) Network-Based Quadratic Boost Converter Suitable for Photovoltaic-Based DC Microgrids

Amir Ghorbani Esfahlan and Kazem Varesi*

Power Electronics Research Lab. (PERL), Faculty of Electrical Engineering, Sahand University of Technology, Tabriz, Iran

Abstract

The Quasi-Z-source (QZS) networks (consisting of an inductor, capacitor, and diode) are added to basic structures to increase the voltage gain. In this paper, a quasi-Z-source network-based DC-DC converter with quadratic voltage conversion ratio is proposed that is applicable for photovoltaic systems and DC microgrids. The advantages of the proposed structure can be summarized as: continuous and low-ripple input current, capability of tracking maximum power point in connection to a solar panel, achievability of high boosting-factors requiring less components, reduced voltage stress on switching devices, simultaneous operation of active switches leading to only two operational modes in continuous conduction operation, simple analysis and easy control of converter. The steady-state calculations of converter for computing gain and other basic parameters have been performed. The suggested topology has been contrasted with similar configurations to validate its supremacy. The validity of theoretical calculations has been verified by simulation results taken out of PSCAD/EMTDC software. Also, the performance of the converter in real cases with consideration of the losses has been investigated, which shows the possibility of practical implementation of the proposed structure in industrial applications or small photovoltaic power plants. The maximum power point tracking ability of the converter has also been certified by simulations.

Keywords: DC-DC converter, high-gain, quasi-z-source, MPPT, voltage stress

**Corresponding author*: kazem83varesi@gmail.com; k.varesi@sut.ac.ir

Nikita Gupta, Mahajan Sagar Bhaskar, Sanjeevikumar Padmanaban and Dhafer Almakhles (eds.) DC Microgrids: Advances, Challenges, and Applications, (325–360) © 2022 Scrivener Publishing LLC

11.1 Introduction

Today, the existence of environmental problems such as air pollution, visual and auditory pollution has led people to think of replacing fossil and pollutant energies with renewable and clean energy sources [1, 2]. Energy sources known as renewable energies are available mostly in the form of wind and solar, where the wind turbines and photovoltaic cells convert these energies into electrical energy. On the other hand, renewable energy sources have features that make them require interfaces between the production unit and the power grid [3, 4]. Solar panels generally produce low voltage and the output of wind turbines is AC power with variable frequency. So, the use of high-gain DC-DC converters is a good method to perform power processing for the power grid [5].

So far, various structures have been presented for high-gain converters, which are generally divided into two categories, insulated and non-insulated. In structures presented in [6, 7], the coupled-inductor is used to increase the conversion ratio. The use of Z-source (ZS) or quasi-Z-source (QZS) networks are other voltage-boosting solutions in converters, where mostly the duty cycle of switches is limited [8, 9]. A bidirectional boost converter with a quadratic voltage-gain is presented in [10, 11]. This structure has an acceptable number of elements concerning the gain it produces, and the voltage stress of its switches is less than the output voltage. These types of converters have two active switches, but the switches work simultaneously and the converter has a simple dynamic and the current stress of each switch is low. A high-gain boost converter with two Z-source networks and an active switch has been presented in [12]. The high current stress in this converter may damage the switch and elements in practical applications. In [13, 14], a combination of an interleaved boost converter with two active switches and a three-stage switched capacitor network is presented. This structure generates high voltage-gain and the voltage stress of semiconductor elements is low, but the active switches do not operate simultaneously, which complicates the dynamics of the converter and its control process.

The multi-stage type of interleaved boost structure with N active switches and N inductors is presented in [15], where the N represents the order of voltage ratio. So, this structure is able to generate high voltage gains. However, with increasing the number of stages, the number of switches and voltage multiplier cells (VMC) increases, which increases the conduction losses and cost of converter. The multi-stage single active switch boost structure presented in [16] produces cubic gain. In multi-stage mode, only the number of Z-source elements is changed, where the number of active switches remains constant. The gain of this converter is higher than similar structures, but the active switch has high current stress. The structures studied above were single-input single-output (SISO) converters. Applications that require several DC sources (such as batteries with different output voltages) require step-up multi-input converters. The [17] presents a high-gain multi-input single-output (MISO) converter that besides generating high gains, has low voltage stress on its switches. A multi-input-multi-output (MIMO) boost structure is also proposed in [18], which is used for high powers.

To improve the voltage-gain, some papers use a combination of quadratic boost converter with Z-source and/or quasi-Z-source source networks. Usually, these types of converters fall into the ultra-high step-up group, and the structure proposed in this paper is part of this type. In [19], a quadratic gain topology with a Z-source network is presented, in which the number of elements is reasonable. The structure presented in [20] is also a quadratic step-up converter with a switched capacitor network that acts as a voltage multiplier cell (VMC). A configuration based on switched-inductor and switched-capacitor is presented in [21], which produces a cubic gain, with a fixed number of active switches while adding extra stages. The [22] presents a converter with *n*-order conversion ratio, where the *n* represents the number of stages. The structure has high gain with an acceptable number of elements. The [23] presents and investigates another Z-source network based step-up structure with cubic voltage ratio. Briefly, the main features aimed in high-gain DC-DC converter are as follows:

- Large voltage gain with reasonable number of elements.
- Continuous and low-ripple input current for tracking maximum power point in solar applications.
- Low loss and high efficiency.
- Low voltage stress on switching devices.

So far, several methods have been presented to solve the problems related to the connection of renewable energy plants to the power grid. The high-gain DC-DC converters can make it possible to connect the solar plants to the grid with an inverter at the end-side. In such a structure, the step-up transformer between the inverter and the low voltage/medium voltage bus of grid can be eliminated. The main advantage of high-gain DC-DC converters over transformers is their MPPT capability in solar or wind applications. Therefore, the presence of these interfaces can eliminate some of the additional costs associated with electromagnetic devices. Also, the high-gain DC-DC converters are the building blocks of DC microgrids.

In this paper, a quasi-Z-source network-based quadratic-gain boost converter is proposed. This converter has only two operating modes in continuous conduction mode (CCM). The two active switches operate simultaneously, which makes the converter dynamic and the control process easier. Due to continuous input current, the suggested converter can effectively track the maximum power point in PV applications. Also, the voltage stress of all semiconductors is less than output voltage.

The configuration of the proposed converter and its operating principles are described in Section 11.2. In Section 11.3, the steady-state analysis regarding boosting-factor, voltage stress across switches and diodes, average inductor current, and other parameters are given. To verify supremacy of proposed converter, it has been compared with existing structures in Section 11.4. The converter performance in discontinuous conduction mode (DCM) is fully described in Section 11.5. The simulation outcomes are delivered in Section 11.6. Section 11.7 covers the real voltage gain and loss calculations of suggested converter. In Section 11.8, the dynamic behavior of proposed converter is evaluated. Section 11.9 investigates the MPPT ability of converter by simulations. Finally, the paper concludes in Section 11.10.

11.2 Proposed Converter

A. Configuration

Figure 11.1 shows an outline of the proposed configuration. Due to continuous source current of converter, the input source can be replaced by solar panel to provide the power required by consumers. The suggested converter consists of two power switches (S_1, S_2), three diodes (D_1, D_2, D_o),

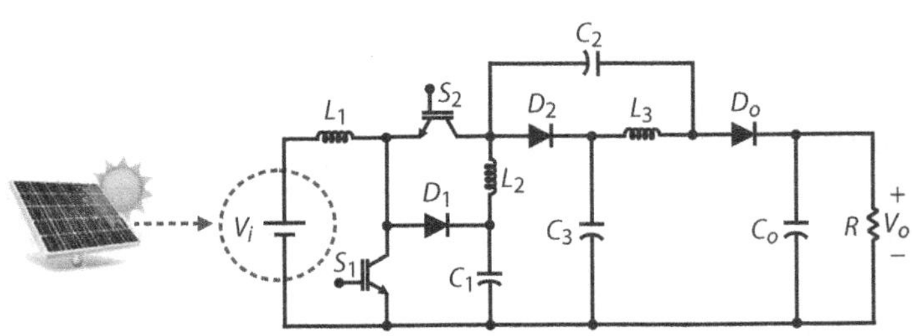

Figure 11.1 Proposed step-up topology.

three inductors (L_1, L_2, L_3), and four capacitors (C_1, C_2, C_3, C_o). The C_2, C_3, L_3, and D_2 belongs to the QZS network that has been employed to enhance the voltage-boosting ability of suggested topology.

B. Operating Principle in Continuous Conduction Mode (CCM)
In continuous conduction mode (CCM), the switches operate simultaneously. So, two operating modes are resulted, which are investigated in steady-state condition with following assumptions:

1. Devices are ideal
2. The voltage ripple of capacitors is neglected due to their large capacitances.

Mode I ($t = 0$ to $t = DT$):
The S_1, S_2 are turned on simultaneously. The diodes are reverse biased and blocked. So, the L_1 and L_2 inductors are energized by input source (V_i) and capacitor C_1, respectively. Also, the capacitor C_3 is discharged and its energy flows to inductor L_3 and capacitor C_2. The demanded energy of load is provided by C_o. In this mode, the voltage of inductors can be calculated from (11.1).

$$v_{L_1} = V_i, v_{L_2} = v_{C_1}, v_{L_3} = v_{C_3} - v_{C_2} \tag{11.1}$$

Mode II ($t = DT$ to $t = T$):
In mode II, the switches are off. The capacitor C_1 is charged by the input source (V_i) and inductor L_1. Similarly, the capacitor C_3 is energized by input source and L_1 and L_2 inductors. In this mode, the energy of inductors, as well as capacitor C_2 flows toward the load. The voltage on inductors is shown in (11.2).

$$v_{L_1} = V_i - v_{C_1}, v_{L_2} = v_{C_1} - v_{C_3}, v_{L_3} = -v_{C_2} = v_{C_3} - V_o \tag{11.2}$$

The equivalent circuits of the proposed converter during CCM are shown in Figure 11.2.

The key waveforms of proposed topology during steady-state conditions are shown in Figure 11.3.

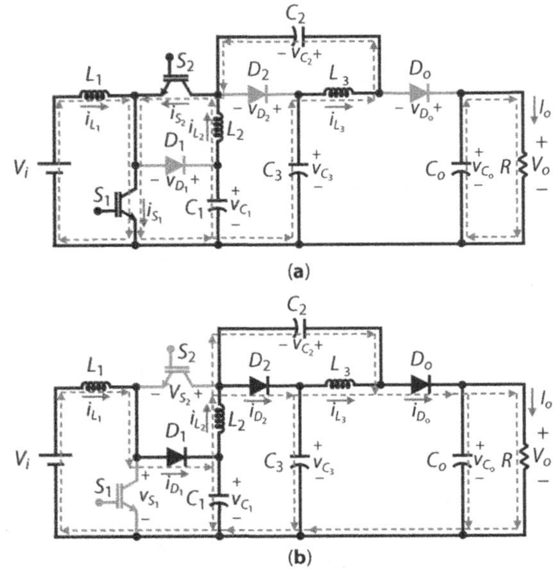

Figure 11.2 Equivalent circuit of proposed converter in CCM, during a) mode I, b) mode II.

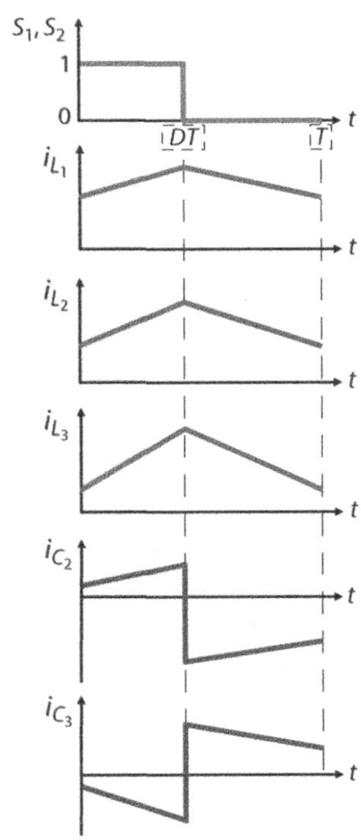

Figure 11.3 Key waveforms of proposed converter during CCM operation.

11.3 Steady-State Analyses

A. Voltage Gain

By applying voltage-second balance principle on inductors, the voltage of capacitors can be calculated as (11.3)–(11.5).

$$\langle v_{L_1}(t)\rangle = 0 \Rightarrow V_{C_1} = \frac{V_i}{1-D} \tag{11.3}$$

$$\langle v_{L_2}(t)\rangle = 0 \Rightarrow V_{C_3} = \frac{V_i}{(1-D)^2} \tag{11.4}$$

$$\langle v_{L_3}(t)\rangle = 0 \Rightarrow V_{C_3} = \frac{DV_i}{(1-D)^2} \tag{11.5}$$

Based on the above-mentioned equations, the gain (boosting factor) of the proposed topology can be calculated from (11.6).

$$G_{CCM} = \frac{V_o}{V_i} = \frac{1+D}{(1-D)^2} \tag{11.6}$$

Where, "D" represents the duty cycle of S_1 and S_2 switches. The (11.6) certifies that the proposed converter can produce large gains even at small duty cycles.

B. Voltage Stress on Semiconductors

The voltage stress on semiconductors is as (11.7), (11.8).

$$V_{S_1} = \frac{1}{1-D}V_i = \frac{1-D}{1+D}V_o,\ V_{S_2} = \frac{D}{(1-D)^2}V_i = \frac{D}{1+D}V_o \tag{11.7}$$

$$V_{D_1} = \frac{1}{1-D}V_i = \frac{1-D}{1+D}V_o,\ V_{D_2} = \frac{1}{(1-D)^2}V_i = \frac{1}{1+D}V_o,$$
$$V_{D_o} = \frac{1}{(1-D)^2}V_i = \frac{1}{1+D}V_o \tag{11.8}$$

It is seen that for $0 < D < 1$, the voltage stress of all semiconductors is less than the V_o, which reduces switching losses.

C. Current Ripple of Inductors
The current ripple of the inductors, which are equal to the difference between the maximum and minimum values of the current passing through them, have been shown in (11.9).

$$\Delta i_{L_1} = \frac{DV_i}{L_1 f}, \Delta i_{L_2} = \frac{DV_i}{(1-D)L_2 f}, \Delta i_{L_3} = \frac{DV_i}{(1-D)L_3 f} \quad (11.9)$$

Where the "f" represents the switching frequency.

D. Average Current of Inductors
From capacitor currents balance law, the average current relations of inductors are obtained in (11.10), where the I_i and I_o denote the input and output currents (in steady-state), respectively.

$$I_{L_1} = I_i = \frac{1+D}{(1-D)^2} I_o, I_{L_2} = (1-D)I_i = \frac{1+D}{1-D} I_o, I_{L_3} = I_o = \frac{(1-D)^2}{1+D} I_i \quad (11.10)$$

E. Maximum and Minimum of Inductor Currents
Using Equation (11.11), the maximum and minimum values of the inductor currents are obtained according to (11.12)–(11.14) in terms of input current, Load resistance (R), inductance value, and switching frequency (f).

$$\begin{cases} I_{L,\max} = I_L + \dfrac{\Delta i_L}{2} \\ I_{L,\min} = I_L - \dfrac{\Delta i_L}{2} \end{cases} \quad (11.11)$$

Not that the "I_L" and "Δi_L" represent the average and ripple of inductor current, respectively.

$$\begin{cases} I_{L_1,\max} = I_i\left[1+\dfrac{R}{2L_1 f}\left(\dfrac{(1-D)^4}{(1+D)^2}\right)\right] \\ I_{L_1,\min} = I_i\left[1-\dfrac{R}{2L_1 f}\left(\dfrac{(1-D)^4}{(1+D)^2}\right)\right] \end{cases} \tag{11.12}$$

$$\begin{cases} I_{L_2,\max} = I_i\left[(1-D)+\dfrac{R}{2L_2 f}\left(\dfrac{D(1-D)^3}{(1+D)^2}\right)\right] \\ I_{L_2,\min} = I_i\left[(1-D)-\dfrac{R}{2L_2 f}\left(\dfrac{D(1-D)^3}{(1+D)^2}\right)\right] \end{cases} \tag{11.13}$$

$$\begin{cases} I_{L_3,\max} = I_i\left[\dfrac{(1-D)^2}{1+D}+\dfrac{R}{2L_3 f}\left(\dfrac{D(1-D)^3}{(1+D)^2}\right)\right] \\ I_{L_3,\min} = I_i\left[\dfrac{(1-D)^2}{1+D}-\dfrac{R}{2L_3 f}\left(\dfrac{D(1-D)^3}{(1+D)^2}\right)\right] \end{cases} \tag{11.14}$$

F. Current Stress on Switches and Diodes

When switches and diodes are conducting, by using the circuit theories, the maximum current passing through them is calculated according to (11.15), (11.16).

$$I_{S_1} = I_{L_1,\max} + I_{L_2,\max} + I_{L_3,\max}, I_{S_2} = I_{L_2,\max} + I_{L_3,\max} \tag{11.15}$$

$$I_{D_1} = I_{L_1,\max}, I_{D_2} = I_{L_2,\max} - \frac{D}{1-D} I_{L_3,\max}, I_{D_o} = I_{L_2,\max} + \frac{D}{1-D} I_{L_3,\max} \tag{11.16}$$

Thus, the current stress of switching devices can be calculated in terms of input current, using equations (11.12)–(11.14).

G. Design of Parameters

a) Inductance Evaluation

Using condition $I_L > \Delta i_L/2$, the inductance values follow the inequalities (11.17)–(11.19).

$$L_1 > \left(\frac{D(1-D)^4}{(1+D)^2}\right)\frac{R}{2f} \tag{11.17}$$

$$L_2 > D\left(\frac{1-D}{1+D}\right)^2 \frac{R}{2f} \tag{11.18}$$

$$L_3 > D\left(\frac{1-D}{1+D}\right)\frac{R}{2f} \tag{11.19}$$

So that the converter to operate in the continuous conduction mode (CCM), the amount of inductance must always be greater than the critical values (determined by Equations (11.17)–(11.19)).

b) Capacitance Evaluation

Assuming that the allowable ripple voltage values of the capacitors are known (Δv_{C1}, Δv_{C2}, Δv_{C3}, Δv_{Co}) the capacitance values are determined as (11.20)–(11.23).

$$C_1 > \frac{V_o}{Rf\Delta v_{C_1}}\left(\frac{D(1+D)}{1-D}\right) \tag{11.20}$$

$$C_2 > \frac{DV_o}{Rf\Delta v_{C_2}} \tag{11.21}$$

$$C_3 > \frac{DV_o}{Rf\Delta v_{C_3}} \tag{11.22}$$

$$C_o > \frac{DV_o}{Rf\Delta v_{C_o}} \tag{11.23}$$

11.4 Comparison with Other Structures

To study the advantages of the proposed converter in more detail, it has been compared with several new structures in terms of voltage gain, the number of elements, and voltage stress across the semiconductors, and its results are shown in Table 11.1. As can be seen from the table, the proposed structure is suitable in terms of the number of switches and diodes. The number of the inductor and the capacitor is also in a better condition, and finally, it can be said that the voltage gain of the proposed converter has a better approach than the compared structures relative to the total number of elements.

To compare the voltage stresses of switches and diodes, three concepts have been used. The total voltage stress of the switches/diodes is the sum of the voltage stresses of the converter switches/diodes, in which the proposed converter is better than similar structures. Normalized voltage stress (NVS) of each switch/diode is switch/diode voltage stress in terms of output voltage, divided by output voltage [30, 31]. For better understanding, the information presented in Table 11.1 has been shown as plots in Figure 11.4. It is observed from Figure 11.4a that for a wide range of duty cycles, the proposed topology produces higher gains than other structures. Also, it is seen from Figure 11.4b-f that for $D>0.5$, the gain per device in the proposed topology is higher than that of other structures. It shows that for achieving the same boosting factor, the proposed topology requires fewer devices. Finally, Figure 11.4g approves the low total voltage stress on switches and diodes of the proposed topology.

11.5 Converter Analyzes in Discontinuous Conduction Mode (DCM)

A. Operating Modes

In discontinuous conduction mode, the converter has five operating modes, so that the first two modes are similar to continuous conduction mode (CCM). Thus:

Mode I [$t = 0$ to $t = DT$] and Mode II [$t = DT$ to $t = D_3T$]:
These states are similar to CCM operation.

Table 11.1 Comparative analyses.

Structure	**[24]**	**[25]**	**[26]**	**[27]**	**[28]**	**[29]**	**Proposed**
Voltage gain	$\frac{2(1+D)}{1-D}$	$\frac{1+3D}{1-D}$	$\frac{2+D}{1-D}$	$\frac{1+D}{1-D}$	$\frac{2}{1-D}$	$\frac{D^2-3D+3}{(1-D)^2}$	$\frac{1+D}{(1-D)^2}$
N_{Switch}	1	2	2	2	1	2	2
N_{Diode}	6	8	3	1	3	3	3
$N_{Inductor}$	2	4	2	2	1	2	3
$N_{Capacitor}$	3	9	6	1	3	3	4
$N_{Component}$	12	23	12	6	8	10	12
The total voltage stress on switches	V_o	$\frac{2}{1+3D}V_o$	$\frac{2}{2+D}V_o$	$\frac{2}{1+D}V_o$	$\frac{1}{2}V_o$	$\frac{2-D}{D^2-3D+3}V_o$	$\frac{1}{1+D}V_o$
The total voltage stress on diodes	[6/(3+D)]Vo	$\frac{2}{1+3D}V_o$	$\frac{3}{2+D}V_o$	$\frac{2}{1+D}V_o$	$\frac{3}{2}V_o$	$\frac{5-3D}{D^2-3D+3}V_o$	$\frac{3-D}{1+D}V_o$
Normalized total voltage stress (switches + Diodes)	$\frac{9+D}{3+D}$	$\frac{4}{1+3D}$	$\frac{5}{2+D}$	$\frac{4}{1+D}$	2	$\frac{7-4D}{D^2-3D+3}$	$\frac{4-D}{1+D}$

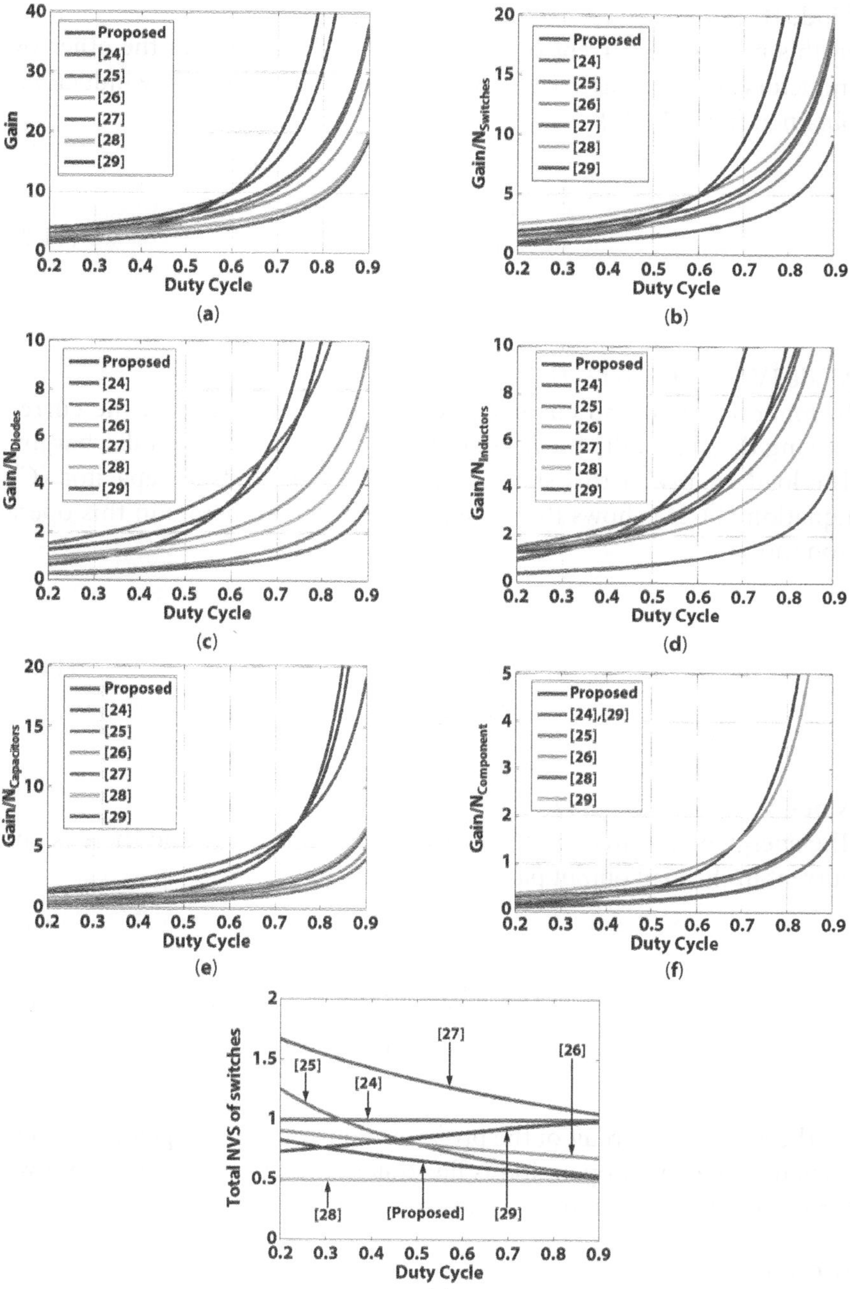

Figure 11.4 Comparison of the proposed structure in terms of (a) gain, (b) number of switches, (c) number of diodes, (d) number of inductors, (e) number of capacitors, (f) number of components, (g) total normalized voltage stress of switches.

Mode III [$t = D_3T$ to $t = D_2T$]:
In this mode, the current of inductor L_3 reaches zero, but the other two inductors are being discharged. The voltage relationship of the inductors is obtained from (11.24).

$$\begin{cases} v_{L_1} = V_i - v_{C_1} \\ v_{L_2} = v_{C_1} - v_{C_3} \\ v_{L_3} = 0 \end{cases} \tag{11.24}$$

Mode IV [$t = D_2T$ to $t = D_1T$]:
In mode IV, while no current passes through inductor L_3, the current passing through inductor L_2 is zero, and inductor L_1 is still discharging. The load energy in this state is supplied by the output capacitor (C_o). Equation (11.25) shows the voltage drop of the inductors in this operation mode.

$$\begin{cases} v_{L_1} = V_i - v_{C_1} \\ v_{L_2} = 0 \\ v_{L_3} = 0 \end{cases} \tag{11.25}$$

Mode V [$t = D_1T$ to $t = T$]:
The energy of all three inductors is empty in this mode and there is only current in the load output part. Thus:

$$\begin{cases} v_{L_1} = 0 \\ v_{L_2} = 0 \\ v_{L_3} = 0 \end{cases} \tag{11.26}$$

The equivalent circuit of the proposed converter in its operating modes related to the DCM operation is illustrated in Figure 11.5 and also waveforms of inductor currents are shown in Figure 11.6.

B. Gain Calculation
To calculate the gain, it is necessary to apply the inductor voltage balance law for relations (11.1), (11.2), and (11.24)–(11.26). In this case,

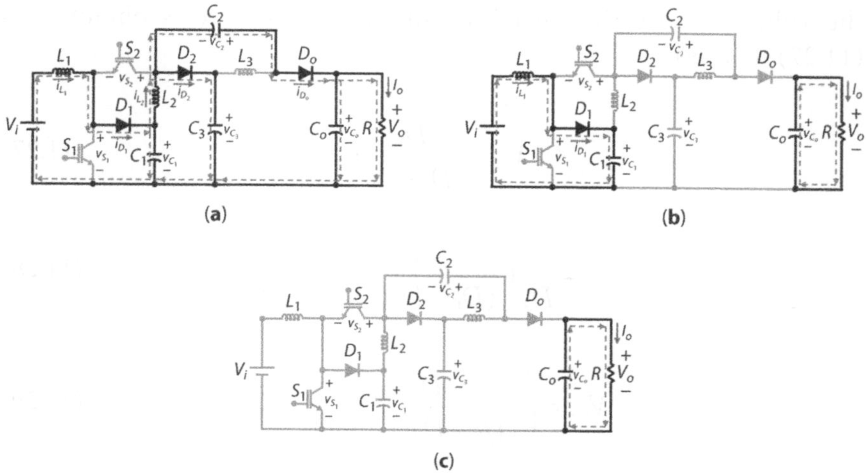

Figure 11.5 The equivalent circuit of the proposed converter during DCM operation in (a) mode III, (b) mode IV, (c) mode V.

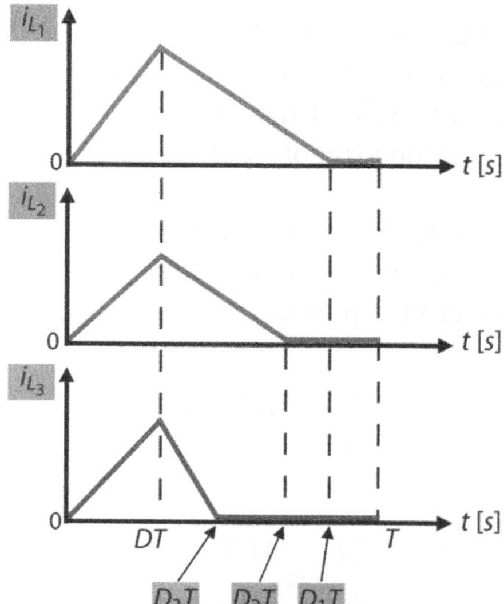

Figure 11.6 The theoretical waveform of inductor currents in DCM.

the voltage across the capacitors in the steady-state is obtained as (11.27)–(11.29).

$$V_{C_1} = \frac{D_1}{D_1 - D} V_i \tag{11.27}$$

$$V_{C_2} = \frac{D}{D_3}\left(\frac{D_1 D_2}{(D_1 - D)(D_2 - D)}\right) V_i \tag{11.28}$$

$$V_{C_3} = \left(\frac{D_1 D_2}{(D_1 - D)(D_2 - D)}\right) V_i \tag{11.29}$$

Therefore, using the above equations and circuit relations, the converter voltage gain in discontinuous conduction mode is calculated according to (11.30).

$$G_{DCM} = \frac{V_o}{V_i} = \left(1 + \frac{D}{D_3}\right)\left(\frac{D_1 D_2}{(D_1 - D)(D_2 - D)}\right) \tag{11.30}$$

According to Figure 11.6, D_1T, D_2T, and D_3T are the times of energy depletion of the inductors L_1, L_2, and L_3, respectively in the "T" period and "D" represents the duty cycle of the converter, which is provided by the power switch driver, in the form of a pulse.

C. Demarcate between CCM and DCM Areas

To ensure that all inductors operate in the CCM region, the following inequalities must govern the performance of the converter:

$$\frac{2L_1}{RT} < \frac{D(1-D)^4}{1+D} \tag{11.31}$$

$$\frac{2L_2}{RT} < \frac{D(1-D)^2}{(1+D)^2} \tag{11.32}$$

$$\frac{2L_3}{RT} < \frac{D(1-D)}{1+D} \tag{11.33}$$

Usually, the right and left side of the above inequalities is represented by "$K_{Critical}(D)$" and "K" respectively [32]. So it can be written:

$$\begin{cases} K_{1-crit}(D) = \dfrac{D(1-D)^4}{1+D} \\ K_{2-crit}(D) = \dfrac{D(1-D)^2}{(1+D)^2} \\ K_{3-crit}(D) = \dfrac{D(1-D)}{1+D} \end{cases}, \begin{cases} K_1 = \dfrac{2L_1}{RT} \\ K_2 = \dfrac{2L_2}{RT} \\ K_3 = \dfrac{2L_3}{RT} \end{cases} \tag{11.34}$$

Therefore, to determine the boundary between the CCM and DCM regions, for all three inductors, it is necessary to plot the "$K_{Critical}(D)$" curve in terms of duty cycle (D) as shown in Figure 11.7.

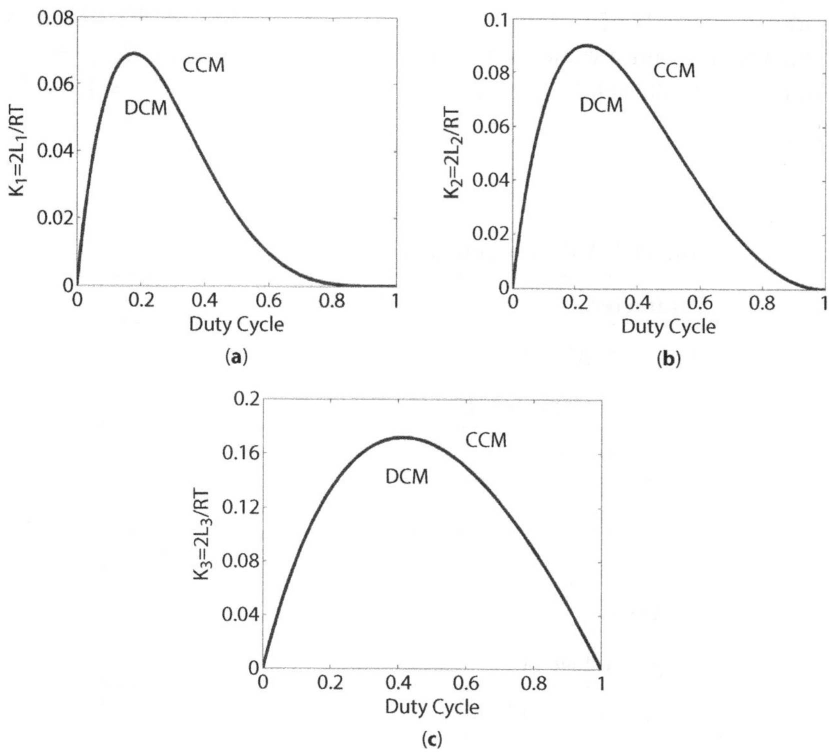

Figure 11.7 The separator curve of the CCM and DCM regions, related to the inductor (a) L_1, (b) L_2, (c) L_3.

11.6 Simulation Results

To assure correct performance of the proposed topology, it has been simulated in PSCAD/EMTDC using the parameter values shown in Table 11.2.

Figure 11.8 shows the output voltage and load current during the steady-state condition. It is seen that the output voltage and load current are 323 [V] and 3.23 [A]. So the voltage boosting factor of the proposed topology is about 13.46 that is verified by (11.6). Figure 11.8c displays the voltage ripple on the load (C_0 capacitor), which is negligible (about 0.25 [V]). It should be noted that this value is much less than the allowable value, i.e., 5% of the output voltage.

Figure 11.9 shows the voltage of capacitors. It is seen that the voltage of C_1, C_2, C_3, and C_o capacitors have been set on 68.57 [V], 127.34 [V], 195.9 [V], and 323 [V], as confirmed by (11.3)–(11.6).

The voltage and current waveform of the inductors in the CCM mode is illustrated in Figure 11.10 and Figure 11.11, respectively. Depending on the figures, in mode I, $v_{L1} = 24$ [V], $v_{L2} = 68.55$ [V], $v_{L3} = 68.60$ [V] and in mode II, $v_{L1} = -44.55$ [V], $v_{L2} = -127.35$ [V], $v_{L3} = -127.38$ [V]. The maximum and minimum values of the inductors L_1, L_2, and L_3 are $IL_{1,max} = 44.60$ [A], $I_{L2,max} = 17.08$ [A], $I_{L3,max} = 4.32$ [A], $I_{L1,min} = 42.60$ [A], $I_{L2,min} = 13.36$ [A], $I_{L3,min} = 2.05$ [A].

Table 11.2 Values of simulation parameters.

Parameter	Value
Input voltage (V_i)	24 [V]
Switching frequency (f)	40 [kHz]
Duty cycle (D)	65 [%]
L_1, L_2, L_3	200, 300, 500 [μH]
C_1, C_2, C_3, C_o	220, 100, 100, 220 [μF]
Load (R)	100 [Ω]
Output voltage (V_o)	323 [V]
Output current (I_o)	3.23 [A]
Gain (V_o/V_i)	13.46

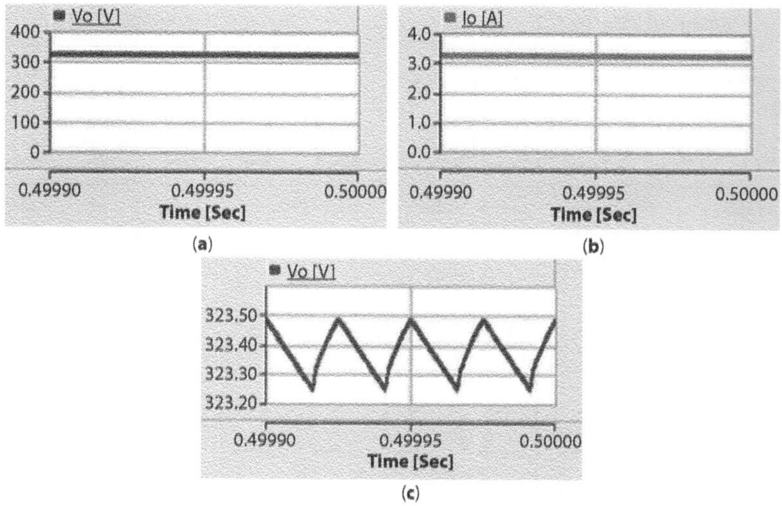

Figure 11.8 Waveforms of (a) load voltage, (b) load current, (c) voltage ripple of the output capacitor (load).

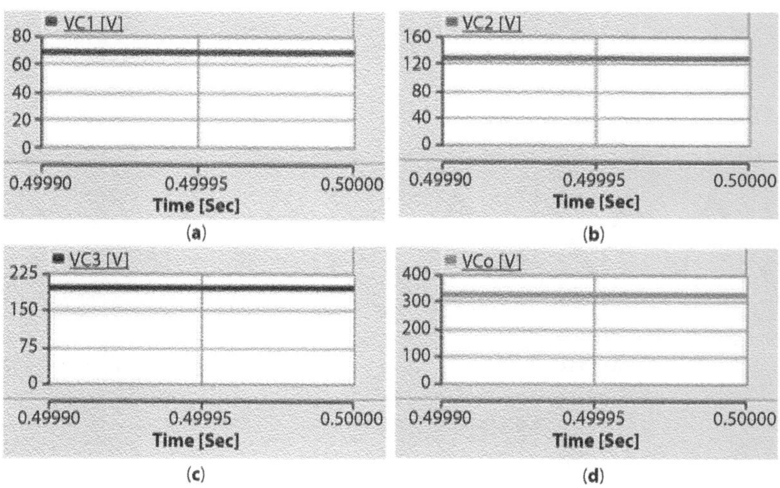

Figure 11.9 Voltage waveform of capacitors (a) C_1, (b) C_2, (c) C_3, (d) C_o.

The waveform of the current passing through the capacitors C_2 and C_3 is shown in Figure 11.12, so that the peak value of the currents is $(IC_2)_{peak} = 4.32$ $[A]$, $(I_{C3})_{peak} = 7.52$ $[A]$ and its minimum values is $(I_{C2})_{min} = -15.34$ $[A]$, $(I_{C3})_{min} = -4.22$ $[A]$.

The blocking voltage (BV) waveform of each of the switches and diodes is shown in Figure 11.13. According to the figure, the voltage stress of the

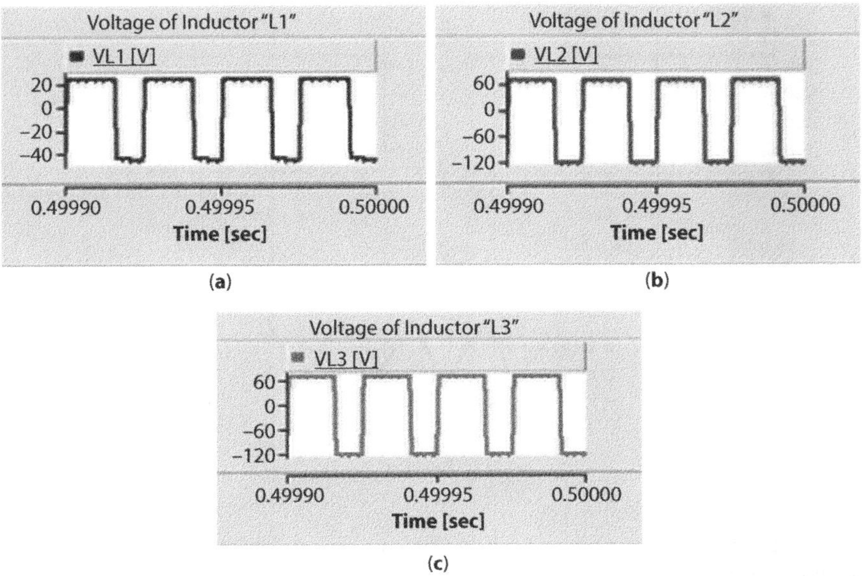

Figure 11.10 The voltage waveform of inductors (a) L_1, (b) L_2, (c) L_3.

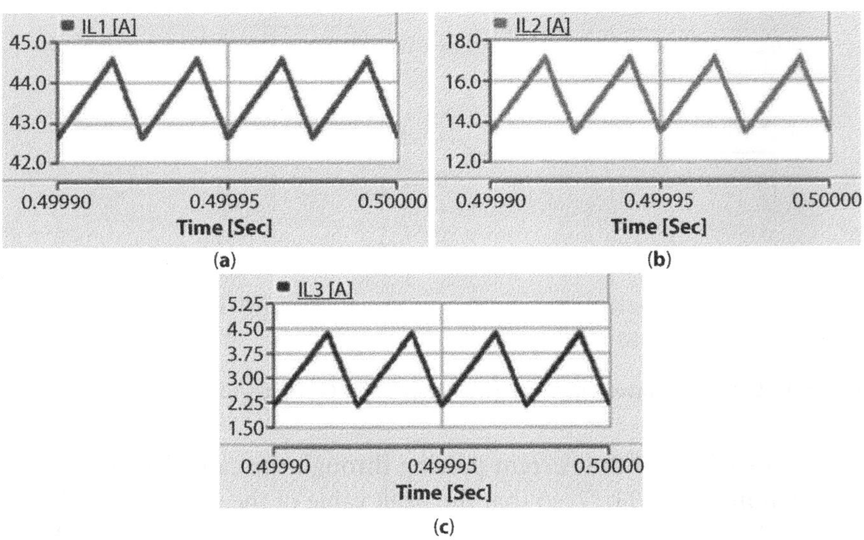

Figure 11.11 The current waveform of inductors (a) L_1, (b) L_2, (c) L_3.

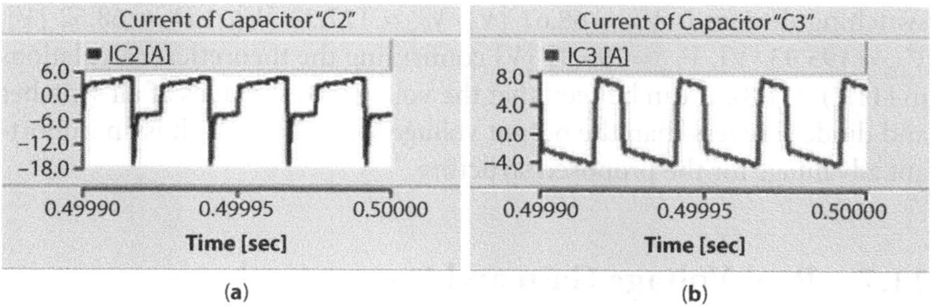

Figure 11.12 The waveforms of the capacitor currents (a) C_2, (b) C_3.

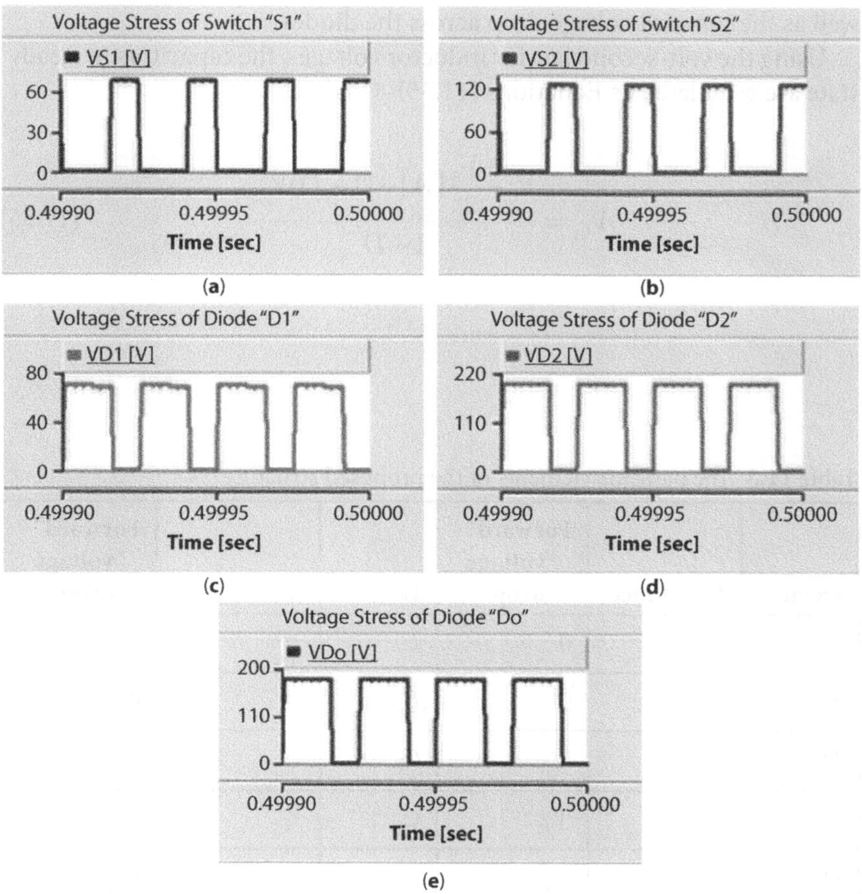

Figure 11.13 The voltage stress waveforms of switches (a) S_1, (b) S_2, and diodes (c) D_1, (d) D_2, e) D_o.

switching devices is $V_{S1} = 68.61$ [V], $V_{S2} = 127.33$ [V], $VD_1 = 68.52$ [V], $V_{D2} = 195.95$ [V], $V_{Do} = 195.91$ [V] confirming the theoretical calculations in (11.7), (11.8). It can be seen that the voltage stress values of all switches and diodes are less than the output voltage (323 [V]), which is an important advantage for the proposed structure.

11.7 Real Voltage Gain and Losses Analyzes

A. Real Gain of the converter
To achieve the real state gain, it is necessary to consider parasitic elements for each of the constituent devices of the structure. Table 11.3 shows the conduction resistance of switches, diodes, inductors, and capacitors, as well as the forward voltage drop across the diodes.

Using the volt-second law for inductor voltages, the capacitors in steady-state are calculated as Equations (11.35)–(11.37).

$$V_{C_1} = \frac{V_i - \frac{V_o}{R}[A] - (1-D)V_{f1}}{1-D} \tag{11.35}$$

Table 11.3 The parasitic elements of the proposed structure.

Device	Resistance	Forward voltage drop	Device	Resistance	Forward voltage drop
S_1	r_{S1}	0	L_2	r_{L_2}	0
S_2	r_{S2}	0	L_3	r_{L_3}	0
D_1	r_{D_1}	V_{f1}	C_1	r_{C_1}	0
D_2	r_{D2}	V_{f2}	C_2	r_{C_2}	0
D_o	r_{D_o}	V_{f_o}	C_3	r_{C_3}	0
L_1	r_{L1}	0	C_o	r_{C_o}	0

$$V_{C_3} = \frac{V_{C_1} - \frac{V_o}{R}[B] - (1-D)V_{f_2}}{1-D} \tag{11.36}$$

$$V_{C_2} = \frac{V_{C2} - \frac{V_o}{R}\left[C + R(1-D)^3\right] - (1-D)V_{f_o}}{D} \tag{11.37}$$

Where:

$$A = \frac{D(3-D)r_{S_1} + D(1-D^2)r_{C_1} + (1+D)r_{L_1} + (1-D^2)r_{D_1}}{(1-D)^2} \tag{11.38}$$

$$B = \frac{D(3-D)r_{S_1} + (1-D^2)r_{L_2} + 2D(1-D)r_{S_2} - 2(1-D^2)r_{C_3} - (1+D)(1-D)^2 r_{D_2}}{(1-D)^2} \tag{11.39}$$

$$C = \frac{D(3-D)r_{S_1} + 2D(1-D)r_{S_2} - D(1-D)^2 r_{C_3} + (1-D)^2 r_{L_3} + (1-D)^3 r_{D_o}}{(1-D)^2} \tag{11.40}$$

The output voltage relationship is also as follows:

$$V_o = V_{C_2} + V_{C_3} + V_{f_2} - V_{f_o} + \frac{V_o}{R}\left[2Dr_{C_3} + \left(\frac{1+D}{1-D}\right)r_{D_2} - r_{D_o}\right] \tag{11.41}$$

Using the basic formulas of circuit theory, the relationship between the input and output voltage of the converter is obtained as (11.42), which according to previous knowledge, in the real case, the voltage gain relationship has the input voltage (V_i).

$$V_o = \frac{(1+D)V_i - (1+D)(1-D)V_{f_1} - (1-D)^2 V_{f_2} - (1-D)^2 V_{f_o}}{D(1-D)^2 + \frac{1}{R}[P]} \tag{11.42}$$

The term " P " is a function of the conduction resistance of the elements, the duty cycle of the converter, and Load (R) and is:

$$P=\frac{D(3-D)^2}{(1-D)^2}r_{S_1}+4Dr_{S_2}+\frac{D(1+D)}{(1-D)^2}r_{L_1}+D(1+D)r_{L_2}+(1-D)^2r_{L_3}$$
$$+\frac{D^2(1+D)}{1-D}r_{C_1}-(1-D)(2-D)(2D^2+D+1)r_{C_3}-2D(1+D)(1-D)r_{D_2}$$
$$+(1-D)^2r_{D_o}+R(1-D)^3 \qquad (11.43)$$

Using the values in Table 11.4, the real gain of the converter is plotted in Figure 11.14. According to this figure, the maximum voltage gain is about 24 and has been adapted from practical values, which, of course, if low-loss semiconductors are used, the maximum value of gain will increase significantly.

B. Efficiency
The efficiency relationship for each DC-DC converter is as follows [33]:

$$\eta=\frac{P_{out}}{P_{in}}=\frac{P_{out}}{P_{out}+P_{loss}} \qquad (11.44)$$

Table 11.4 The values of the losses factors for the plot in Figure 11.14.

Parameter	Value	Parameter	Value
r_{S1}, r_{S2}	0.05 [Ω]	r_{C1}, r_{C2}	0.01 [Ω]
r_{L1}	0.05 [Ω]	r_{C3}, r_{Co}	0.01 [Ω]
r_{L2}	0.08 [Ω]	V_{f1}, V_{f2}	0.7 [V]
r_{L3}	0.1 [Ω]	V_{fo}	0.7 [V]
r_{D1}, r_{D2}	0.1 [Ω]	R_{Load}	200 [Ω]
r_{Do}	0.2 [Ω]	V_i	24 [V]

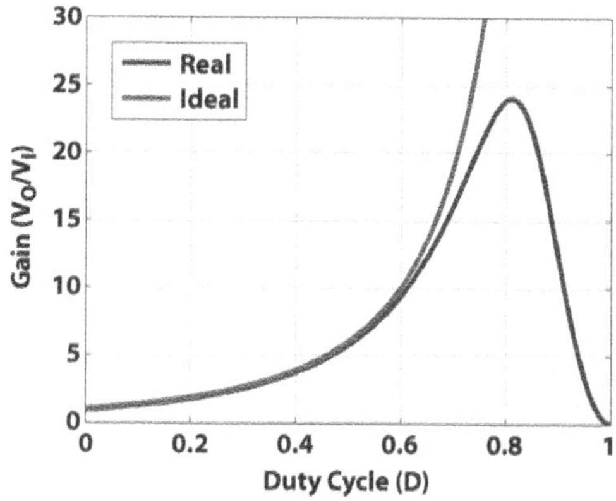

Figure 11.14 The real gain of the proposed converter versus Ideal gain.

Thus, the efficiency of the converter is calculated using Equations (11.42)–(11.44), which are shown in terms of output voltage, parasitic elements, duty cycle, and output load resistance in relation (11.45).

$$\eta = \frac{1}{\left[D + \frac{1}{R(1-D)^2}\left[P\right]\right] + \frac{1}{V_o}\left[\frac{1+D}{1-D}V_{f_1} + V_{f_2} + V_{f_o}\right]} \tag{11.45}$$

It should be noted that the term " P " is the same as (11.43).

Using the values in Table 11.4, the efficiency of the proposed converter in terms of the duty cycle is plotted in Figure 11.15a. According to the figure, the maximum efficiency of the proposed structure is about 96%, which occurs at $D = 0.45$. To improve the voltage gain and adapt the converter for renewable energy applications, the duty cycle value is considered to be 0.60, in which case the converter has an efficiency of about 94%, which according to the number of elements, are very convenient. In Figure 11.15b, the efficiency curve of the proposed structure in terms of output power is shown. The highest efficiency is related to 50 watts load and its value is equal to 96.01%. At the operating point for the converter, the load power is about 240 watts, so at this point, the efficiency is 93.8%. It is observed that the converter also has a good efficiency at higher powers. Figure 11.15c shows the efficiency changes with increasing input voltage from 12 to 60 volts. Also, to examine the performance of the converter more closely, the efficiency curves in terms of output power have been scared for different values of the input voltage in Figure 11.15d. In other words, if 5 solar panels with a

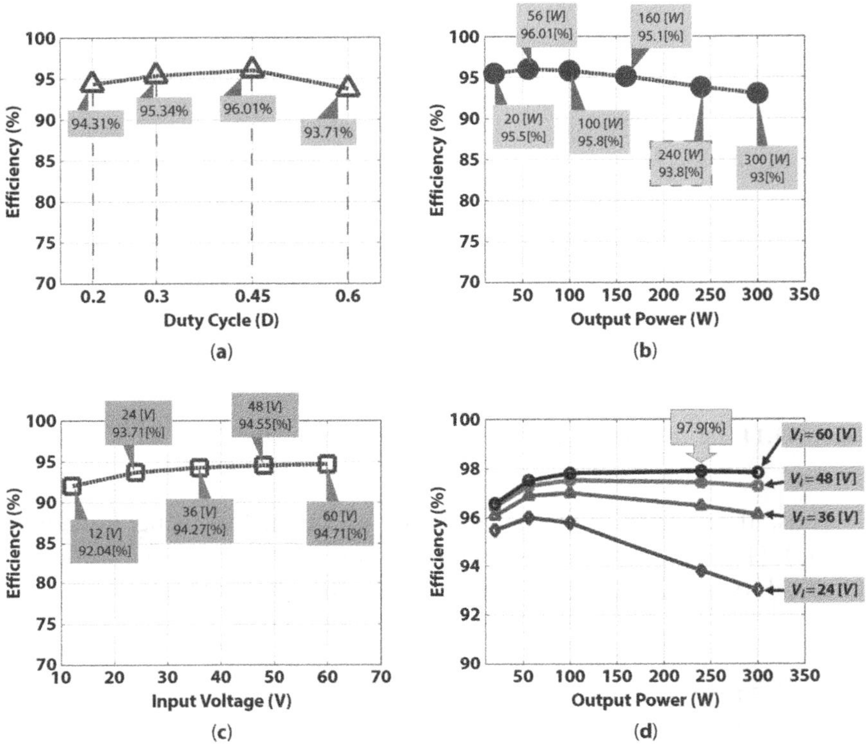

Figure 11.15 The efficiency curve of the proposed converter in terms of (a) Duty cycle for $0.2 < D < 0.6$, (b) Output power (load), (c) Input voltage, (d) Output power for the various value of input voltage.

nominal voltage of 12 volts are connected in series and the output power is equal to 240 watts, 97.9% efficiency can be achieved.

C. Numerical Losses Calculation

In numerical calculations of power losses, it is necessary to determine the effective current of each element. Using the information in Table 11.4, the effective value of current passing through the elements of the structure are given in Table 11.5 in $D = 60$ [%].

The amount of power that each element lose and converts to heat is calculated in watts as (11.46)–(11.49).

The switches losses:

$$\begin{cases} (P_{Loss})_{S_1} = r_{S_1} \times (I_{rms})_{S_1}^2 = 0.05 \times 13.01^2 = 8.46 \text{ [W]} \\ (P_{Loss})_{S_2} = r_{S_2} \times (I_{rms})_{S_2}^2 = 0.05 \times 4.48^2 = 1.00 \text{ [W]} \end{cases} \tag{11.46}$$

Table 11.5 RMS value of the converter devices.

Device	RMS current	Device	RMS current
S_1	13.01 $[A]$	L_2	4.47 $[A]$
S_2	4.48 $[A]$	L_3	1.19 $[A]$
D_1	6.91 $[A]$	C_1	5.50 $[A]$
D_2	1.73 $[A]$	C_2	1.43 $[A]$
D_o	1.71 $[A]$	C_3	1.41 $[A]$
L_1	11.01 $[A]$	Input source	11.01 $[A]$

The diodes losses:

$$\begin{cases}(P_{Loss})_{D_1} = r_{D_1} \times (I_{rms})_{D_1}^2 = 0.1 \times 6.91^2 = 4.77 \text{ [W]} \\ (P_{Loss})_{D_2} = r_{D_2} \times (I_{rms})_{D_2}^2 = 0.1 \times 1.73^2 = 0.29 \text{ [W]} \\ (P_{Loss})_{D_o} = r_{D_o} \times (I_{rms})_{D_o}^2 = 0.2 \times 1.71^2 = 0.58 \text{ [W]}\end{cases} \tag{11.47}$$

The power losses related to inductors:

$$\begin{cases}(P_{Loss})_{L_1} = r_{L_1} \times (I_{rms})_{L_1}^2 = 0.05 \times 11.01^2 = 6.06 \text{ [W]} \\ (P_{Loss})_{L_2} = r_{L_2} \times (I_{rms})_{L_2}^2 = 0.08 \times 4.47^2 = 1.59 \text{ [W]} \\ (P_{Loss})_{L_3} = r_{L_3} \times (I_{rms})_{L_3}^2 = 0.1 \times 1.19^2 = 0.14 \text{ [W]}\end{cases} \tag{11.48}$$

The power losses of capacitors:

$$\begin{cases}(P_{Loss})_{C_1} = r_{C_1} \times (I_{rms})_{C_1}^2 = 0.01 \times 5.5^2 = 0.30 \text{ [W]} \\ (P_{Loss})_{C_2} = r_{C_2} \times (I_{rms})_{C_2}^2 = 0.01 \times 1.43^2 = 0.02 \text{ [W]} \\ (P_{Loss})_{C_3} = r_{C_3} \times (I_{rms})_{C_3}^2 = 0.01 \times 1.41^2 = 0.019 \text{ [W]}\end{cases} \tag{11.49}$$

Thus, the total power dissipated is:

$$\begin{aligned}P_{Loss} &= (P_{Loss})_{Switches} + (P_{Loss})_{Diodes} + (P_{Loss})_{Inductors} + (P_{Loss})_{Capacitors} \\ &= 23.23 \text{ } [W]\end{aligned} \tag{11.50}$$

The output power of the converter is calculated as follows:

$$P_{out} = P_{in} - P_{loss} = V_i I_i - P_{loss} = 24 \times 11.01 - 23.23 = 241.01\ [W] \quad (11.51)$$

Finally, the converter efficiency is calculated according to equation (11.52):

$$\eta = \frac{P_{out}}{P_{out} + P_{loss}} = \frac{241.01}{241.01 + 23.23} = 91\% \quad (11.52)$$

It should be noted that if low-loss semiconductors are used, this value will be much higher.

11.8 Dynamic Behavior of the Proposed Converter

In this section, using PSCAD software, the proposed converter is simulated as a closed-loop system with a PI controller. Details of the control mechanism are given in the block diagram in Figure 11.16.

It should be noted that the speed of dynamic response in tracking an input control signal or eliminating the effect of disturbance in a short time is one of the most important features of closed-loop electrical systems. The simulation is performed by applying the data in Table 11.6, which includes the values of the converter structure parameters and the carrier wave frequency. Also, according to the table, the converter should be able to follow the changes of the reference signal (from 200 to 100) and reach its final value and the steady-state error is zero. The value of proportional gain and integral factor related to the controller is also specified in this table according to the design.

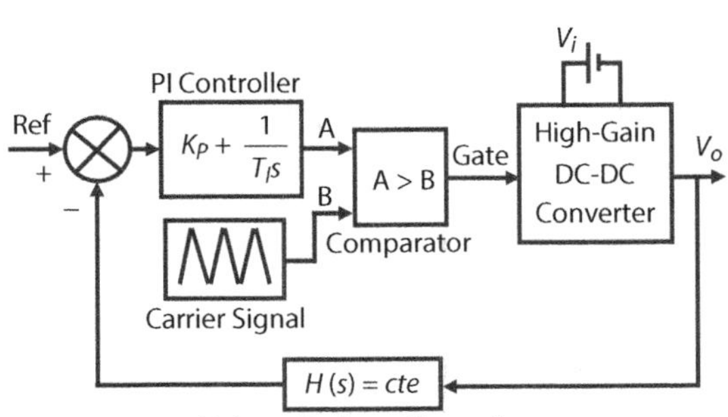

Figure 11.16 Block diagram of control mechanism to better utilization of the proposed converter.

Table 11.6 The parameter values of closed-loop simulation.

Parameter	Value
Input voltage (V_i)	24 V
Switching frequency (f)	40 kHz
L_1 ,L_2 ,L_3	200 ,300 ,500 μH
C_1 ,C_2 ,C_3 ,C_o	50 ,100 ,50 ,220 μF
Load (R)	200 Ω
Reference signal	200/100 V
Output voltage (V_o)	200/100 V
Output current (I_o)	1/0.5 A
K_P	0.0018
T_I	5.9 [sec]

A. Reference Signal Tracking

Figure 11.17a shows the response of the closed-loop system in response to changes in the reference signal. It can be seen that the system reaches its steady-state with just one overshoot and provides a voltage of 200 volts. But in $t = 0.3$ [sec], when the reference signal value decreases to 100 in step form, the system output also reaches 100 with an undershoot. Therefore, it can be understood that in this case, the converter shows good dynamic behavior. Also, during the system output changes, the converter operation duty cycle will be subject to changes as shown in Figure 11.17b and varied between 50% and 32%.

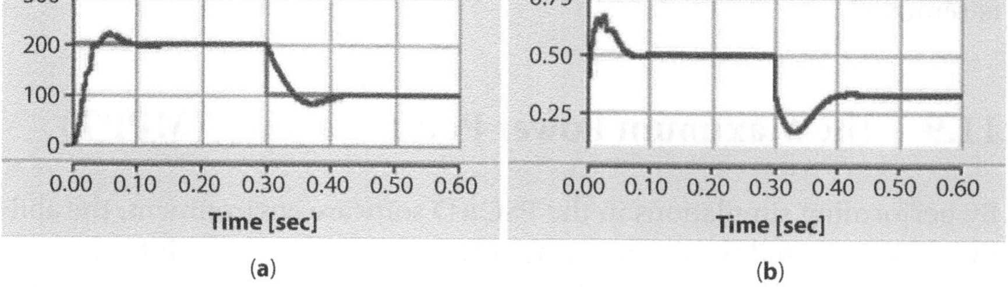

Figure 11.17 Variation of (a) output voltage and (b) duty cycle versus reference signal changes.

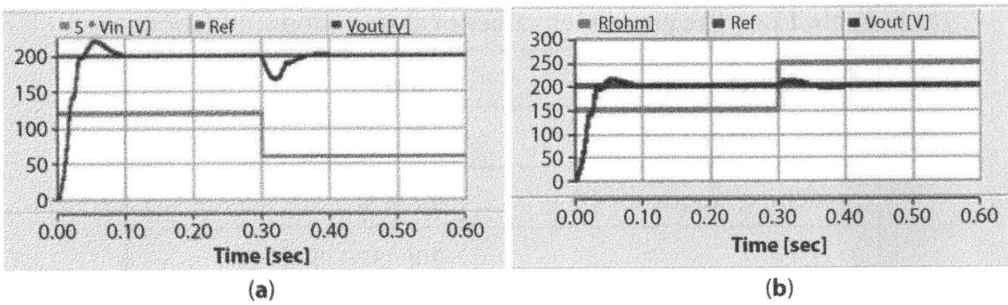

Figure 11.18 Dynamic behavior of the proposed converter due to the changes of (a) input voltage, (b) Load.

B. Eliminate the Effect of Disturbance

The major disturbances that may occur for high-gain DC-DC converters are input voltage fluctuations or output load changes. Usually, the DC voltage that is connected to these converters is the output of a solar panel or the rectified mode of the power grid. Climate and temperature changes lead to fluctuations in the output voltage of the solar panels [34]. Also, the voltage of the power grid is constantly fluctuating. Therefore, the high-gain converter with its closed-loop system must be able to eliminate the effects of these types of changes in a short time and return to the desired initial value. Output load changes occur when, for example, the converter supplies a DC motor. It is known that as a result of increasing and decreasing the mechanical load of the motor, the amount of output current of the DC-DC converter and consequently the amount of the load (V_o/I_o) changes [35].

Figure 11.18 shows how to eliminate the disturbance in the converter. In Figure 11.18a, the input voltage is initially 24 volts but decreases to 12 [V] in $t = 0.3$ [sec], and as can be seen, the output of the system first changes and after a short time it reaches its desired value. In Figure 11.18b, the output load changes from 150 [Ω] to 250 [Ω], and the output voltage returns to the desired value (200 [V]) after the disturbance effect is removed.

11.9 The Maximum Power Point Tracking (MPPT)

By performing simulations in the PSCAD software environment, the ability of the proposed converter to track the maximum power point is demonstrated. Each PV (solar panel) has a *P-V* and *P-I* characteristic curves that indicates the amount of voltage and current for which the maximum power is obtained. Under standard ambient conditions ($T = 25°C$, G = 1000[W/m^2]), a 120 watts solar panel produces its maximum power (120 [W]) at $V_{PV} = 30$ [V] and $I_{PV} = 4$ [A]. The P-V and P-I characteristic curves for this panel are shown in Figure 11.19.

To achieve the maximum power point, the perturb and observe (P&O) algorithm is implemented on the proposed converter, and its block diagram is shown in Figure 11.20.

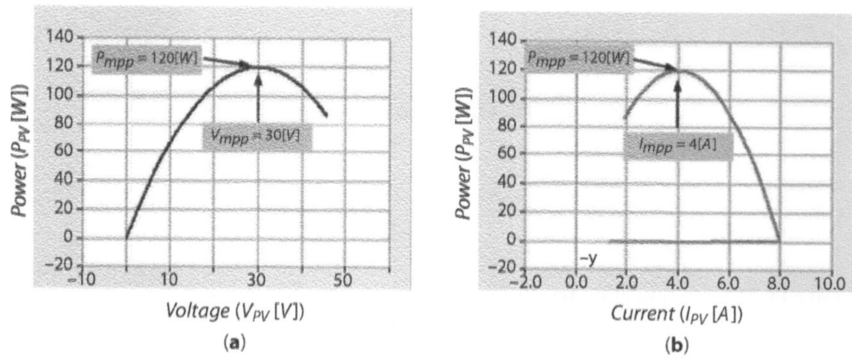

Figure 11.19 (a) P-V characteristic, and (b) P-I characteristic of a 120 [W] solar panel (PV).

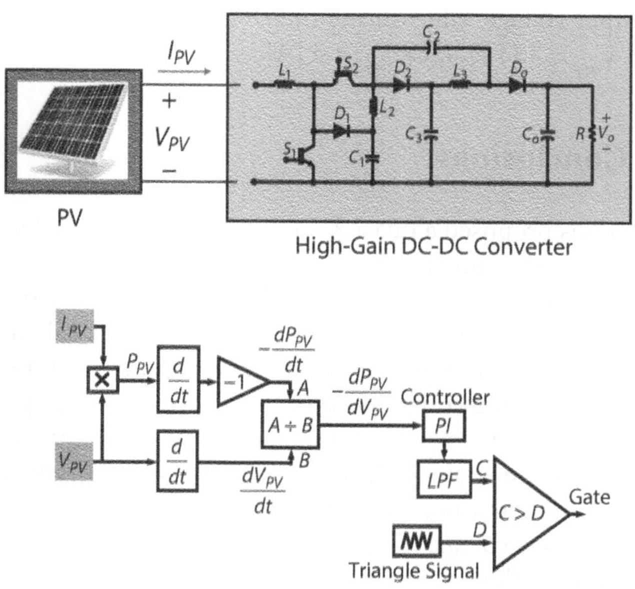

Figure 11.20 The connection between a photovoltaic panel and proposed structure with block diagrams of the P&O algorithm for MPP tracking.

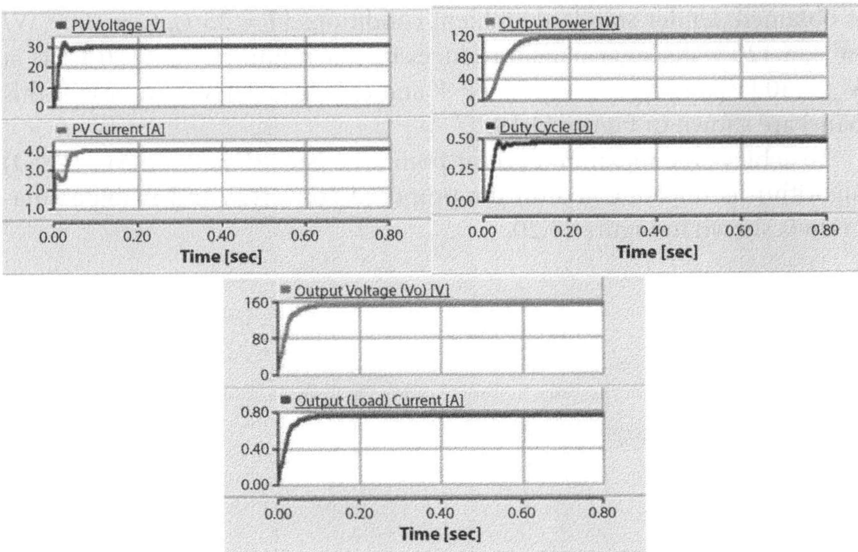

Figure 11.21 The simulation results for MPP tracking by the converter.

The simulation results are shown in Figure 11.21 in terms of time. *PI* controller coefficients are considered ($K_p = 1.5$, $K_i = 0.95$). As Figure 11.21 shows, the output power is approximately equal to 120 [*W*] (PV power) and the output voltage value during the $D = 45$ [%] is about 187 [*V*] and also: $V_{PV} = V_{mpp} = 30$ [*V*], $I_{PV} = I_{mpp} = 4$ [*A*].

11.10 Conclusions

This chapter has proposed a quasi Z-source-based quadratic DC-DC converter with large step-up ability. The suggested structure has continuous (low-ripple) input (source) current, which leads to its better performance in renewable energy applications. Using the steady-state calculations, the static gain of the converter as well as other key parameters of the converter have been obtained and compared with that of similar existing structures. The comparisons have been done through boosting capability, device count and voltage stress viewpoints. The outcomes certify the competitiveness of the proposed configuration. The voltage stress of the switching elements is less than the output-voltage which reduces switching losses. Also, the two-mode operation of the converter eases the control process. The operating principles of the converter in CCM and DCM have been fully investigated.

The theoretical equations have been approved by simulation (PSCAD/EMTDC) results. The real-gain has also been computed considering the losses of elements. The highest efficiency (96%) of the converter occurs at a load of approximately 240 $[W]$ ($D = 45\%$). This low duty-cycle moderates the losses. It should be noted that to confirm the equations of losses and efficiency, the numerical calculation section has been included. Dynamic analysis and load step-change response of closed-loop system has also been studied. The results verify appropriate dynamic performance (fast operation with low overshoot/undershoot) of the converter in tracking reference signal and eliminating possible disturbances. Finally, the MPPT ability of the converter has been certified by simulation.

11.11 Appendix

Definition of Symbols and Variables

V_i, V_o	Symbols of Input Voltage and Output Voltage
R	Load Resistance
L_1, L_2, L_3	Symbol of Inductors and Inductances
S_1, S_2	Switches of Converter
D_1, D_2, D_o	Diodes of Converter
C_1, C_2, C_3, C_o	Symbol of Capacitors and capacitances
t	Time [second]
D	Duty Cycle (a present of period time)
T	Period Time of switching [second]
$\Delta i_{L_1}, \Delta i_{L_2}, \Delta i_{L_3}$	Current Ripple of Inductors L_1, L_2, L_3
$I_{L_2}, I_{L_2}, I_{L_3}$	Average Current of Inductors L_1, L_2, L_3
f	Switching Frequency
$\Delta v_{C_1}, \Delta v_{C_2}, \Delta v_{C_3}, \Delta v_{C_o}$	Voltage Ripple of Capacitors C_1, C_2, C_3, C_o
r_{S_1}, r_{S_2}	The on-state Resistance of Switches S_1, S_2
$r_{D_1}, r_{D_2}, r_{D_o}$	The on-state Resistance of diodes D_1, D_2, D_o
$r_{L_1}, r_{L_2}, r_{L_3}$	The Winding Resistance of Inductors L_1, L_2, L_3
$r_{C_1}, r_{C_2}, r_{C_3}, r_{C_o}$	The series Resistance of Capacitors C_1, C_2, C_3, C_o
$V_{f_1}, V_{f_2}, V_{f_o}$	Forward Voltage Drop of Diodes D_1, D_2, D_o
η	Symbol of Converter Efficiency
K_P, T_I	Proportional and Integral Coefficients of *PI* Controller

V_{PV}, I_{PV}	Voltage and Current of solar panel
V_{mpp}, I_{mpp}, P_{mpp}	Voltage, Current, and Power of maximum power point
LPF	Low Pass Filter
V_{in}, V_{out}	Input and Output Voltage of Converter in Simulation

References

1. T. Taghiloo, K. Varesi, and S. Padmanaban, "A Large-Gain Continuous Input-Current DC-DC Converter Applicable for Solar Energy Systems," *Green Energy: Solar Energy, Photovoltaics, and Smart Cities,* pp. 345–367, 2020.
2. H. Mostafapour, K. Varesi, and S. Padmanaban, "A Developed Large Boosting Factor DC-DC Converter Feasible for Photovoltaic Applications," *Green Energy: Solar Energy, Photovoltaics, and Smart Cities,* pp. 515–548, 2020.
3. S. A. Gorji, A. Mostaan, H. T. My, and M. Ektesabi, "Non-isolated buck–boost dc–dc converter with quadratic voltage gain ratio," *IET Power Electronics,* vol. 12, pp. 1425–1433, 2019.
4. K. Varesi, S. H. Hosseini, M. Sabahi, and E. Babaei, "A high-voltage gain nonisolated noncoupled inductor based multi-input DC-DC topology with reduced number of components for renewable energy systems," *International Journal of Circuit Theory and Applications,* vol. 46, pp. 505–518, 2018.
5. P. Ma, W. Liang, H. Chen, Y. Zhang, and X. Hu, "Interleaved high step-up boost converter," *Journal of Power Electronics,* vol. 19, pp. 665–675, 2019.
6. W. Hassan, D. D.-C. Lu, and W. Xiao, "Single-Switch High Step-Up DC–DC Converter With Low and Steady Switch Voltage Stress," *IEEE Transactions on Industrial Electronics,* vol. 66, pp. 9326–9338, 2019.
7. M. Forouzesh, Y. Shen, K. Yari, Y. P. Siwakoti, and F. Blaabjerg, "High-efficiency high step-up DC–DC converter with dual coupled inductors for grid-connected photovoltaic systems," *IEEE Transactions on Power Electronics,* vol. 33, pp. 5967–5982, 2017.
8. Z. Xiaoquan and B. Zhang, "High step-up quasi-Z-source DC–DC converters with single switched capacitor branch," *Journal of Modern Power Systems and Clean Energy,* vol. 5, pp. 537–547, 2017.
9. J. Liu, J. Wu, J. Qiu, and J. Zeng, "Switched Z-source/quasi-Z-source DC-DC converters with reduced passive components for photovoltaic systems," *IEEE Access,* vol. 7, pp. 40893–40903, 2019.
10. H. Ardi, R. R. Ahrabi, and S. N. Ravadanegh, "Non-isolated bidirectional DC–DC converter analysis and implementation," *IET Power Electronics,* vol. 7, pp. 3033–3044, 2014.

11. H. Ardi, A. Ajami, F. Kardan, and S. N. Avilagh, "Analysis and implementation of a nonisolated bidirectional DC–DC converter with high voltage gain," *IEEE Transactions on Industrial Electronics,* vol. 63, pp. 4878–4888, 2016.
12. S. Ghabeli Sani, F. Mohammadi, M. R. Banaei, and M. Farhadi-Kangarlu, "Design and implementation of a new high step-up DC-DC converter for renewable applications," *International Journal of Circuit Theory and Applications,* vol. 47, pp. 464–482, 2019.
13. A. Alzahrani, M. Ferdowsi, and P. Shamsi, "High-voltage-gain DC–DC step-up converter with bifold dickson voltage multiplier cells," *IEEE Transactions on Power Electronics,* vol. 34, pp. 9732–9742, 2019.
14. G. Lin and Z. Zhang, "Low Input Ripple High Step-Up Extendable Hybrid DC-DC Converter," *IEEE Access,* vol. 7, pp. 158744–158752, 2019.
15. B. Zhu, S. Liu, D. M. Vilathgamuwa, and Y. Li, "High step-up SVMC-based DC/DC converter for offshore wind farms," *IET Power Electronics,* vol. 12, pp. 1445–1454, 2019.
16. G. Zhang, H. Chen, S. Yu, B. Zhang, Y. Chen, H. Iu, *et al.*, "An extendable single-switch n-cell boost converter with high voltage gain and low components stress for renewable energy," *International Journal of Circuit Theory and Applications,* vol. 48, pp. 817–831, 2020.
17. K. Varesi, S. H. Hosseini, M. Sabahi, E. Babaei, S. Saeidabadi, and N. Vosoughi, "Design and Analysis of a Developed Multiport High Step-Up DC–DC Converter With Reduced Device Count and Normalized Peak Inverse Voltage on the Switches/Diodes," *IEEE Transactions on Power Electronics,* vol. 34, pp. 5464–5475, 2018.
18. P. Mohseni, S. H. Hosseini, M. Sabahi, T. Jalilzadeh, and M. Maalandish, "A new high step-up multi-input multi-output DC–DC converter," *IEEE Transactions on Industrial Electronics,* vol. 66, pp. 5197–5208, 2018.
19. K. Varesi and A. G. Esfahlan, "An Enhanced-Gain Quadratic-Boost DC-DC Configuration," in *2020 28th Iranian Conference on Electrical Engineering (ICEE)*, 2020, pp. 1–5.
20. Y. Zhang, H. Liu, J. Li, M. Sumner, and C. Xia, "DC–DC boost converter with a wide input range and high voltage gain for fuel cell vehicles," *IEEE Transactions on Power Electronics,* vol. 34, pp. 4100–4111, 2018.
21. K. Varesi and M. Ghorbani, "A generalized common-ground single-switch continuous input-current boost converter favourable for DC microgrids," *International Journal of Circuit Theory and Applications,* vol. 48, pp. 1658–1675, 2020.
22. K. Varesi, A. A. Gandomi, and S. Saeidabadi, "A new multi-phase high step-up DC-DC converter appropriate for PV applications," in *2019 Iranian Conference on Renewable Energy & Distributed Generation (ICREDG)*, 2019, pp. 1–6.
23. K. Varesi, "A Novel Extra High Gain Non-Isolated DC-DC Converter With Continuous Input/Output Current," in *2019 Iranian Conference on Renewable Energy & Distributed Generation (ICREDG)*, 2019, pp. 1–6.

24. L. Liu, D. Li, and L. Yao, "Non-isolated high step-up DC-DC conversion circuits for photovoltaic system," *International Transactions on Electrical Energy Systems*, vol. 30, p. e12192, 2020.
25. H. Ye, G. Jin, W. Fei, and N. Ghadimi, "High step-up interleaved dc/dc converter with high efficiency," *Energy Sources, Part A: Recovery, Utilization, and Environmental Effects*, pp. 1–20, 2020.
26. Y. Zhang, Q. Liu, Y. Gao, J. Li, and M. Sumner, "Hybrid switched-capacitor/switched-Quasi-Z-source bidirectional DC–DC converter with a wide voltage gain range for hybrid energy sources EVs," *IEEE Transactions on Industrial Electronics*, vol. 66, pp. 2680–2690, 2018.
27. X. Wu, W. Shi, and J. Du, "Dual-Switch Boost DC–DC Converter for Use in Fuel-Cell-Powered Vehicles," *IEEE Access*, vol. 7, pp. 74081–74088, 2019.
28. M. Karthikeyan, R. Elavarasu, P. Ramesh, C. Bharatiraja, P. Sanjeevikumar, L. Mihet-Popa, ., "A Hybridization of Cuk and Boost Converter Using Single Switch with Higher Voltage Gain Compatibility," *Energies*, vol. 13, p. 2312, 2020.
29. K. Varesi, N. Hassanpour, and S. Saeidabadi, "Novel high step-up DC–DC converter with increased voltage gain per devices and continuous input current suitable for DC microgrid applications," *International Journal of Circuit Theory and Applications*, 2020.
30. K. Varesi, S. H. Hosseini, M. Sabahi, and E. Babaei, "A multi-port high step-Up DC-DC converter with reduced normalized voltage stress on switches/diodes," in *2018 9th Annual Power Electronics, Drives Systems and Technologies Conference (PEDSTC)*, 2018, pp. 1–6.
31. K. Varesi, A. A. Ghandomi, S. Hosseini, M. Sabahi, and E. Babaei, "An improved structure for multi-input high step-up DC-DC converters," in *2017 8th Power Electronics, Drive Systems & Technologies Conference (PEDSTC)*, 2017, pp. 241–246.
32. R. W. Erickson and D. Maksimovic, *Fundamentals of power electronics*: Springer Science & Business Media, 2007.
33. M. K. Kazimierczuk, *Pulse-width modulated DC-DC power converters*: John Wiley & Sons, 2015.
34. A. G. Esfahlan, K. Varesi, "A New High Step-Up DC-DC Converter Based on Impedance Network," in *2021 12th Annual Power Electronics, Drives Systems and Technologies Conference (PEDSTC)*, 2021 (In Press).
35. N. Hassanpour and K. Varesi, "A New Non-Isolated High Gain DC-DC Converter Suitable for Renewable Energies," in *Power Electronics, Drives Systems and Technologies Conference (PEDSTC), 2019 10th Annual, Shiraz, Iran*, 2019, pp. 1–6.

12

Research on Protection Strategy Utilizing Full-Scale Transient Fault Information for DC Microgrid Based on Integrated Control and Protection Platform

Shi Bonian* **and Sun Gang**

R&D Center, Beijing Sifang Automation Company, Beijing, China

Abstract

With the increasing demand for better power quality and more reliable power supply, a Voltage Source Converter (VSC) based DC microgrid with simple structure, high efficiency, and high reliability has gained much attention in recent years. To guarantee the safe and reliable operation of a DC microgrid, control and protection is the key problem, which need to be solved urgently. In this chapter, a typical multi-terminal VSC DC microgrid for study is proposed, with careful consideration of topology design and grounding mode selection. Then the DC fault characteristics of a multi-terminal DC microgrid is analyzed, including DC unipolar fault and DC bipolar fault, and is utilized as the theoretical basis for protection configuration. Further, the protection zone division principle and protection configuration method for the studied multi-terminal DC microgrid are provided, which can help accurately locate the faults on the DC microgrid and achieve fast fault isolation and recovery. An integrated control and protection platform is introduced, on which a layered setting of protection action delay and some supplementary control measures, such as deblocking the VSCs and reclosing the DC breakers in the non-fault area, can be implemented. This integrated control and protection strategy can guarantee the selectivity of protection action, minimize the affected range by fault and quickly recover the power supply of a DC microgrid. A large amount of test results under the RT-LAB environment have verified the effectiveness of the integrated control and protection strategy.

Keywords**:** DC microgrid, Voltage Source Converter (VSC), unipolar fault, bipolar fault, fault location, integrated control and protection platform

**Corresponding author*: shibonian@sf-auto.com

Nikita Gupta, Mahajan Sagar Bhaskar, Sanjeevikumar Padmanaban and Dhafer Almakhles (eds.) DC Microgrids: Advances, Challenges, and Applications, (361–396) © 2022 Scrivener Publishing LLC

12.1 Introduction

With the increasing demand for better power quality and more reliable power supply, a Voltage Source Converter (VSC) based DC microgrid with simple structure, high efficiency, and high reliability has gained attention in recent years [1–3]. Incorporating Distributed Generators (DGs) and DC loads such as electric vehicles (EVs) to a DC microgrid can save lots of converters and a large amount of power loss.

Compared with the HVDC system, a DC microgrid may include different types of converters, AC distribution systems, DC lines, distributed generations, energy storage and various kinds of loads. Failure of any part will affect the operation reliability of the whole DC microgrid and the safety of relevant equipment. The performance of protection for the DC microgrid is the key to its safe and reliable operation [4, 5].

Recently Modular Multilevel Converter (MMC) has been developed as a novel topology of VSC [6]. Since MMC technology has advantages such as low harmonics, high efficiency, low step voltage change rate, low switching frequency and modularity, an MMC-based DC microgrid will gain more and more recognition in future application. Although the half-bridge modular multi-level converter (HB-MMC) has many advantages, it cannot interrupt DC fault current due to the free-wheeling diodes within the submodules. The full-bridge modular multi-level converter (FB-MMC) has the DC faults isolation capability, but it needs double switching devices compared with the HB-MMC and leads to high project investment. Therefore, if the HB-MMC structure is selected, to solve the problem that HB-MMC is incapable of isolating DC fault current, DC circuit breakers are necessary to be adopted. Since the fault characteristics of a DC microgrid displays fast change speed, high amplitude and strong nonlinearity, it is urgent to study the fault mechanism analysis, fault identification and location, as well as the setting coordination of all levels of protection [7].

The remainder of this chapter will be organized as follows: in Section 12.2, the topology and grounding mode of the studied DC microgrid are introduced. Then the DC fault types and characteristics are presented in Section 12.3. After that, DC fault clearing and system restoration strategies are elaborated in Section 12.4. Performance verification via RT-LAB simulation tests is illustrated in Section 12.5. Finally, conclusions are provided in Section 12.6.

12.2 Topological Structure and Grounding Model of Studied Microgrid

12.2.1 Proposed DC Distribution Network Topology

Generally, the topological structure of a DC microgrid can be divided into three types: radial, hand-in-hand and ring network [8]. In the radial network, there is only one line to supply power for any load, so the reliability of the power supply is relatively low. For the hand-in-hand and ring topology, any load has two lines of the power supply, so the power supply reliability is higher. However, it is relatively difficult to identify faults and coordinate control and protection. A DC microgrid with ring topology is more generalized, since it can be divided into two terminal or radial structure. Therefore, this chapter takes the ring DC microgrid topology as the study object to conduct research. The ring topology of the studied multi-terminal DC microgrid is shown in Figure 12.1.

As shown in Figure 12.1, converters VSC1~VSC3 are connected to the 10kV AC distribution network. Converter VSC4 is connected to a microgrid composed of AC load and a synchronous generator. DC transformers DCT1~DCT3 are connected to the AC load, wind power and energy storage, photovoltaic power generation and DC load in sequence. Compared with two- or multiple-level voltage source converters, MMC have advantages in technology development trends, AC and DC harmonics, losses, reliability and flexibility [9, 10]. Therefore, the converters VSC1~VSC4 adopt the form of MMC. The voltage parameters and capacity of the major equipment connected to each converter are shown in Table 12.1.

The converters shown in Figure 12.1 are further divided into the following four types:

1. Type 1, such as VSC1, VSC2 and VSC3, which connects with the medium-voltage AC power grid or AC microgrid. The device can complete the AC/DC or DC/AC conversion with bi-directional power flow capability. It can operate in DC voltage control mode, active power and reactive power control mode.
2. Type 2, such as VSC4, which connects with the independent AC power generation equipment like some small synchronous generators. It can complete AC/DC conversion and only have unidirectional power flow capability.

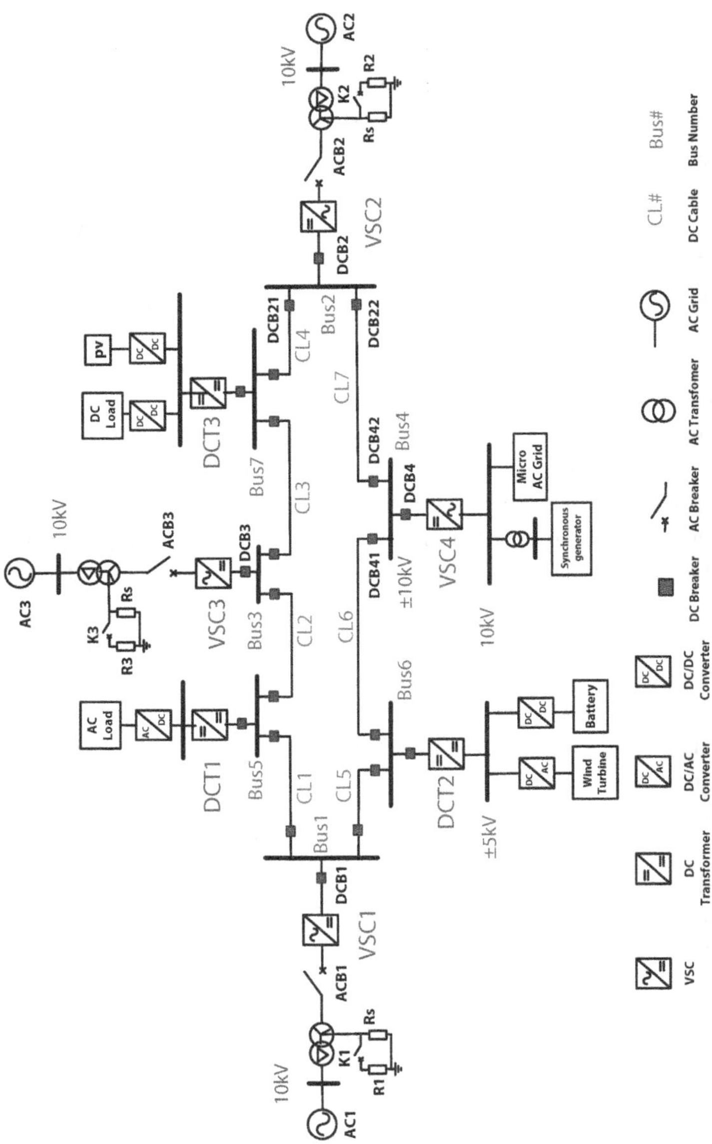

Figure 12.1 Ring topology of multi-terminal DC microgrid.

Table 12.1 Parameters of major components for the studied microgrid.

No.	Name of the converters	Voltage on both sides of the VSCs	Capacity	Equipment connected to the converters
1	VSC1	10kVac/±10kVdc	10MVA	AC distribution network
2	VSC2	10kVac/±10kVdc	10MVA	AC distribution network
3	VSC3	10kVac/±10kVdc	10MVA	AC distribution network
4	VSC4	380Vac/10kVac 380Vac/10kVac	5MW 1MW	AC microgrid/ Synchronous generator
5	DCT1	380Vac/±5kVdc	8MW	AC load
6	DCT2	0.6kVac/±5kVdc 1.5kVdc/±5kVdc	1MW 2MW*1h	Wind power/ battery energy storage
7	DCT3	400Vdc/±5kVdc 1kVdc/±5kVdc	7MW 6MW	DC Load/ Photovoltaic

3. Type 3, such as DCT1, which connects with the DC microgrid. It can have unidirectional power flow capability, and usually operates in Low DC side constant voltage control mode and active power control mode.
4. Type 4, such as DCT2 and DCT3, which connects with wind turbines, photovoltaics and storage battery to the DC microgrid. It has bi-directional power flow capability, and can operate in High DC side constant voltage control mode, Low DC side constant voltage control mode and active power control mode.

As shown in Figure 12.1, the lengths of DC cable lines CL1~CL7 are 2km, 1.5km, 2km, 4km, 1km, 2km and 5km, respectively. Both terminals of each line and converters are configured with hybrid DC circuit breakers [11]. The zero-crossing point of DC fault current limits the application of DC circuit breakers in the field of high-voltage and large-capacity DC transmission [12]. However, it has been partially applied in medium and low-voltage DC microgrids [13]. The tripping time of medium-voltage hybrid DC circuit breakers is around 3ms [14, 15].

12.2.2 Neutral Grounding Model

12.2.2.1 Grounding Position Selection

The grounding point of a DC microgrid usually has two choices: DC side grounding and AC side grounding, as shown in Figure 12.2. Two-level or three-level VSC has a concentrated capacitor in DC side, so it can be used to directly lead out the grounding branch. However, for the system with MMC topology, there is no centralized capacitance and no natural neutral point; the grounding point needs to be constructed artificially.

If there is a neutral point on the valve side of the converter transformer, it can ground directly through a resistance, i.e., AC side grounding mode 1, as shown in Figure 12.2. If there is no neutral point on the valve side, it can ground through a resistance by configuring a star-shaped reactance, i.e., AC side grounding mode 2. In DC side grounding mode, large resistance can be used to lead out the grounding branch, as shown in Figure 12.2. Considering the recovery time of the single-pole short-circuit fault and steady-state power loss, AC side grounding mode 1 is recommended.

12.2.2.2 Grounding Mode Selection

If high resistance grounding mode is selected, the unbalanced current of the DC microgrid system is small, but overvoltage will be generated. When a single pole to ground fault occurs, the fault current is effectively limited due to the large grounding resistance (equivalent to open circuit). However, since the VSC still controls the DC voltage to the rated voltage, the maximum non-fault pole to ground voltage will increase to twice the normal working voltage.

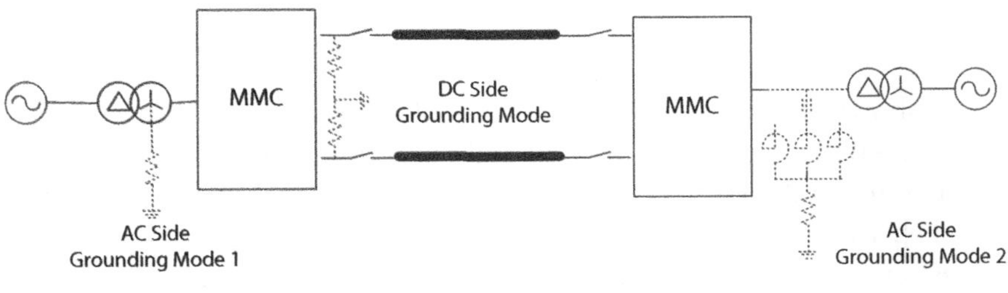

Figure 12.2 Schematic diagram of MMC-based DC microgrid grounding position.

When a single-pole grounding short-circuit fault occurs, and the grounding resistance is very small (equivalent to a short circuit here), large current will be generated. Since this high-current grounding system does not generate overvoltage, and the insulation monitoring devices may not be installed, then the circuit breaker must be quickly tripped to remove the fault.

Limited by the development level of DC circuit breakers, most flexible DC transmission systems in actual projects currently choose high resistance grounding methods to limit the level of fault short-circuit current. In order to save the cost of insulation coordination, there is no short-time operation mode with faults. Once a fault occurs, the entire DC system will be immediately shut down to prevent the system from over-voltage operation. However, for the medium-voltage DC microgrid, the DC voltage level is low and the insulation coordination cost is not high. High-resistance grounding can be used, so that the system can still operate for a short period of time in the case of a single-pole fault to improve system reliability.

Based on the above analysis, this chapter selects the AC side grounding mode 1, as shown in Figure 12.2, with the grounding resistance $R_S = 2000\Omega$ for the following study.

12.3 Fault Characteristics of DC Microgrid

The possible faults occurring on the DC microgrid may include external AC fault, DC line fault, DC bus fault, and load-side equipment fault [16]. The common fault types are listed as follows:

i. Short-circuit faults of converter transformer, AC connection line and start-up circuit;
ii. Voltage source converter internal fault, including bridge arm short circuit, etc.;
iii. Failure of DC side equipment, such as DC breaker and DC reactor;
iv. DC line and DC bus fault, such as DC line or DC bus unipolar fault, DC line or DC bus bipolar fault, open circuit fault, etc.;
v. Fault on load side equipment, such as DGs and EVs connected with the microgrid.

Due to the small damping of the DC microgrid, the fault propagation speed is fast and the amplitude is large. In order to protect the safety of power electronics and other related equipment in the DC microgrid, stricter requirements are put forward for the reliability and operation speed of protection.

12.3.1 DC Unipolar Fault Characteristics

When DC unipolar fault occurs, the magnitude of fault current is related to DC voltage amplitude, system grounding resistance and transition resistance and so on [17]. There is no centralized capacitor on the DC side of MMC, so the capacitor of sub-module (SM) will not produce large discharge surge current during the fault, and the capacitor voltage is basically stable. Fault current is the superposition of AC fault current injected through bridge arm, capacitance of bridge arm sub-module and equivalent capacitance discharge current of lines on both sides of fault point. Fault current path is shown in dotted line in Figure 12.3, where: u_s is the phase voltage value of AC system; L is the equivalent inductance value of the system considering the bridge arm reactors; R_s is the system grounding resistance value; R_f is the fault transition resistance; SMn is the converter sub-module (SM) and n is the number of each bridge arm module; C_0 is the capacitance value in a sub-module; C_1 is the equivalent capacitance value taking into account the DC line ground capacitance and stray capacitance; T_1 and T_2 are the converter valve; D_1 and D_2 are the freewheeling diodes; U_c is the voltage at both ends of the capacitor

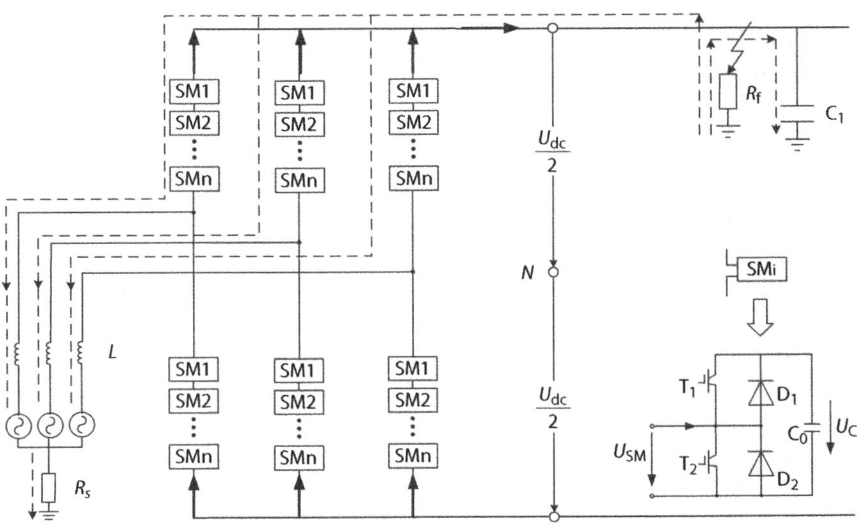

Figure 12.3 Fault current path at MMC side under unipolar fault.

of the sub module; U_{SM} is the voltage of converter sub-module; N is the neutral reference point; U_{dc} is the voltage between the poles of DC line.

The equivalent circuit for fault current path is shown in Figure 12.4, where m is the number of sub-modules put into the upper bridge arm at the time of fault.

If the DC line impedance is neglected, the fault transient discharge current can be expressed as the superposition of the capacitor discharge current of the bridge arm sub-module $i_f(t)$, and set $i_{c1}(t)$ as the stray capacitor discharge current, then

$$i_f(t) = i_{c0}(t) + i_{c1}(t) \tag{12.1}$$

Where, $i_{c0}(t)$ is the discharge current of RLC second-order circuit, and is derived from

$$\begin{cases} L\dfrac{C_0}{m}\dfrac{d^2U_c(t)}{dt^2} + R_0\dfrac{C0}{m}\dfrac{dU_c(t)}{dt} + U_c(t) = 0 \\ i_{c0}(t) = \dfrac{C_0}{m}\dfrac{dU_c(t)}{dt} \end{cases} \tag{12.2}$$

Where, the total resistance of the discharge circuit $R_0 = R_s + R_f$.
For the discharge current of RC circuit $i_{c1}(t)$, it is satisfied by

$$\begin{cases} R_fC_1\dfrac{dU_{c1}(t)}{dt} + U_{c1}(t) = 0 \\ i_{c1}(t) = C_1\dfrac{dU_{c1}(t)}{dt} \end{cases} \tag{12.3}$$

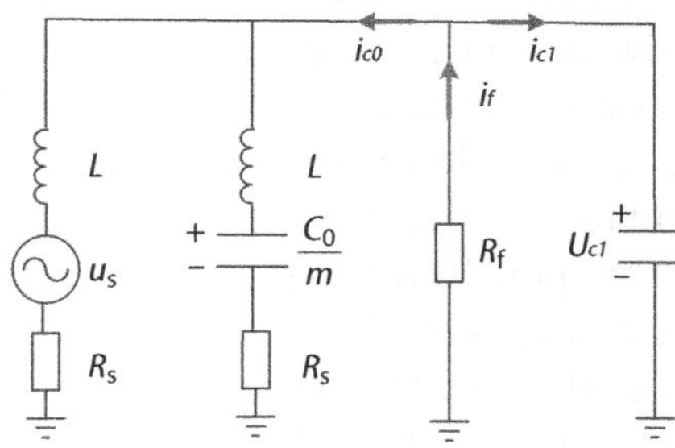

Figure 12.4 Equivalent circuit at MMC side under unipolar fault.

Where, $U_{c1}(t)$ is the equivalent voltage of stray capacitor under fault.

After the capacitor discharge process is finished, the fault steady-state current is approximately expressed as follows:

$$I_{fst} = \frac{U_{\mathrm{fDC}}}{R_{\mathrm{s}} + R_{\mathrm{f}}} \tag{12.4}$$

Where, U_{fDC} is the equivalent voltage of the bridge arm of the fault pole and can be approximately taken as $U_{\mathrm{fDC}} = 0.5U_{\mathrm{dc}}$. In case of metal short-circuit fault, set $R_{\mathrm{f}} = 0$.

Due to the clamping effect of the large ground resistance, the unipolar fault only changes the position of the DC system voltage reference point. The imbalance of the upper and lower bridge arm capacitance will cause the faulty polar voltage to be biased, and the DC voltage between the electrodes remains unchanged. Under this case, the power still can be delivered between MMCs, but the voltage of the pole under fault drops to zero and the voltage of the non-fault pole doubles. Therefore, if the insulation requirement is satisfied, the DC microgrid can run under fault for a short period. After the fault is cleared, because the converter transformer with Δ/Yn connection provides the charging circuit after the fault, the voltage between the positive and negative poles of the converter can quickly return to normal. This feature is very beneficial to the transient fault recovery of the DC line.

12.3.2 DC Bipolar Fault Characteristics

DC bipolar short circuit is a very serious fault in a DC microgrid. Fault overcurrent will trigger the protection of the converter itself and turn off the IGBT. The DC fault current will still exist even if the converter valve is locked due to the freewheeling effect of diodes in the sub-module of MMC [18]. The short-circuit current path of a typical DC power distribution DC cable when the positive and negative poles are short-circuited, take AC phase A and phase B as an example, as shown in Figure 12.5. The blue arrow indicates the discharge current flow of the sub-module capacitor before the IGBT is turned off, and the red arrow indicates the current flow injected by the AC system to the fault point through the freewheeling diode.

Before the converter valve is blocked, the fault current is composed of the discharge current of the parallel capacitor of the sub-module and the current fed into the AC system. After the converter is blocked, the fault current is mainly the steady-state fault current fed from the AC side.

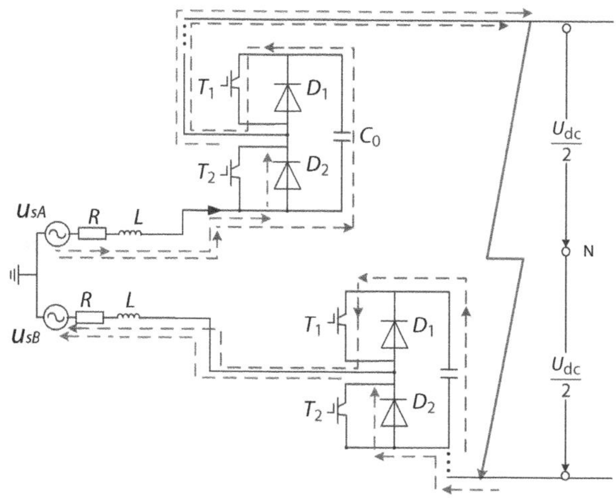

Figure 12.5 Fault current path under bipolar fault.

The discharge process of blocking the front bridge arm capacitance is equivalent to the second-order oscillation process formed by the bridge arm capacitance and the equivalent impedance of the system. The magnitude of the fault current depends on the number of sub-modules of the bridge arm, the parameters of the bridge arm capacitance, and the equivalent impedance of the system. Detailed analysis is given as follows.

1. Transient discharge process of DC capacitor before MMC being blocked
The DC capacitor forms an RLC discharge circuit through the fault point, as shown in Figure 12.6. Usually the line resistance is much smaller than the inductance and capacitance value. The discharge process can be regarded as a second-order oscillating discharge process. The capacitor discharge current $i_c(t)$ is expressed as [19]:

$$\begin{cases} i_c(t) = e^{-\frac{t}{\tau}}\left[\sqrt{\frac{C_0}{nL}U_{dc}^2 + I_0^2}\,\sin(\omega' t + \beta)\right] \\ \beta = \arctan\left(\frac{I_0}{U_{dc}}\sqrt{\frac{nL}{C_0}}\right) \end{cases} \tag{12.5}$$

Where: τ is the decay time constant of discharge current; ω' is the oscillating discharge angular frequency; n is the number of bridge arm module blocks; I_0 is the current value at the beginning of discharge in the circuit.

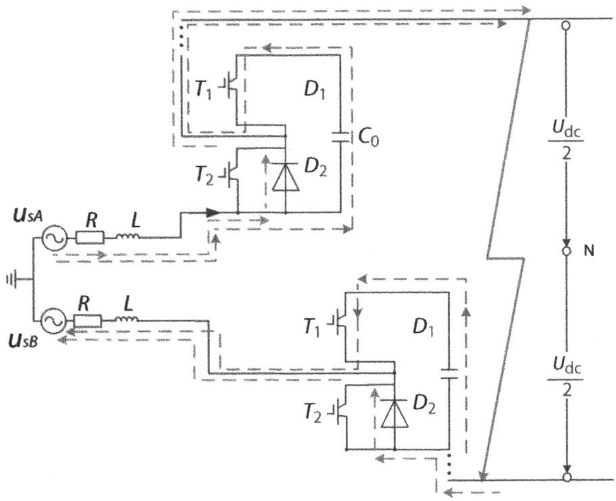

Figure 12.6 Fault current path under bipolar fault before MMC being blocked.

It can be seen from the above equation that there is a maximum value of the discharge current, which is directly related to the capacitance C_0 of the sub-module, the equivalent reactor of the circuit L and the circuit current I_0 at the fault time. The amplitude of the discharge current can be reduced by increasing it, and the arrival time of the maximum value of the discharge current can be delayed.

2. AC steady-state fault current feeding process after MMC being blocked
As long as the AC circuit breaker does not trip, the steady-state fault current fed from the AC grid through flywheel diode will always exist, the fault characteristics are equivalent to three-phase short circuit of AC system. The fault path circuit after MMC being blocked is shown in Figure 12.7.

Set the voltage expression of AC grid phase A as $u_{sA}(t) = U_m \sin(\omega t + \phi_0)$, where U_m is the peak value of the AC phase voltage, ω is the rated angular frequency of AC system, and ϕ_0 is the phase angle of the AC system at the time of the fault. Then, the steady-state fault phase current provided by the AC grid can be approximately expressed as:

$$i_{sA}(t) = \frac{U_m}{\sqrt{R^2 + (\omega L)^2}} \sin(\omega t + \phi_1) \tag{12.6}$$

Where, R and L are the equivalent resistance and reactance of the external AC grid; ϕ_1 is the phase angle of the AC system after the DC microgrid begins the new steady-state operation with the still-existing fault.

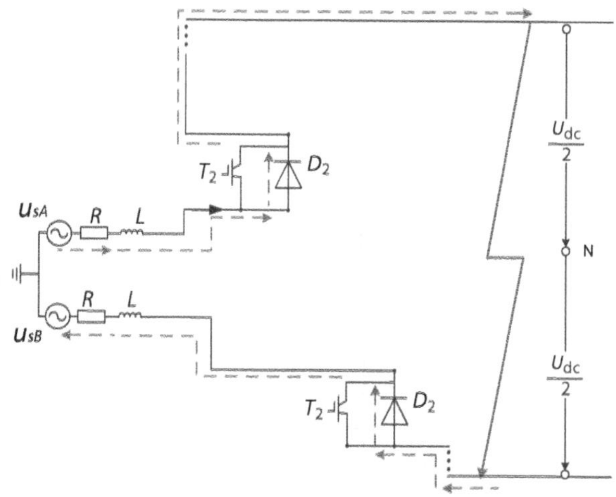

Figure 12.7 Fault current path under bipolar fault after MMC being blocked.

It can be seen from the above equation that the amplitude of short-circuit current can be limited by increasing the system impedance, equivalent impedance of converter transformer or inductance of bridge arm reactor.

In summary, the bipolar fault current rises rapidly and has large amplitude before blocking. Even after the converter is blocked, there is still fault current equivalent to three-phase short circuit in the circuit. If the fault is not isolated in time, the whole DC microgrid will be seriously damaged.

12.4 DC Microgrid Protection Strategy

12.4.1 Protection Zone Division and Protection Configuration

12.4.1.1 Protection Zone Division

The division of the Protection zone must be realized only if the protection function can reflect all the faults in all operating modes and realize the protection function without dead zone. The protection zone is divided according to the function of the protected object and the electrical bay of the protection zone. In the area with similar structure and components, the protection zone should be reproducible as much as possible to facilitate the configuration of protection.

Generally, the fault areas can be divided into AC power grid fault area, converter internal fault area, DC line side fault area and DC microgrid load area. The specific division of protection zone for the studied DC microgrid is shown in Figure 12.8.

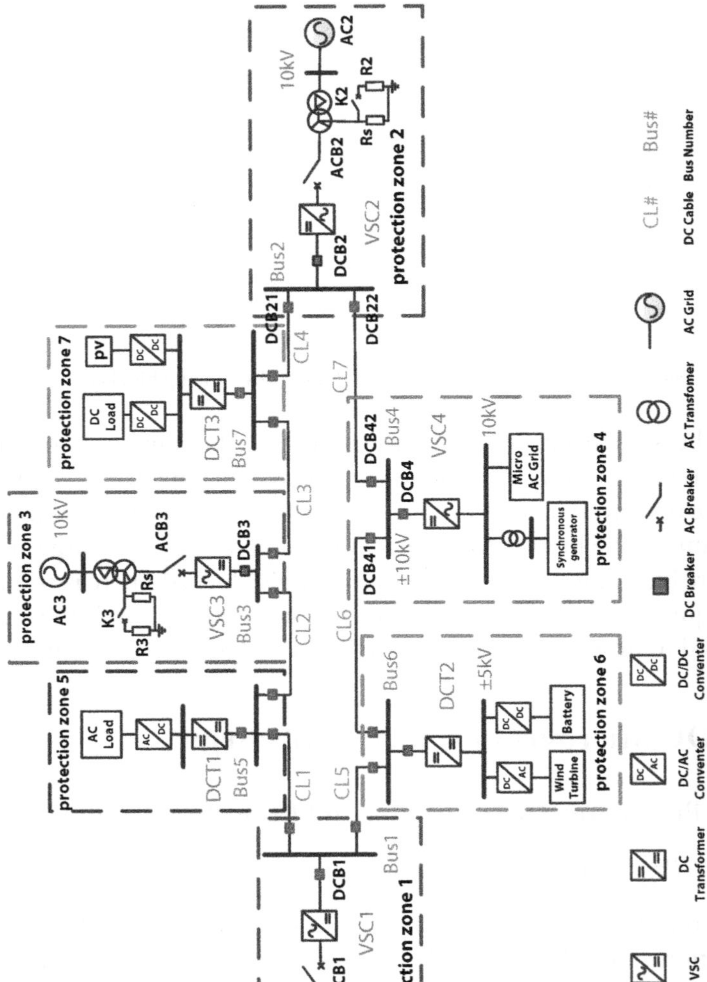

Figure 12.8 Schematic diagram of protection zone division.

According to the above principles, with the goal of simplifying the protection configuration and having high reproducibility in the protection zone, the protection zones are divided as follows:

1. The main VSC and its associated AC system are set as a single protection zone, like the protection zone 1, protection zone 2 and protection zone 3, as shown in Figure 12.8. Each of these protection zones respectively consists of converter VSC1/VSC2/VSC3 and the auxiliary switchgear, as well as the associated AC system components (connecting transformer, AC line, AC line connecting DC side of transformer to converter, etc.).
2. The protection zone 4 is set for the VSC4. It includes converter VSC4, DC bus, DC circuit breaker, synchronous generator, transformer and other auxiliary equipment, etc.
3. The protection zone 5 is set for the unidirectional DC transformer DCT1. It includes DC transformer DCT1, DC bus, DC circuit breaker, AC load and other ancillary equipment, etc.
4. The protection zone 6 is set for the bidirectional DC transformer DCT2. It includes DC transformer DCT2, high-voltage DC bus and switches, low-voltage DC bus, wind turbine, energy storage and other auxiliary equipment, etc.
5. The protection zone 7 is set for the bidirectional DC transformer DCT3. It includes DC transformer DCT3, high-voltage DC bus and switch, low-voltage DC bus, photovoltaic (PV), DC load and other auxiliary equipment, etc.

12.4.1.2 Protection Configuration

With the above divided protection zones for the studied DC microgrid, the basic principal protection configuration is developed as: After meeting the requirements for the safety of equipment, according to specific application requirements and power grid conditions, protection schemes are determined to be suitable for different operating modes, topologies and power transfer levels. It should minimize the scope of outages, shorten the outage time, and improve the reliability of the DC microgrid.

1. The protection configuration for the DC converter, such as VSC1~VSC4, is mainly used to protect the safety of internal converter equipment, and is set as:

i. bridge arm over-current protection;
ii. bridge arm differential protection;
iii. valve DC over-current protection;
iv. bridge arm reactor differential protection;

2. The protection configuration for the DC transformer, such as DCT1~DCT3, is mainly over-current protection. The over-current protection can serve as backup protection for the load and subsidiary microgrid which are connected with the DC transformer.
3. The protection configuration for the DC line and DC bus is set as:
   i. DC voltage unbalance protection;
   ii. DC low voltage over-current protection;
   iii. DC line differential protection;
   iv. DC bus differential protection.

With the above protection zone division and protection configuration scheme, the DC microgrid with different topologies can be flexibly adapted. The equipment within the DC microgrid is controlled by its own protection system to complete the protection function according to the withstanding electrical capability of the equipment.

12.4.2 Integrated Control and Protection Platform

As will be introduced in the following sections, the fast fault isolation and recovery in each protection zone relies on the control measures, such as switching on some parallel resistors, blocking and deblocking VSCs, and reclosing of the DC circuit breakers. So, a novel integrated control and protection platform for the studied DC microgrid is proposed. Figure 12.9 illustrates the architecture of integrated control and protection platform for the DC microgrid, which consists of one control master station and some control & protection IEDs (Intelligent Electronic Devices). Each control & protection IED is deployed for one protection zone, as shown in Figure 12.8. All the control & protection IED communicate with the control master station, which is the core part of the platform.

Taking the DC bipolar fault as the example, the integrated control & protection IED detects the fault, and then sends a command to the VSC control system to block the VSC valve. The control & protection IED will also trip the DC breakers within its protection zone. Then the control & protection IED will send the protection action information and the status of adjacent VSC and DC breakers to the control master station. The control

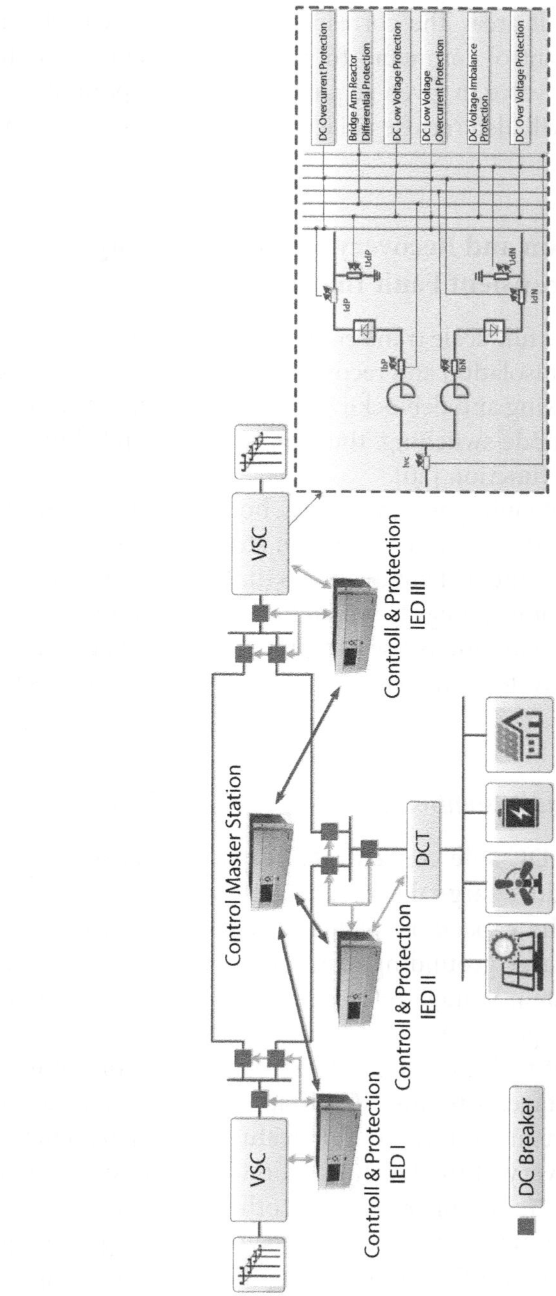

Figure 12.9 Schematic architecture of integrated control and protection platform for the VSC-based DC microgrid.

master station will further judge the fault area and non-fault area, and then issue the command of deblocking certain VSCs and reclosing certain DC breakers in the non-fault area. The corresponding control & protection IED will transfer the control command to adjacent VSC control system for deblocking the VSC valve in given sequence. The corresponding control & protection IED will also reclose adjacent DC breakers with a given sequence for recovery.

12.4.3 Fault Isolation and Recovery Strategy Utilizing Full-Scale Transient Fault Information

The strategy utilizing the full-scale transient information is adopted in this chapter for fast DC fault isolation and recovery by integrated control measures, such as VSC blocking and deblocking, DC circuit breaker switching and system operation mode switching, thus improving the reliability and sensitivity of protection function [20].

According to different fault type and severity, both slow and fast protection strategies are derived. The slow protection strategy is used to isolate slight short-circuit faults, and its fault diagnosis time is less than few milliseconds. The fast protection strategy based on FPGA (Field-Programmable Gate Array) hardware technique is used to protect serious faults such as DC bipolar fault on converter and the fault diagnosis time can be less than 100 microseconds.

12.4.3.1 Unipolar Fault Isolation and Recovery of DC Line/Bus

Because MMC has no DC side concentrated capacitor, its sub-module capacitor will not produce large discharge surge current when unipolar fault occurs, and the capacitor voltage is stable. Moreover, due to the clamping effect of large grounding resistance, the fault only changes the position of DC system voltage reference point, and the DC bipolar voltage remains unchanged, MMC can normally transfer active power. However, the voltage of fault pole drops to zero, and the non-fault pole increases to normal voltage. After the fault disappears, its pole-to-ground voltage can quickly return to normal. This feature is very beneficial to the transient fault recovery of the DC line. By rationally designating the insulation level of DC lines, converters and other DC equipment, the converter can run continuously without emergency blocking, as long as the fault is eliminated before the AC/DC equipment withstanding voltage is exceeded.

For the unipolar fault of DC line/bus, the identification method based on the full-scale fault current is adopted. Firstly, the differential protection based on transient fault current is used to identify the fault location. Then, if the system operation mode changes or the transition resistance is too large to accurately identify the fault, the steady-state current differential protection by small parallel resistance switching is adopted to determine the fault position. By closing the switches K1~K3, as illustrated in Figure 12.8, the small parallel resistors R1~R3 are connected to the DC microgrid. As a result, the system resistance is reduced and the amplitude of steady-state differential current is increased for the steady-state differential current calculation. After certain given period, such as 1s, these small parallel resistors are switched off; even the fault location may not be successfully identified under very large fault transition resistance condition.

The above scheme is described in detail as follows:

1. When the DC microgrid is in normal operation, the neutral point resistance is large. The DC fault current amplitude is limited below 10A to meet the requirements of system voltage clamping and the safety of personnel and equipment near the fault point.
2. When the DC voltage unbalance protection function acts, the control & protection IED will issue alarm message to the control master station.
3. With collected fault information and after certain fixed time delay, the control master station will send command to the control & protection IEDs to decrease microgrid grounding resistance, i.e., closing the switches K1~K3 in turn with 200ms interval. When the fault current amplitude becomes greater than 10A, which is the threshold for fault location by the differential protection function, the fault position is identified.
4. After 1s period with the K1~K3 being closed, the control & protection IED will switch off the K1~K3. Then, with the derived fault location information, the integrated control and protection system will execute the fault isolation and system operation mode change program, and the DC microgrid will recover its normal operation again.

The DC voltage unbalance protection is configured as the main protection for this kind of unipolar fault, and the protection criterion is set as:

$$\begin{cases} \left|U_{\mathrm{dp}} + U_{\mathrm{dn}}\right| > U_{\mathrm{set}} \\ U_{\mathrm{set}} = k_{\mathrm{u}} U_{\mathrm{DCBase}} \end{cases} \tag{12.7}$$

Where, U_{dp} and U_{dn} are the voltage vector values of the positive and negative electrodes of the DC line to ground, respectively; U_{set} is the threshold value of the DC voltage unbalance protection action; U_{DCBase} is the rated DC unipolar voltage and $U_{DCBase} = 10kV$; k_u is the sensitivity coefficient, and $0 < k_u \leq 1$.

Since the voltage unbalance protection function only reflects unbalance fault of DC voltage, to consider the reliability of the protection and other factors, the DC unipolar rated voltage with the action threshold value k_u is adopted. After a fixed time delay T_{d1}, the protection function delivers an alarm message, which will be used as the startup signal for the control master station to apply the scheme of closing switches K1~K3. In this chapter, T_{d1} is set as 200ms. Then the K1~K3 will be switched off again after 1 s respectively by the control & protection IEDs.

The unipolar fault location function of DC line CL1 ~ CL7 is realized by the low fixed value section of DC line differential protection, the protection criterion is set as:

$$\begin{cases} \left|I_{\mathrm{dpL}} - I_{\mathrm{dpR}}\right| > \max(I_{\mathrm{c\_set\_L}}, k_{\mathrm{set\_L}} \times I_{\mathrm{res\_L}}) \\ or \\ \left|I_{\mathrm{dnL}} - I_{\mathrm{dnR}}\right| > \max(I_{\mathrm{c\_set\_L}}, k_{\mathrm{set\_L}} \times I_{\mathrm{res\_L}}) \end{cases} \tag{12.8}$$

Where, differential current threshold value $I_{c\_set\_L} = 0.03 \times L_{DCBase}$; restraint current $I_{res\_L} = \max(I_{dL}, I_{dR})$; restraint coefficient $k_{set\_L} = 0.03$; rated DC line current $I_{DCBase} = 1.25$ kA; I_{dpL} and I_{dpR} represent the measured values of current at both ends of the DC positive pole line; I_{dnL} and I_{dnR} represent the measured values of current at both ends of the DC negative pole line.

After a certain fixed time delay T_{d2} with the above criterion satisfied, the control & protection IED will isolate the faulty line. In this chapter, T_{d2} is set as 300ms.

12.4.3.2 Bipolar Fault Isolation and Recovery of DC Line/Bus

In order to quickly identify and isolate DC bipolar faults and realize rapid recovery in a multi-terminal DC microgrid, a strategy based on the rapid reclosing of DC circuit breakers is proposed. Specifically, once the fast protection function detects the fault, the control & protection IEDs immediately block the inverters and trip the DC circuit breakers to isolate the fault. Later, the control master station with collected fault information from different control & protection IEDs, will confirm the fault location and decide the scheme for recovery. With the recovery command received from the control master station, the control & protection IED will reclose the DC breaker and deblock the VSC in the non-fault area to achieve rapid recovery under a new topology.

The flow chart of the above-introduced control and protection processing is shown in Figure 12.10.

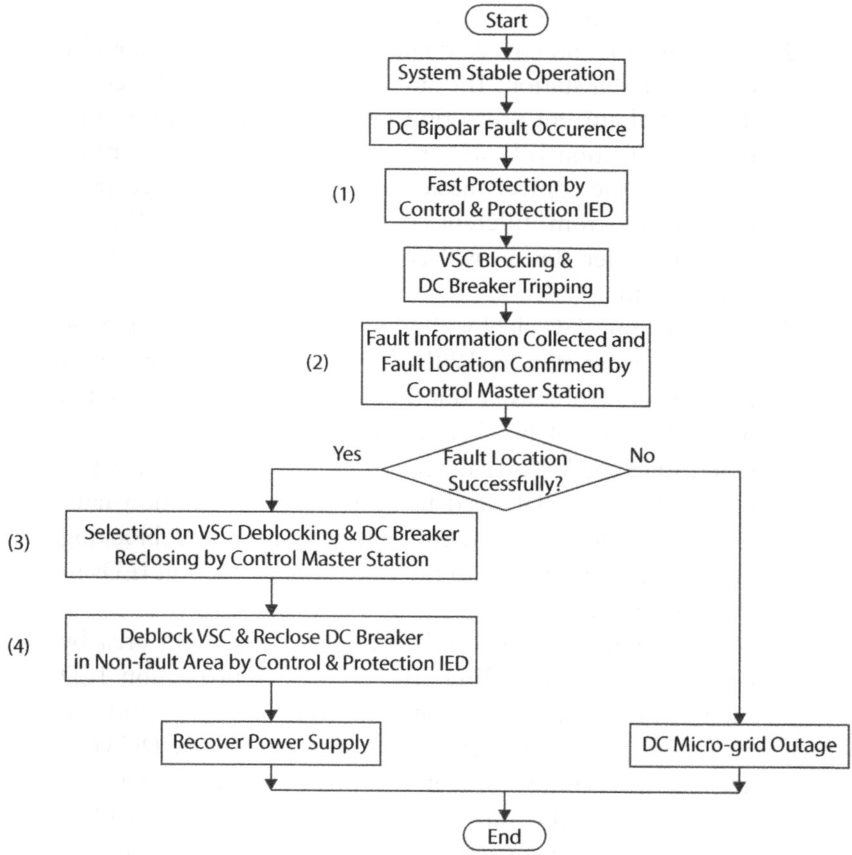

Figure 12.10 Flow chart of DC bipolar fault isolation and recovery.

Some key procedures as illustrated in Figure 12.9 are introduced in detail as follows:

1. Fast protection by control & protection IED. Once the bipolar fault occurs, the fast protection function offered by control & protection IED in each protection zone detects the characteristic value of the fault, and acts in hundreds of microseconds. The control & protection IED will send command to the VSC control system and the VSC valve will be blocked in less than one millisecond. The control & protection IED will also trip the DC circuit breaker in less than 3ms and thus the fault will be isolated. This process may lead to the protection action in non-fault area and will expand the affected range by the fault.
2. Fault information collected and fault location confirmed by control master station. The fault location is derived based on differential current protection principle, and then the control & protection IED will transfer the fault location information to the control master station. With collected fault information from different control & protection IEDs, the control master station will confirm the fault type and the fault location.
3. Selection on VSC deblocking & DC breaker reclosing by control master station. With the confirmed fault information, the control master station further compares the status of the DC circuit breakers before and after the fault, and determines the VSCs that need to be deblocked and the DC circuit breakers that need to be reclosed in the non-fault area. Then the control master station issues the deblocking or reclosing commands to the control & protection IEDs in the non-fault area.
4. Deblock VSC & reclose DC breaker in non-fault area by control & protection IED. The control & protection IED then sends command to the VSC control system and the VSC valve will be deblocked in turn to rebuild the DC voltage. Later the control & protection IED recloses the DC circuit breakers in the non-fault area and power supply will be gradually resumed. Finally, the DC microgrid will operate normally under a new network topology.

The detection of DC bipolar fault can be realized by DC low voltage over-current protection function. Considering the reliability of protection and other factors, the voltage criterion acts when the DC voltage between the poles is lower than 1.5 times of the rated voltage of the single pole U_{DCBase}, and the current criterion acts when the line current is greater than 2 times of the rated value I_{DCBase}. Then, the protection criteria is set as

$$\begin{cases} \left|U_{dp} - U_{dn}\right| > U_{set\_S} \wedge \left|I_{dp}\right| > I_{set\_S} \\ or \\ \left|U_{dp} - U_{dn}\right| > U_{set\_S} \wedge \left|I_{dn}\right| > I_{set\_S} \end{cases} \tag{12.9}$$

Where, the voltage action threshold $U_{set\_S} = 1.5 \times U_{DCBase}$ and the current action threshold $I_{set\_S} = 2 \times I_{DCBase}$.

When the voltage and current criteria are satisfied simultaneously, the control & protection IED starts and acts to trip the corresponding DC circuit breakers.

In order to protect DC equipment, the MMC valve should be blocked by the DC control system as soon as possible once the bipolar fault is detected. However, after the valve is blocked, the current, voltage and other electrical quantities necessary for fault location will drop rapidly, which may lead to failure of fault location. As a result, the blocking time of the valve should be configured comprehensively, which not only meets the requirements of the fault judgment, but also does not damage the DC equipment.

The fast line differential protection function is adopted in this chapter. Based on above analysis, the line differential protection function must complete the fault line location before the useful fault information disappears. The bipolar fault location can be determined by the high-set value section of the DC line differential protection. The protection criteria are set as follows:

$$\begin{cases} \left|I_{dpL} - I_{dpR}\right| > \max(I_{c\_set\_H}, k_{set\_H} \times I_{set\_H}) \\ \text{or} \\ \left|I_{dnL} - I_{dnR}\right| > \max(I_{c\_set\_H}, k_{set\_H} \times I_{set\_H}) \end{cases} \tag{12.10}$$

Where, differential current threshold $I_{c\_set\_H} = 0.5 \times I_{DCBase}$, restraint current $I_{res\_H} = \max(I_{dpL}, I_{dpR})$ and restraint coefficient $k_{set\_H} = 0.03$.

12.5 Simulation Verification

In order to verify the performance of the control and protection strategy, an RT-LAB hardware in loop simulation laboratory was built based on the studied multi-terminal DC microgrid[21]. In the studied DC microgrid, the VSC converter adopts MMC topology, which is composed of six bridge arms. Each bridge arm is composed of 25 sub-modules in series. Each sub-module adopts half bridge structure and switching frequency of 800 Hz. DC transformer adopts high and low voltage DC voltage of 10kV and 5kV respectively, and switching frequency is set as 20 kHz. The testing setting is shown in Figure 12.11.

F1~F4 fault positions, as shown in Figure 12.12, are selected to study the single and double pole faults of the DC line and bus. For F1 at DC cable line CL1 and F2 at DC Bus 1, the fault type is unipolar grounding short circuit. The fault type of F3 at DC cable line CL7 and F4 at DC Bus 2 is bipolar short circuit. Because the fault characteristics between DC line and DC bus are similar, this chapter adopts F1 and F3 for unipolar and bipolar fault study, respectively.

Figure 12.11 Laboratory setting for the DC microgrid tests.

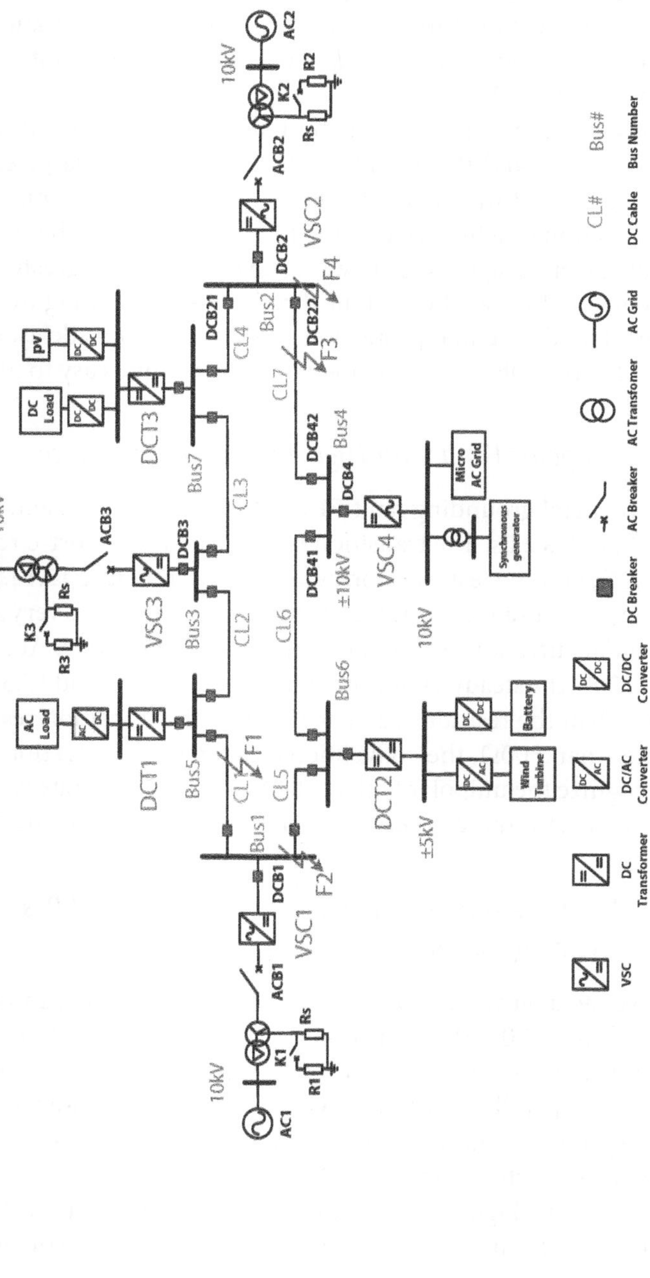

Figure 12.12 Fault sets selected on the studied DC microgrid.

12.5.1 Verification under DC Unipolar Fault

12.5.1.1 Metal Short Circuit Fault of DC Line

The positive pole metal grounding fault F1 is set on DC line CL1, as in Figure 12.12, the starting time of fault is 2.0s and the fault duration is 0.5 s. The restraint coefficient is taken as $k_i = 0.3$. The recorded fault waveforms are shown in Figure 12.13.

The transient current polarities of both terminals are opposite as shown in Figure 12.13(a), and the amplitude of fault current is large enough as shown in Figure 12.13(b), so the fault can be judged by the transient fault current protection function. According to Equation (12.8), the steady-state differential current is approximately 205A, which is much greater than the restraint current 25A, as the current threshold shown in Figure 12.13(c). The steady-state differential protection acts after a 300ms delay, as shown in Figure 12.13(d). Obviously, the metallic fault is quite easy to identify.

12.5.1.2 Unipolar Fault with High Transition Resistance

The positive metal grounding fault F1 is set in DC line CL1, and the starting time of fault is 2.0s. The transition resistance of the short-circuit point is set as 300Ω, and the test waveforms are shown in Figure 12.14.

As shown in Figure 12.14, the fault transient component is very small and the differential current protection function does no operate. According to Equation (12.8), the steady-state differential current is around 17.5A~18.0A and greater than the restraint current 17.0A. If the transition resistance is even greater than 300Ω, the conventional differential protection function cannot recognize the unipolar fault at all. As a result, the strategy proposed in section 12.4.3.1 is quite necessary, and will be tested in the next section.

12.5.1.3 High Resistance Unipolar Fault with Parallel Resistance Switching Strategy

The positive metal grounding fault F1 is set in DC line CL1, and the starting time of fault is 2.0s. The transition resistance of the short-circuit point is set as 300Ω, and restraint coefficient is set as $k_{i2} = 0.15$. According to section 12.4.3.1, parallel resistance switches K1~K3 are closed at 2.4s, 2.6s and 2.8s respectively, and later opened at 3.4s, 3.6s and 3.8s, respectively. Test waveforms are shown in Figure 12.15.

As illustrated in Figure 12.15(a), the DC voltage is biased due to the unipolar fault. According to Equation (12.7), the logic signal of DC voltage

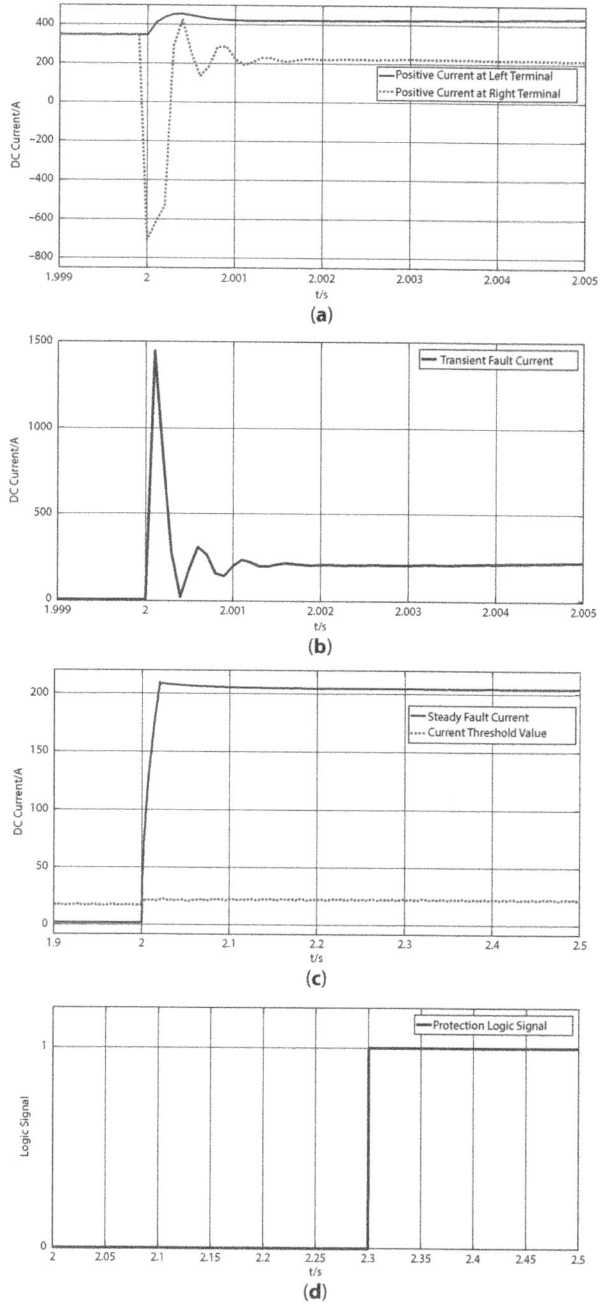

Figure 12.13 Simulation of positive pole metallic fault on DC line CL1. (a) Fault current at both terminals of line CL1. (b) Transient fault current at fault point. (c) Steady fault current at fault point. (d) Differential protection logic signal at left terminal of CL1.

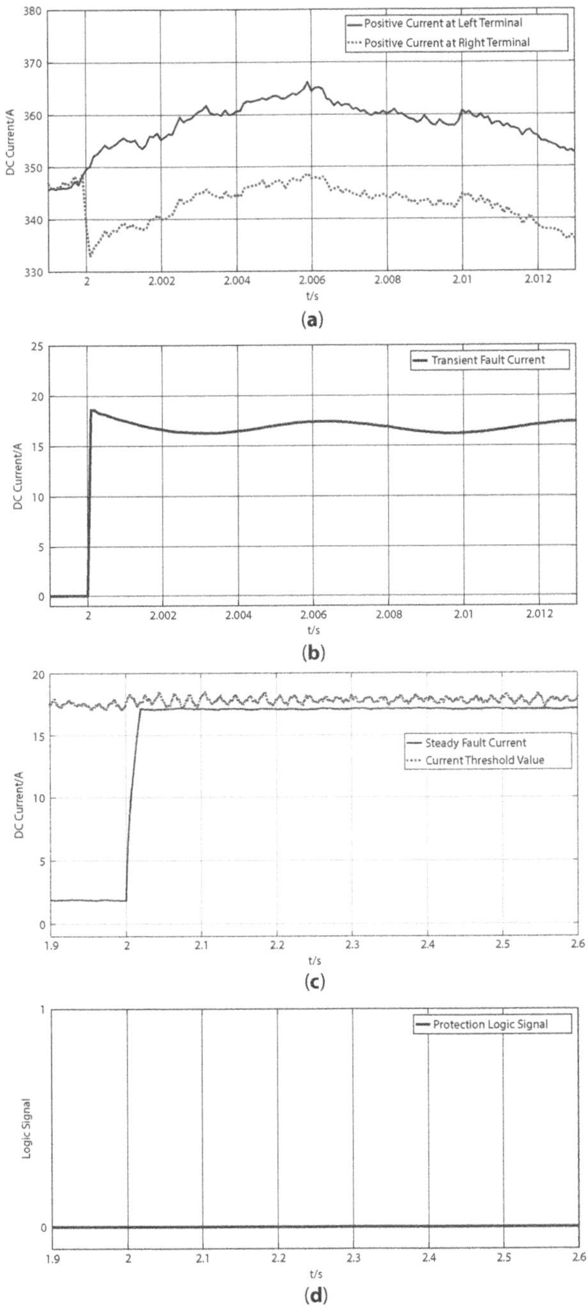

Figure 12.14 Positive pole fault through 300 Ω transition resistance on DC line CL1. (a) Fault current at both terminals of CL1. (b) Transient fault current at fault position. (c) Steady fault current at fault position. (d) Protect logic signal of left terminal of CL1.

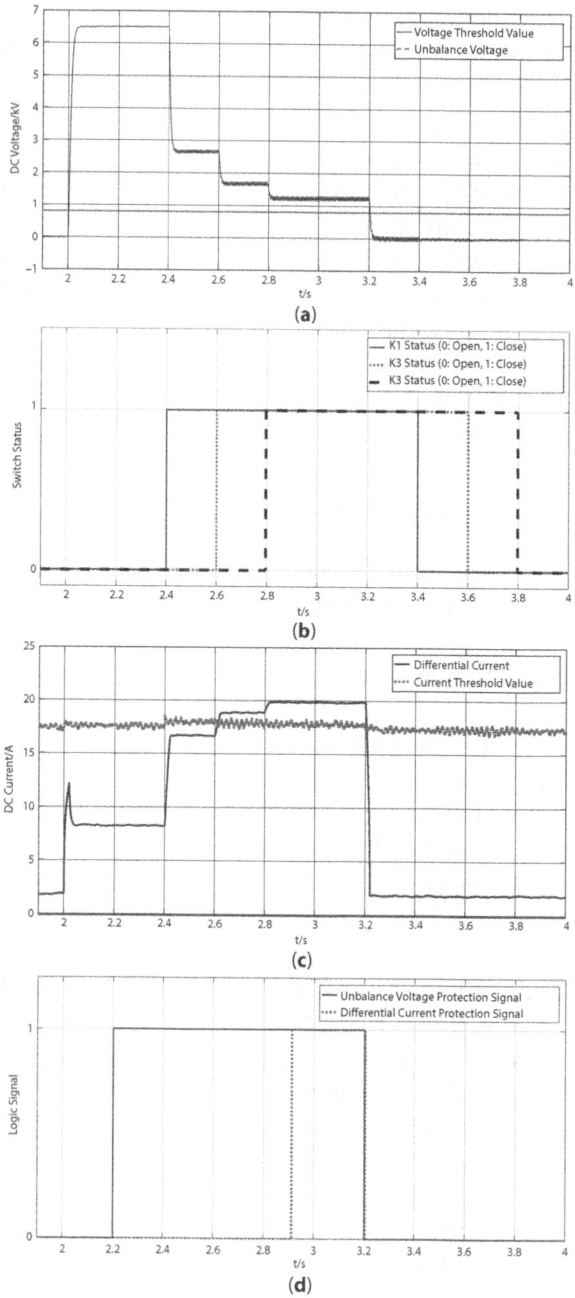

Figure 12.15 Positive pole fault through 300 Ω transition resistance on DC line CL1 with parallel resistance switching operation. (a) DC unbalance voltage at left terminal of CL1. (b) Statuses of the switches K1~K3. (c) Differential current at fault position. (d) Protection logic signals of left terminal of CL1.

unbalance protection function is derived as the blue line in Figure 12.15(d). This output with jump from 0 to 1 at 2.2s will be used as the trigger condition for the closing of switches K1 ~ K3 by the control master station.

Then the switches are closed successively at an interval of 200ms until the differential current is greater than the restraint current, as shown in Equation (12.8). The parallel switch K_1 at the VSC1 grounding resistance R_1 is closed at 2.4s, and then K_2 and K_3 after 200ms interval each. The steady-state current value of the protection becomes to be greater than the protection action threshold value at 2.6s, as shown in Figure 12.15(c). After 300ms delay, i.e., at 2.9s, the protection output control & protection IED will jump from 0 to 1, as shown in the red line of Figure 12.15(d). Later, after the fault is cleared at 3.2s, the unbalance voltage disappears, as displayed in Figure 12.15(a).

12.5.2 Verification under DC Bipolar Fault

Bipolar fault F3 is set on in DC line CL7, as shown in Figure 12.12. The fault start time is 2.0s, and the fault duration is 100ms. Fast DC overcurrent protection function and DC line differential protection function are used as the main protection and fault line selection protection for DC bipolar fault.

Around 700 μs after the fault occurrence, the currents at both terminals of the fault line are greater than the action threshold value of the overcurrent protection 1.0kA, as shown in Figure 12.16(a), so the control & protection IEDs at VSC2 and VSC4 act according to Equation (12.9). This process is shown in the procedure (1) of Figure 12.10. Then the VSC1~VSC4 are quickly blocked, and the status of VSC2 is displayed in Figure 12.16(c). The DC breakers, including DCB2, DCB21 and DCB22, trip around 2.5ms after the fault occurrence. As shown in Figure 12.12, under this fault case, DCB22 is located on the fault line CL7, and DCB2 and DCB21 are located in the non-fault area. As shown in Figure 12.16(c), the DC breaker DCB21 in the non-fault area also trips at 2.0025s.

Also around 700 μs after the fault occurrence, the differential current is greater than the threshold value 2.0kA, as shown in Figure 12.16(a), the control & protection IEDs at both terminals of DC line CL7 act according to Equation (12.10), and then the fault position can be derived. The protection output of different control & protection IEDs, together with the statuses of VSCs and DC breakers, are collected by the control master station. The control master station further confirms the fault information. This process is shown in the procedure (2) in Figure 12.10.

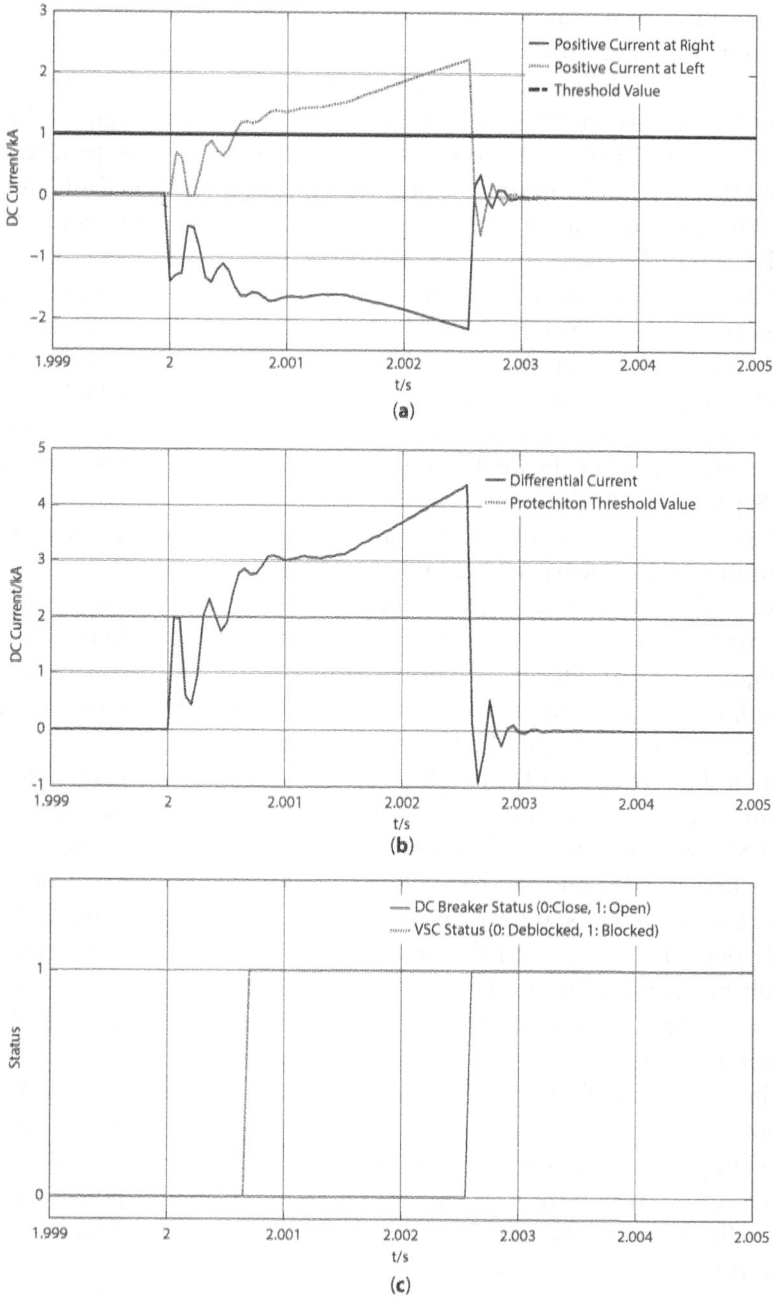

Figure 12.16 Simulation of bipolar fault on DC line CL7. (a) Fault current at both terminals of CL7. (b) DC differential current at fault position. (c) Statuses of VSC2 and DCB21.

The fault current waveforms and statuses of the VSC and DC breaker are shown in Figure 12.16.

After the DC breakers tripping, the fault currents decrease to almost zero around 3ms after the fault occurrence, as shown in Figure 12.16(a) and Figure 12.16(b), and the bipolar fault is isolated. This process also causes the VSC blocking and DC breaker tripping in the non-fault area. To maintain the power supply, it is necessary to deblock the VSCs and reclose the DC circuit breakers in the non-fault area.

With the confirmed fault information, the control master station determines the VSCs that need to be deblocked and the DC circuit breakers that need to be reclosed in the non-fault area, as the procedure (3) in Figure 12.10. Under this fault case, VSC1~VSC4 need to be deblocked, and the DC breakers outside the fault line CL7 need be reclosed. Then the control master station issues the deblocking and reclosing commands to the control & protection IEDs.

Take the inverter VSC2 as an example; its voltage and current waveforms during the recovery process are shown in Figure 12.17.

After the VSC2 blocking and the DC circuit breakers tripping, the DC voltage returns to the uncontrolled rectifier charging level as shown in Figure 12.17(a) during the period from 2.0007s to 2.25s. During this period, the VSC2 bridge arm current and VSC2 output current both drop to zero, as shown in Figure 12.17(b) and Figure 12.17(c).

Then the procedure (4) in Figure 12.10 is executed. The control & protection IED at VSC2 sends a command to the VSC control system and the VSC valve will be deblocked to rebuild the DC voltage. The VSC2 is deblocked at 2.25s and the DC voltage and bridge arm current of VSC2 during 2.25s to 2.35s are shown in Figure 12.17(a) and Figure 12.17(b). And 100ms later, the control& protection IED at VSC2 recloses the DC circuit breakers, including DCB2 and DCB21. Both the bridge arm current and output current of VSC2 increase to the normal value gradually, as shown in Figure 12.17(b) and Figure 12.17(c).

The responses of VSC1, VSC3 and VSC4 are similar to that of VSC2. About 250ms after the fault occurrence, the VSCs are deblocked and the DC voltages begin to rebuild. About 350ms after the fault occurrence, the DC circuit breaker at the non-fault area is reclosed, and the power transfer gradually returns to the level before the fault. The whole time used for fault isolation and recovery is less than 1s.

A large amount of tests under different operation modes, VSC control modes, load levels, and fault conditions are performed. All the test results demonstrate that the integrated control and protection strategy based on full-scale transient information can quickly identify and isolate serious

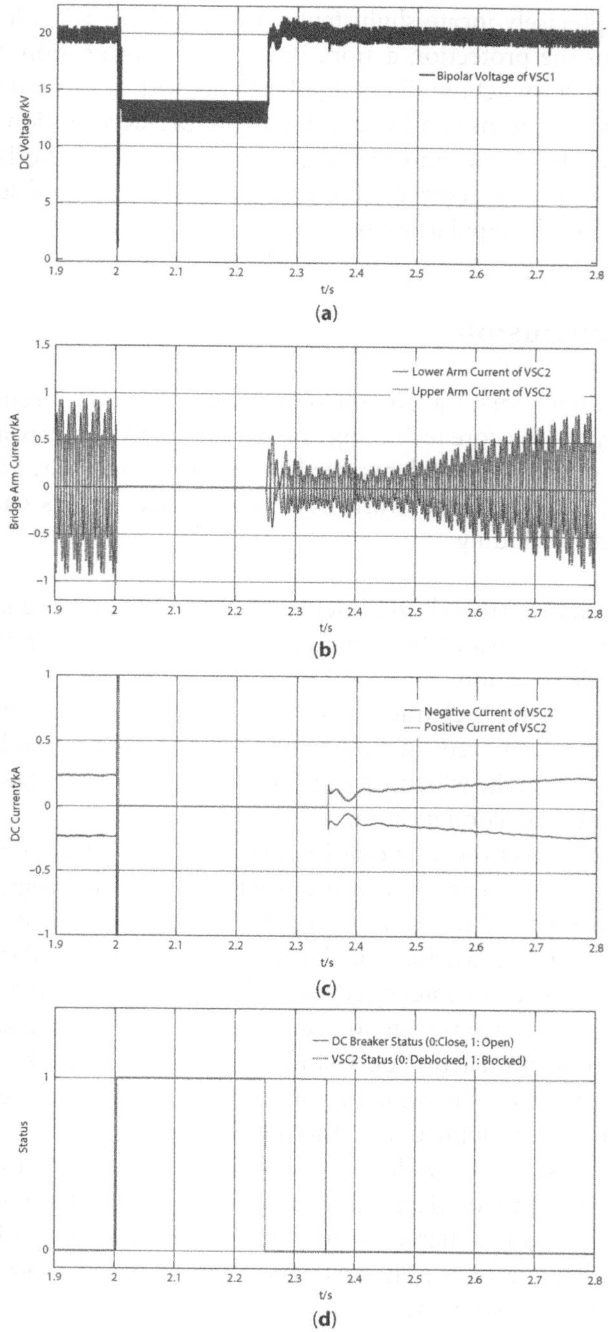

Figure 12.17 Output of VSC2 under DC line CL7 bipolar fault. (a) Bipolar DC voltage. (b) Bridge arm current. (c) Bipolar DC current. (d) Protect action logic signal.

faults, and precisely locate slight faults, thus ensuring the selectivity and sensitivity of the protection action. Based on the protection criterion of small resistance switching, DC unipolar faults with transition resistances greater than 300 ohms can be identified. The pre-acceleration protection based on the full-scale transient fault information and the DC breaker reclosing scheme can achieve millisecond-level isolation and second-level recovery under the bipolar faults.

12.6 Conclusion

For the past few years, the DC microgrid application has been developed very quickly. As the core technology of DC microgrid, control and protection is one of the key problems need to be solved urgently. For this new technology, some research work has been studied in this chapter from many aspects, including:

1. A multi-terminal DC microgrid for test is proposed, with careful consideration on topology design and grounding mode selection.
2. The DC fault characteristics of multi-terminal DC microgrid is analyzed, including typical DC unipolar fault and bipolar fault, which provides theoretical research basis for protection configuration.
3. The protection zone division principle and protection configuration method of a multi-terminal DC microgrid are provided, which can accurately locate the fault of a DC microgrid and achieve fault isolation and recovery. Based on the strategy of integrated control and protection, the selectivity of protection action can be ensured by layered setting of protection action delay and reclosing of DC breaker, so as to minimize the influence of the fault, and quickly recover the power supply of DC microgrid.
4. The test results under RT-LAB environment are consistent with the theoretical analysis, which can reflect the steady and dynamic characteristics of multi-terminal DC microgrid, and verify the effectiveness of the proposed control and protection strategy.

The work of this chapter lays a foundation for the in-depth study and practical operation of a DC microgrid. The coordination and optimization between different control and protection strategies should be studied in the future.

References

1. M. E. Baran, N. R. Mahajan. DC Distribution for Industrial Systems: Opportunities and Challenges. *IEEE Trans. Industrial Application*, 2003, 39(6):1596–1601.
2. Guidelines for Typical Configuration and Power Supply Schemes of Middle Voltage DC Distribution Network, T/CEC 166-2018. Beijing: China Electricity Council, 2018.
3. S. K. Chaudhary, J. M. Guerrero, R. Teodorescu. Enhancing the capacity of the AC distribution system using DC interlinks - A step toward future DC grid. *IEEE Trans. Smart Grid*, 2015, 6(4):1722–1729.
4. K. Jia, Z. Xuan, T. Feng, C. Wang, T. Bi, D. W. P. Thomas. Transient High-Frequency Impedance Comparison-Based Protection for Flexible DC Distribution Systems. *IEEE Trans. Smart Grid*, 2020, 11(1):323–333.
5. P. Ruffing, N. Collath, C. Brantl, A. Schnettler. DC Fault Control and High-Speed Switch Design for an HVDC Network Protection Based on Fault-Blocking Converters. *IEEE Trans. Power Delivery*, 2019, 34(1):397–406.
6. A. Lesnicar, R. Marquardt. A New Modular Voltage Source Inverter Topology. *2003 European Conference on Power Electronics and Applications, Toulouse, France,* 2003:1–10.
7. L Tang, B-T Ooi. Locating and Isolating DC Faults in Multi-Terminal DC Systems. *IEEE Trans Power Delivery*, 2017, 22(3):1877–1884.
8. G. Jing, A. Zhang, H. Zhang, S. Sun. Network topology and operation control of DC distribution network with AC DC converter. 2019 *Chinese Control and Decision Conference (CCDC)*, 2019:6216–6220.
9. G. Li, J. Liang, F. Ma, C. E. Ugalde-Loo, H. Liang. Analysis of Single-Phase-to-Ground Faults at the Valve-Side of HB-MMCs in HVDC Systems. *IEEE Trans. Industrial Electronics*, 2019, 66(3):2444–2453.
10. S. Cui, S.-K. Sul. A Comprehensive DC Short-Circuit Fault Ride-Through Strategy of Hybrid Modular Multilevel Converters (MMCs) for Overhead Line Transmission. *IEEE Trans. Power Electronics*, 2016, 31(11):7780–7796.
11. N. Ahmed, L. Angquist, S. Mahmood. Efficient modeling of an MMC-based Multi terminal DC System Employing Hybrid HVDC Breakers. *IEEE Trans. Power Delivery*, 2015, 30(4):1792–1801.
12. C. M. Franck. HVDC Circuit Breakers: A Review Identifying Future Research Needs. *IEEE Trans. Power Delivery*, 2011, 26(2):998–1007.
13. R. Lazzari, L. Piegari. Design and Implementation of LVDC Hybrid Circuit Breaker. *IEEE Trans. Power Electronics*, 2019, 34(8):7369–7380.
14. Y. J. Hafner. Interface arrangement between AC and DC systems for reliable opening of the circuit breaker in time. Patent US8971070B2, Oct. 27, 2011.

15. C. Meyer, S. Schroder, R.W. DeDoncker. Solid-state circuit breakers and current limiters for medium-voltage systems having distributed power systems. *IEEE Trans. Power Electronics*, 2004, 19(5):1333–1340.
16. L. Qi, A. Antoniazzi, L. Raciti. DC distribution fault analysis, protection solutions, and example implementations. *IEEE Trans. Industrial Application*, 2018, 54(4):3179–3186.
17. Sun Gang, Shi Bonian, Zhao Yuming. Research on the fault location method and protection configuration strategy of MMC based DC distribution grid. *Power System Control and Protection*, 2015, 43(22):127–134.
18. G. Zhang, X. Qi, Y. Chen, L. Qi, C. Yue. Characteristic analysis of modular multilevel converter under DC pole-to-pole short circuit fault. *Automation of Electrical Power System*, 2016, 40(12):151–157.
19. Bin Li, Jiawei He, Jie Tian, Yadong Feng, Yunlong Dong. DC Fault Analysis for MMC- based System. *Journal of Modern Power Systems and Clean Energy*, 2017, 5(2):275–282.
20. Shi Bonian, Li Yan, Sun Gang. Research on DC Fault Protection Strategy for Multi-terminal Flexible DC Distribution Network based on Full Fault Current Information. *High Voltage Engineering*, 2019, 10(5): 3076–3083.
21. Opal-RT. RT-LAB Solution for Real-time Application-OP104: Artemis and RT-events. Technical Support, 2016.

13

A Decision Tree-Based Algorithm for Fault Detection and Section Identification of DC Microgrid

Shankarshan Prasad Tiwari and Ebha Koley*

Department of Electrical Engineering, N.I.T. Raipur, C.G., India

Abstract

The unprecedented development in domestic and commercial infrastructure has led to an increase in power demand. Further, the continuous stress on the reduction of the carbon emission to decrease the air pollutants in the atmosphere has motivated research towards the integration of green energy sources to fulfill the energy requirement in the distribution system. However, integration with the microgrid presents a great challenge to the adoption of traditional overcurrent-based protection schemes due to the weather-dependent intermittent behavior of nonconventional energy sources such as photovoltaic and wind energy. Also, the dependence of the fault current magnitude on the operating mode of the microgrid enhances the challenge in achieving reliable protection. The reliability of any power network demands a protection strategy which can easily and accurately detect the fault. In this regard, in the present chapter, a decision tree-based protection scheme is designed for fault detection/classification and faulty section identification of DC microgrids during faulty condition. The simple structure of the decision tree allows for achieving high accuracy in executing the protection task. The proposed algorithm is based on the acquisition of the voltage and current signals from the relaying buses. For evaluating the proposed scheme, various test cases have been considered to quantify the accuracy of the decision tree-based classification module designed for mode detection, fault detection/classification and faulty section for both the operating modes, i.e., islanded and grid connected. The numerical simulations reflect the efficacy of the proposed scheme in effectively performing the protection task under widely varying operating scenarios.

**Corresponding author*: ekoley.ele@nitrr.ac.in

Nikita Gupta, Mahajan Sagar Bhaskar, Sanjeevikumar Padmanaban and Dhafer Almakhles (eds.)
DC Microgrids: Advances, Challenges, and Applications, (397–420) © 2022 Scrivener Publishing LLC

Keywords: DC microgrid protection, fault detection/classification, DC fault, DERs (distributed energy resources), decision tree, section identification

Acronyms

DERs	Distributed energy resources
WTG	Wind turbine generator
PV	Photovoltaic
SDG	Synchronous diesel generator
BS	Battery storage
UG	Utility grid
DT	Decision tree
AC	Alternating current
DC	Direct current
PG	Pole to ground
PP	Pole to pole
NF	No fault
TP	True positive
TN	True negative
FP	False positive
FN	False negative
FD/C	Fault detection/classification
FSI	Fault section identification
GC	Grid connected
ISM	Islanded mode
ROC	Receiver operator characteristic

Symbols

Ω	Ohm
R_f	Fault resistance
L_f	Length of fault
D1-D6	Sections of DC microgrid
L1, L2	Loads

13.1 Introduction

In recent years the pressure toward reducing carbon emission is continuously increasing due to the hazardous effects of air pollutants. Also, the demand for power is drastically rising due to developments in domestic and commercial infrastructure at the distribution end. The reduction in the availability of nonrenewable sources in nature has moved us towards adoption of the alternate and green energy sources such as solar, biomass, tidal, geothermal and wind-based sources to fulfill the growing energy requirement [1–3]. The adoption of these green sources of energy has propelled the development of the microgrid [2–4]. The microgrid is a part of the electrical power distribution network [5], which can be defined as the interconnection of the low rating renewable DERs (distributed energy resources), energy storage device, power electronic converters and load with clearly defined electrical boundaries [6, 7]. The microgrid imparts smartness to the power distribution network by the efficient utilization of the distributed energy resources [8]. As compared to transmission of DC power at high voltage, the DC power distribution network is a relatively new approach [9]. On the basis of output power the microgrid can be categorized as AC, DC and hybrid microgrid [10]. Microgrids can be operated in GC or in ISM if any contingency occurs in the utility grid [11]. Recently, DC microgrids have found wide application in different areas like lighting appliances, power system control devices, data centers, protection systems, shipboard power systems and residential loads [12–14]. A DC microgrid offers numerous benefits over the AC microgrid system such as absence of skin effect, no need of synchronization, safer for human bodies, less power conversion stages, improved reliability and voltage quality and higher power transfer capability [15–19]. The wide usages of DC power in various types of electrical and electronics apparatus due to the rapid development in semiconductor technology [20] has also contributed to the popularity of DC microgrid. In spite of the various benefits of the DC system over the AC system, protection of a DC-based microgrid system is a challenging task due to the unipolar current behavior, limited fault current capacity of converters and variation in the magnitude of fault current during grid-connected and isolated mode. Faults in DC microgrids are mainly classified into pole to ground (PG) and pole to pole (PP) fault [21]. A numbers of protection schemes for DC microgrids have been reported by researchers in literature regarding the protection and analysis of a DC microgrid system. Some of the significant works in this regard include differential current-based fault detection [22], unit protection scheme using superimposed current [23], combined time-frequency domain-based efficient protection scheme for low-voltage DC microgrid [24], centralized unit base protection scheme [25], multi-sample differential protection [26], probe power unit based non-iterative approach for estimation of fault location [27], short circuit fault analysis [28], differential protection for [29], adaptive load shedding [30], poverty severity index-based protection strategy [31], and event-based classification approach [32]. Though the above-reported techniques have been well validated for different fault scenarios, achieving reliable protection in DC microgrid under stressed and dynamic operating conditions is still a challenge. Motivated by the above-mentioned challenges, in this chapter a decision tree-based protection algorithm has been proposed for a DC microgrid. The proposed algorithm utilizes the post-fault voltage and current samples for executing the protection tasks under dual modes of operation. The major contribution of the work can be summarized as:

1. Modeling and simulation of a DC microgrid system for different fault levels and varying degree of weather intermittency.
2. Development of decision tree-based algorithm for detection of fault and faulty section identification.
3. Comparison of the DT-based algorithm with SVM and stand-alone kNN-based techniques for reliability analysis.

13.2 DC Test Microgrid System

The 350V, 500 KW DC ring microgrid test model used in the present work is illustrated in Figure 13.1 [33]. The test microgrid system consists of three DERs (photovoltaic array, SDG and WTG). The DERs fed power to the grid through DC-DC converters. In the microgrid system, the sensors at the six buses (A1 to A6) are used to execute the protection tasks. The utility grid is connected at bus A1 through a bidirectional converter. To increase the

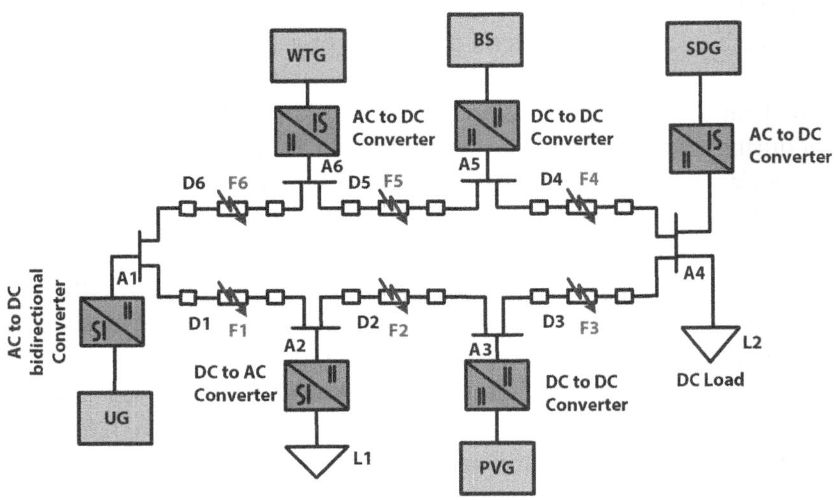

Figure 13.1 DC microgrid system under study.

reliability of the system, a battery management system is used for improved power quality and load leveling [34]. The model is simulated in MATLAB/Simulink software for implementing the proposed decision tree-based protection algorithm. The microgrid has been divided into six sections, namely (D1, D2), (D2, D3), (D3, D4), (D4, D5) and (D5, D6) for designing the protection scheme. The entire length of the system is 6 kilometers, in which each section is extended over the length of 1 kilometer. To facilitate the proper interfacing between the grid and DERs, power electronics converters are used [35]. The SDG and WTG connects with the microgrid at bus A1, A4 and A6. Similarly, PV array, Load L2 and battery are connected through bus A3, A4 and A5.

13.3 Overview of Decision Tree-Based Proposed Scheme

In recent times, the decision tree has emerged as a significant machine learning technique for the analysis of transient conditions in a power distribution network [36, 37]. As compared to other data mining tools, the decision tree possesses a simple structure which allows for easier implementation. The classification accuracy is high as compared to SVM and stand-alone kNN-based machine learning algorithms. It is a non-parametric supervised machine learning tool where the targets are in a supervised manner [38]. The decision tree algorithm works on the principle of the pattern reorganization. In literature, DT has been used for FD/C, security assessment, transient stability and islanding detection of distributed energy resources-based power network [39–41]. The basic structure of a decision tree comprises root node, branches and leaf nodes [42]. The root node processes the entire dataset to arrive at the preliminary decision regarding the appropriate class. The decision process initiates at root node. Branches emanating from the root node demonstrate the conjunctions of features to lead the classification. Leaf nodes are subnodes of the internal node (Figure 13.2). In this algorithm a tree is formed by the division of source set which has the various subsets of root node. The variables in the classification process can be expressed as:

$$\bar{Z} = \{Z_1, Z_2, \ldots\ldots\ldots\ldots\ldots Z_m\}^{T_r}$$

$$Z_i = \left\{Z_{i1}, Z_{i2} \ldots\ldots\ldots\ldots\ldots Z_{in}\right\}$$

$$P = \left\{P_1, P_2 \ldots\ldots\ldots\ldots\ldots P_m\right\}^{T_r} \tag{13.1}$$

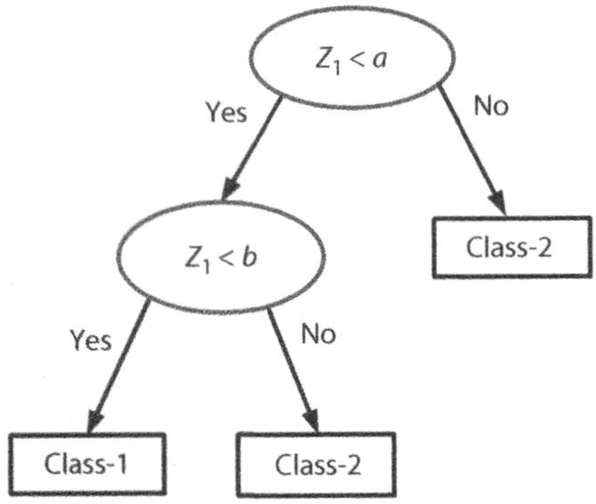

Figure 13.2 Structure of decision tree for binary classification.

where $\bar{Z}$ represents the feature dataset of m feature vectors Z_i. Each feature vector in $\bar{Z}$ is represented in the feature space with n components. The vector P denotes the m dimensional vector of dependent variables.

The DT-based classifier processes the input data to predict the value of P based on the sensor information. The algorithm aims at growing a DT with the help of local optimum decisions about the system parameters for partitioning the dataset $\bar{Z}$. In its basic form, the decision tree is the binary tree which executes the classification task by segregating the feature space by a set of partitions in two-dimensional space, represented as [43]

$$T = \{(n, q, m, n^L, n^R)\ T^L, T^R\} \tag{13.2}$$

In equation (13.2) n, q, n^L, n^R, T^L, and T^R respectively indicate the label of the partition, feature axis, threshold value of partition, labels of the nodes for the left partition and labels of the nodes for the right partition, respectively. The partitioning of the two-dimensional feature space into two classes is depicted in Figure 13.3.

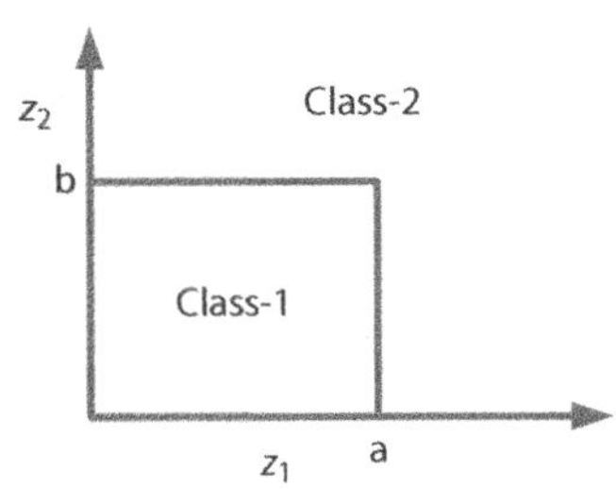

Figure 13.3 Classes of two-dimensional classification in the feature space.

13.4 DC Microgrid Protection Using Decision Tree Classifier

A decision tree-based scheme for performing the protection functions under diversified dynamics of a DC microgrid network has been dealt with in this section. The faulty condition of the system creates an interruption in the power supply. So, designing a protection scheme which can easily detect and classify different faults is necessary for improved resiliency of the power network. To meet these challenges, a protection scheme has been developed to operate under islanded and grid-connected mode. The proposed algorithm utilizes the post-fault voltage and current samples measured at bus A1. The procedure adopted for executing the scheme is illustrated in the form of a flowchart in Figure 13.4, which outlines the steps involved in the design of the protection scheme. First, in order to

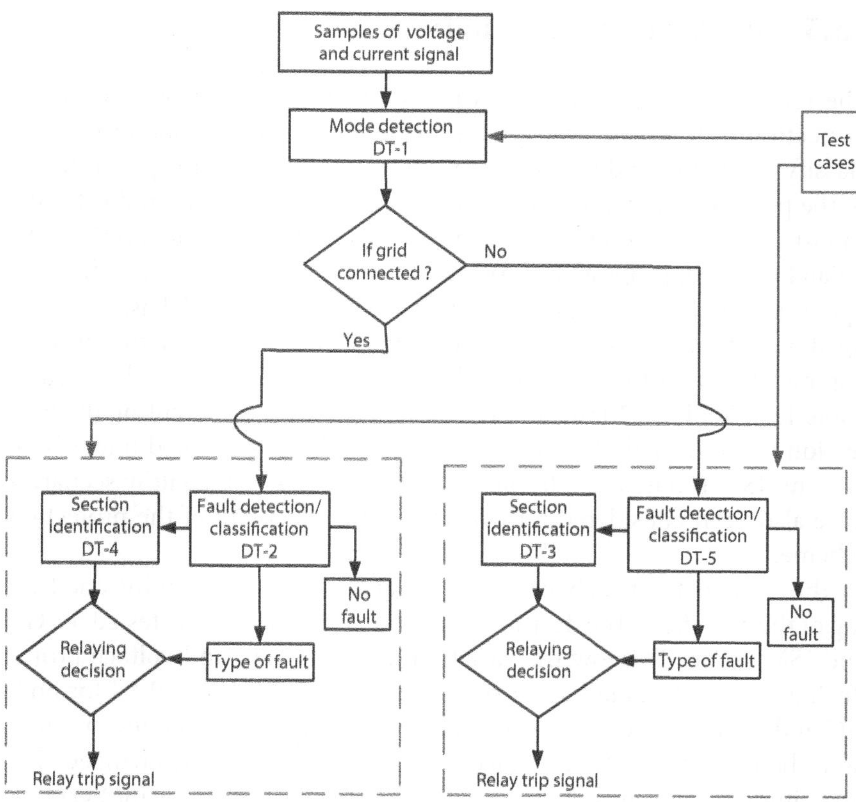

Figure 13.4 Flow chart of decision tree-based algorithm.

identify the operational modes, i.e., GC and ISM, DT-1 is developed. Once the mode is detected successfully, then the algorithm is triggered to execute other decision trees for detecting/classifying the fault and identify the effected section.

In the proposed scheme DT-2 and DT-5 have been designed for FD/C, while DT-3 and DT-4 have been developed to perform the task of FD/C and FSI, respectively. For evaluating the proposed decision-tree modules, various test cases of a DC microgrid under stressed conditions have been considered.

Once the faulty section and type of fault (PG, PP) is identified by the protection scheme, a signal is communicated to operate the circuit breaker and isolate the faulty section. If the system is isolated from the unhealthy section, then the extended duration of the power interruption is avoided. Also, costly electrical equipments are protected, thereby saving the assets of the power supply chain.

13.5 Performance Evaluation

The reliability of any power network is linked to the performance of its protection system. The protection scheme must be fast enough to detect/classify the fault condition in few milliseconds. A superior performance of the protection scheme under various operating condition indicates the robustness and reliability of the protection system. In this section, the reliability and appropriateness of the scheme is examined under different operating conditions. The effectiveness of the proposed DT-based scheme for different faults has been analyzed. The details of the variation in fault parameter during the training and testing of the classifier is depicted in Table 13.1. Table 13.2 indicates DT parameters used in executing the protection tasks. A total of 1,584 fault cases have been generated under both GC and ISM mode. In addition to the faulty cases, 450 healthy scenarios have also been considered for evaluating the selectivity of the protection scheme.

The entire dataset is divided into 70% and 30% for the training and testing of the classifiers. The decision tree modules are separately tested for GC and ISM mode. To show the variation in the current and voltage profile during fault, a PG fault has been created in section D1 at t=0.4 s for both GC and ISM (Figure 13.5). It can be observed that with the inception of fault the magnitude of the voltage and current are suddenly changed. The magnitude of fault current during GC mode is higher than the ISM grid for

Table 13.1 Details of the data for training and testing of the algorithm.

Fault parameters varied during training of classifier	**Number of variation in fault parameters during training**	**Total cases**	**Total training cases**	**Total testing cases**
Types of mode	2	1,584+450 no fault cases=2,034	1,258	776
Section	6			
Types of fault	2			
Fault resistance	11			
Length of fault	11			
Inception of fault	1			

Table 13.2 Parameters during the training of the classifier.

	Decision tree				
Parameters	**DT-1**	**DT-2**	**DT-3**	**DT-4**	**DT-5**
Maximum number of splits	4	4	4	4	4
Split criterion	Gini diversity index	Gini diversity index	Gini diversity index	Gini diversity index	Gini diversity index
Maximum surrogates per node	10	10	10	10	10

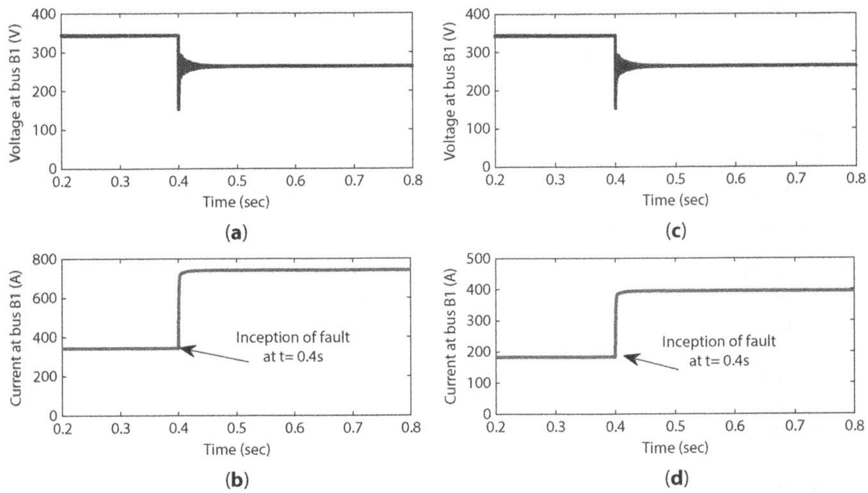

Figure 13.5 Fault current and voltage during a PG fault in section D1 for GC ((a) and (b)) and ISM ((c) and (d)).

the same fault scenario. The same demands separate settings of the protection scheme for both the modes.

After successful training of the decision tree network the confusion matrix has been plotted in Figure 13.6, to determine the effectiveness of the trained classifiers. The superior performance of the classifier pertains to the lesser value of the off-diagonal elements in the confusion matrix. A high accuracy for both the operating modes reflects the appropriateness of

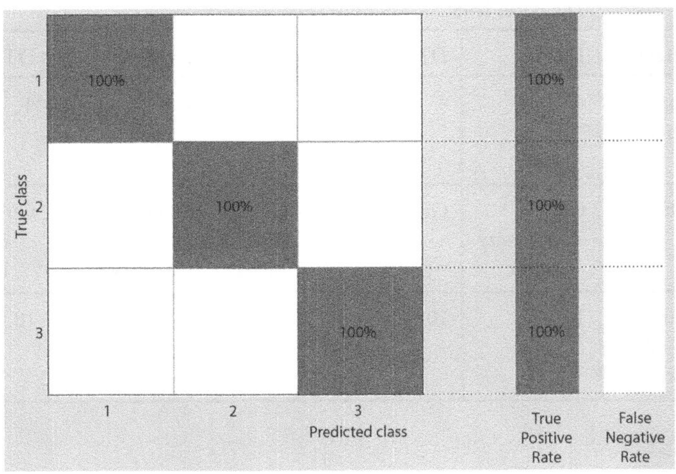

Figure 13.6 Confusion matrix of decision tree-based classifier.

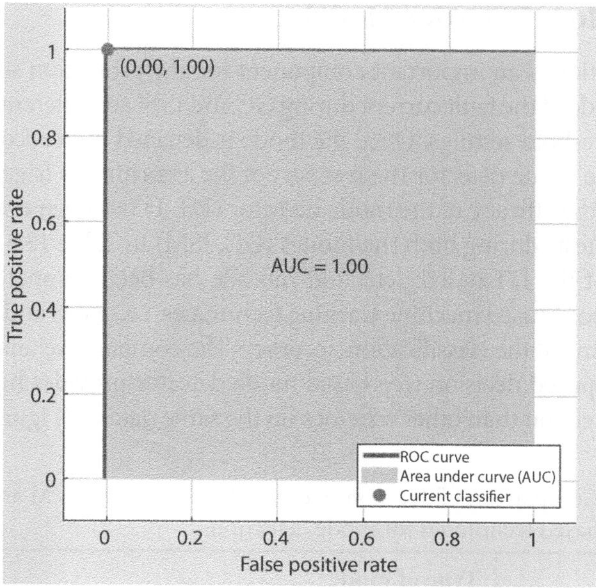

Figure 13.7 ROC curve of the trained DT classifier.

the DTs in performing accurate mapping between the feature vector and the predicted classes.

In addition to the confusion matrix, the effectiveness of the DT-based classifiers has also been evaluated in terms of receiver operator characteristic (ROC). The ROC curve for the datasets used during the training is depicted in Figure 13.7. The curve provides information regarding the instances of correctly detected faults (TPR) and healthy scenarios detected as fault (FPR). The TPR and FPR are defined in terms of classes predicted truly, i.e., true positive (TP) and falsely, i.e., false negative (FN) as:

$$TPR = \frac{TP}{FN + TP}$$

$$FPR = \frac{TP}{FN + TP}$$

An area under the curve (AUC) closer to one indicates the ability of the classifier to avoid true negative and false negative classes in prediction.

13.5.1 Mode Detection Module

Mode detection is an important component for any protection scheme, since the magnitude of the fault current during GC and ISM are different and require different threshold settings. Once the mode is detected then according to the output of the mode detector the rest part of the algorithm is triggered. In this subsection, the efficacy of the mode detector (DT-1) has been evaluated for a total of 640 tests during both the modes (GC, ISM) in Table 13.3. The appropriateness of the DT-based detection module has been compared with two more commonly used machine learning techniques, i.e., SVM and stand-alone kNN, in terms of the classification accuracy. The comparative analysis reveals that the proposed decision tree-based mode detector provides high accuracy in mode detection than other schemes on the same dataset (Figure 13.8).

Table 13.3 Comparison of decision tree-based scheme with SVM and stand-alone kNN-based techniques for mode detection.

Technique	Type of mode		
	Accuracy % grid connected	**Accuracy % islanded**	**Overall accuracy %**
DT	100	100	100
SVM	98.89	97.56	98.22
Stand-alone kNN	97.89	97.13	97.51

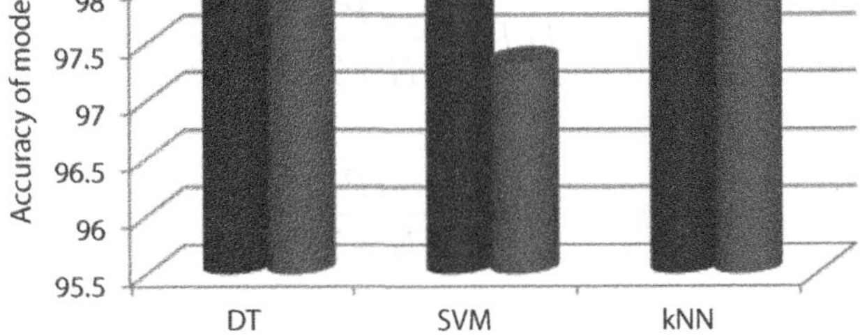

Figure 13.8 Comparison of mode detection accuracy (%) of DT-1 with other machine learning-based schemes on the same feature dataset.

13.5.2 Fault Detection/Classification

In this subsection the performance of the proposed FD/C modules have been analyzed under the GC and ISM mode.

The reliability of any protection scheme is related to its ability to promptly detect fault condition at the earliest. The faster fault detection enhances the immunity of the system against possible shutdown. To test the effectiveness of FD/C modules, various diverse test cases have been considered—a total of 712 test cases under GC and ISM modes along with 136 no fault (healthy) cases. The classifiers DT-2 and DT-5 have been examined in terms of dependability and security. Dependability refers to the possibility of misdetection of faults and security relates to the generation of false alarm signal.

A higher value of dependability and security is desirable for any protection scheme to adopt in a power distribution network. The proposed decision tree-based classifier modules have been compared with other machine learning-based classifiers in Table 13.4. The higher value of dependability and security reveals that the DT-based classifiers are robust and effective for the detection of the fault.

The magnitude of fault current varies significantly under variation in the fault resistance. Therefore, to analyze the response of DT under varying fault resistance, the response of the protection scheme is examined for PP and PG faults. If the conductor is grounded through some weak materials such as branches of tree, sand, building, the behavior of the fault current varies in accordance with the resistance. To validate the immunity of the proposed scheme under varying resistance, both types of faults under the GC and ISM has been considered. The response of FD/C module with dissimilar values of resistance is illustrated in Table 13.5. In the case of PP fault, the variation in the fault resistance is much less, but in the case of PG fault the fault resistance varies over a wider range [44, 45]. The response of the DT-based classification modules to varying levels of fault resistance is depicted in Table 13.5. The test results highlight the robustness of the DT-based scheme to fault resistance (Figure 13.9).

13.5.3 Section Identification

Resiliency in the operation of a power distribution network demands the proper identification of the faulty section to restore the system to the pre-fault scenario at the earliest. If following the detection of fault, the effected section is properly identified and a trip signal is sent by the relay to operate the circuit breaker, then the faulty section can be isolated. The

Table 13.4 Comparison of the proposed scheme with other algorithms.

Technique	Mode of operation					
	GC			ISM		
	Dependability %	**Security %**	**Accuracy %**	**Dependability %**	**Security %**	**Accuracy %**
DT	99.51	100	99.75	98.94	100	99.47
SVM	98.76	98.16	98.46	98.51	97.89	98.2
Stand-alone kNN	97.89	98.43	98.16	97.12	97.86	97.49

Table 13.5 Response of DT against the variation of fault resistance.

Types of mode	Type of fault	Fault resistance	Faulty section	Response of DT
Grid connected	PG	25	D1	PG
	PP	0.4	D1	PP
	PG	50	D3	PG
	PP	0.4	D3	PP
	PG	75	D5	PG
	PP	0.4	D5	PP
	PG	100	D2	PG
	PP	0.4	D2	PP
Islanded mode	PG	25	D1	PG
	PP	0.4	D1	PP
	PG	50	D3	PG
	PP	0.4	D3	PP
	PG	75	D5	PG
	PP	0.4	D5	PP
	PG	100	D2	PG
	PP	0.4	D2	PP

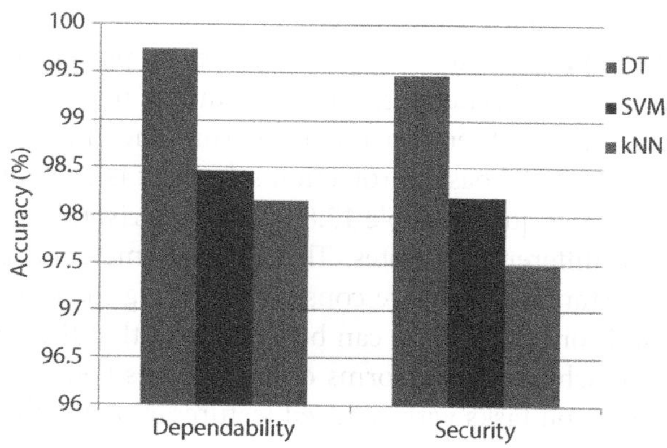

Figure 13.9 Comparison of DT-based scheme in terms of reliability indices.

same minimizes the impact of the fault. In this subsection the performance of DT-based section identifiers (DT-3 and DT-4) have been analyzed. The microgrid network has been divided into six sections from D1 to D6. Similar to mode detection and fault detection/classification, both grid-connected and islanded modes have been considered for section identification.

To examine the performance of section identifier for PG and PP faults, a total number of 504 cases have been considered; the results are detailed in Table 13.6 and Table 13.7. The same datasets have also been utilized to test the support vector machine and stand-alone kNN-based algorithm, and the accuracy during section identification has been compared. The classification accuracy results ascertain that the DT-based section identifier efficiently performs the intended task section identification task, irrespective of the varying system dynamics. Table 13.6 illustrates the overall accuracy of DT, SVM and stand-alone kNN during PG. The same task for PP fault has also been carried out and detailed in Table 13.7. For both types of fault, the proposed DT-based scheme provides the highest classification accuracy, followed by SVM and stand-alone kNN. Further, in order to illustrate the relay operation time, the faults in different sections are simulated. The voltage and current profile for a fault at section D3 are depicted in Figure 13.10. The corresponding trip signal is shown in Figure 13.11. Similarly for a fault in section D2, the post-fault waveforms and the trip signal are shown in Figures 13.12 and 13.13 respectively.

13.5.4 Comparative Analysis of the Proposed Scheme with other DC Microgrid Protection Techniques

As detailed in section 13.1, the diverse operating conditions of a power distribution network is often a hindrance to the efficient operation of power networks. The varying nature of the DERs creates power quality disturbances. Any protection scheme should be immune to such disturbances, generally caused by weather-dependent intermittency in DER operation. Thus to evaluate the DT-based protection scheme it is compared with the other reported techniques in Table 13.8. The comparison has been carried out in terms of different attributes. The fault parameters such as length of fault and resistance of fault are considered during the mode detection, FD/C and FSI. From the table it can be concluded that the decision tree-based proposed scheme outperforms other schemes ([9, 11, 14]) by executing the protection tasks with reduced features, along with maintaining robustness to fault parameter variation.

Table 13.6 Performance analysis of section identifier (DT-3 and DT-4) under PG fault.

Mode of operation	Section	DT (accuracy %)	SVM (accuracy %)	Stand-alone kNN (accuracy %)
Grid-connected mode	D1	99.45	97.23	94.85
	D2	99.56	95.54	95.43
	D3	99.63	97.67	94.95
	D4	99.21	95.13	97.46
	D5	99.35	96.65	95.66
	D6	99.54	97.86	96.78
	Overall accuracy	**99.45**	**96.68**	**95.85**
Islanded mode	D1	99.28	97.46	95.74
	D2	99.36	96.96	94.23
	D3	99.21	96.11	96.12
	D4	99.05	96.22	95.66
	D5	99.16	96.16	95.46
	D6	99.32	98.55	95.48
	Overall accuracy	**99.23**	**96.91**	**95.44**

Table 13.7 Performance analysis of section identifier (DT-3 and DT-4) under PP fault.

Mode of operation	Section	DT (accuracy %)	SVM (accuracy %)	Stand-alone kNN (accuracy %)
Grid-connected mode	D1	99.21	96.35	94.26
	D2	99.61	97.33	95.43
	D3	98.69	96.44	95.65
	D4	99.33	95.63	95.16
	D5	99.22	96.36	95.36
	D6	99.45	97.81	96.48
	Overall accuracy	**99.25**	**96.65**	**95.39**
Islanded mode	D1	99.35	95.69	95.14
	D2	98.86	96.91	94.23
	D3	98.96	97.33	95.12
	D4	99.13	97.11	96.16
	D5	98.91	94.22	95.26
	D6	99.16	97.28	95.18
	Overall accuracy	**99.06**	**96.42**	**95.18**

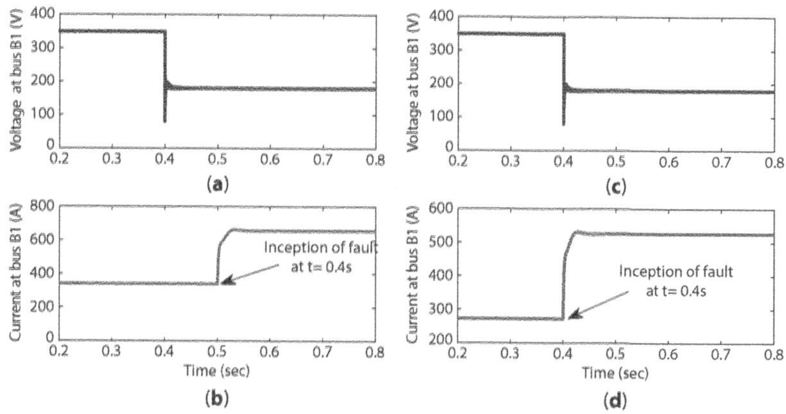

Figure 13.10 Fault current and voltage during a PG fault in section D3 for GC ((a) and (b)) and ISM ((c) and (d)).

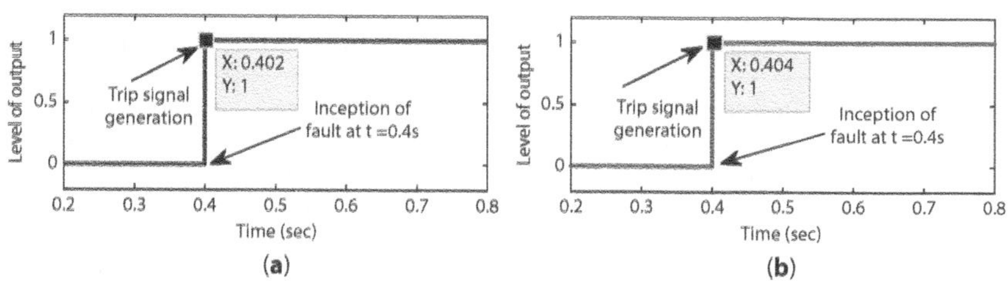

Figure 13.11 Generation of trip signal for the fault case depicted in Figure 13.10 (a) for GC (b) and ISM.

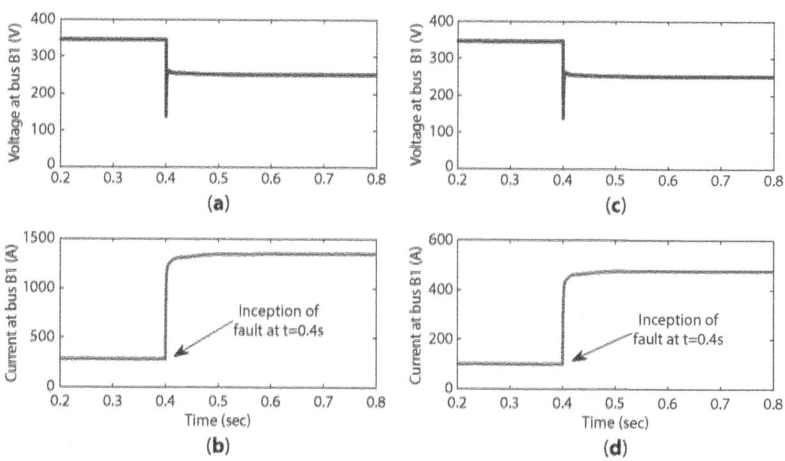

Figure 13.12 Fault current and voltage during a PG fault in section D2 for GC ((a) and (b)) and ISM ((c) and (d)).

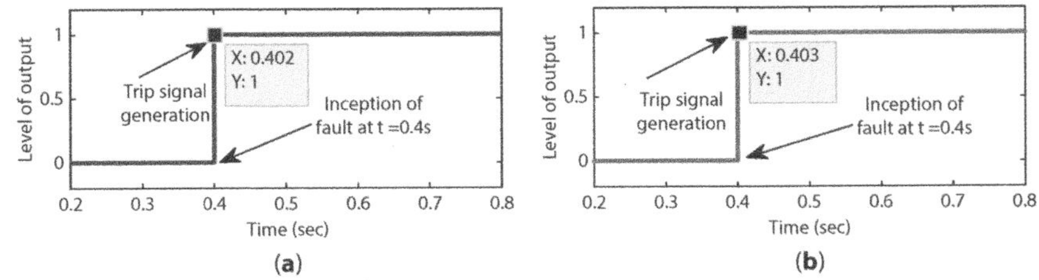

Figure 13.13 Trip signal for the fault case depicted in Figure 13.12 (a) for GC (b) and ISM.

Table 13.8 Comparative analysis of the proposed scheme with other reported techniques.

Parameters of comparison	Microgrid protection schemes			
	[24]	[25]	[9]	**Proposed scheme**
Input features	Wavelet coefficient	Current only	Voltage and current	Voltage and current
Fault parameters	Resistance, and location of fault	Location of fault	Resistance, and location of fault	Resistance and location of fault
Tasks performed by the reported and proposed algorithm	Fault detection and classification	Fault detection, faulty zone and location	Fault detection and location	Mode detection, Fault detection/ classification and section identification
Reliability analysis	Not carried out	Not carried out	Not carried out	Carried out

13.6 Conclusion

The proliferation of energy demand, reduced availability of fossil fuels and easier implementation has propelled the adoption of DC micro-grids. The penetration of renewable sources of energy has also increased in modern power systems to fulfill the peak demand of power. However, the randomness in DER operation due to weather intermittency often poses a serious challenge in DC microgrid protection. The scenario demands a reliable and efficient protection strategy which can properly differentiate the healthy and faulty section. The present chapter proposes a decision tree-based FD/C and FSI scheme to attain the above objectives in a DC microgrid. To evaluate the effectiveness of the algorithm, various test cases have been considered during both GC and ISM. The decision tree-based classifier modules have been compared with SVM and stand-alone kNN-based machine learning techniques. The comparative analysis confirms the effectiveness of the DT-based classification modules in maintaining a high degree of reliability and resiliency in DC microgrid operation.

References

1. Adefarati T, Bansal, RC. Integration of renewable distributed generators into the distribution system: a review. *IET Renew Pow. Gener.*, 2009; 10(7):873–884.
2. M. Manohar, E. Koley and S. Ghosh, "Reliable protection scheme for PV integrated microgrid using an ensemble classifier approach with real-time validation," in IET Science, Measurement & Technology, vol. 12, no. 2, pp. 200–208, 3 2018, doi: 10.1049/iet-smt.2017.0270..
3. M. Manohar, E.Koley, and S. Ghosh, "Stochastic weather modeling-based protection scheme for hybrid PV–wind system with immunity against solar irradiance and wind speed," *IEEE Systems Journal,* vol.14(3), 3430–3439, 2020.
4. Planas E, Andreu J, Gárate JI, De Alegría IM, Ibarra E. AC and DC technology in microgrids: a review. *Renew. and Sust. Energy Reviews*, 2015; 43: 726–749.
5. Lasseter RH. Microgrids. *IEEE Power Engine Society Winter Meeting. Conference Proceedings,*2002 1, 305–308. (Cat. No. 02CH37309).
6. Xu X, Mitra J, Wang T, Mu L. An evaluation strategy for microgrid reliability considering the effects of protection system. *IEEE Trans on Power Deliv.*, 2015; *31*(5):1989–1997.
7. Akhtar, Z., & Saqib, M. A. Microgrids formed by renewable energy integration into power grids pose electrical protection challenges. *Renew. Energy*, 2016; *99*: 148–157.
8. Murli Manohar, Ebha Koley, Subhojit Ghosh, Microgrid protection under weather uncertainty using joint probabilistic modeling of solar irradiance and wind speed, *Computers & Electrical Engineering,* Vol. 86, 2020.
9. Park JD, Candelaria J, Ma L, Dunn, K. DC ring-bus microgrid fault protection and identification of fault location. *IEEE Trans on Power Deliv.*, 2013; *28*(4):2574–2584.
10. Sahoo SK, Sinh AK, Kishore NK.Control techniques in AC, DC, and hybrid AC–DC microgrid: a review. *IEEE Journal of Emerging and Selected Topics in Power Electronics*, 2017; *6*(2): 738–759.

11. Li, Yunwei, Nejabatkhah F. Overview of control, integration and energy management of microgrids. *Journal of Modern Power Syst. and Clean Energy*, 2014; *2*(3):212–222.
12. Salomonsson D, Sannino A. Load modelling for steady-state and transient analysis of low-voltage DC systems. *IET Electr. Power Appli.*, 2007; *1*(5): 690–696.
13. Cuzner RM., Venkataramanan G. The status of DC micro-grid protection. *IEEE Industry Appli Society Annual Meeting*, 2008; 1–8.
14. Salomonsson D, Soder L, Sannino A. Protection of low-voltage DC microgrids *IEEE Trans on Power Deliv.*, 2009; *24*(3): 1045–1053.
15. Dahale S, Das A, Pindoriya NM, Rajendran S. An overview of DC-DC converter topologies and controls in DC microgrid. *International Conference on Power Syst (ICPS)*, 2017; 410–415.
16. Amamra SA, Ahmed H, El-Sehiem RA. Firefly algorithm optimized robust protection scheme for DC microgrid. *Electric Power Compo. and Syst.*, 2017;*45*(10): 1141–1151.
17. M. Sagar Bhaskar, M. Meraj, A. Iqbal, S. Padmanaban, P. Kiran Maroti and R. Alammari, "High Gain Transformer-Less Double-Duty-Triple-Mode DC/DC Converter for DC Microgrid," *IEEE Access*, vol. 7, pp. 36353–36370, 2019, doi: 10.1109/ACCESS.2019.2902440.
18. Hakala T, Lähdeaho T, Järventausta P. Low-voltage DC distribution utilization potential in a large distribution network company. *IEEE Trans on Power Deliv.*, 2015; *30*(4):1694–1701.
19. Azizi S, Sanaye Pasand M, Abedini M, Hasani A. A traveling-wave-based methodology for wide-area fault location in multiterminal DC systems. *IEEE Trans on Power Deliv.*, 2014; *29*(6): 2552–2560.
20. M. S. Bhaskar, R. Alammari, M. Meraj, S. Padmanaban and A. Iqbal, "A New Triple-Switch-Triple-Mode High Step-Up Converter with Wide Range of Duty Cycle for DC Microgrid Applications," *IEEE Trans on Industry Applications*, vol. 55, no. 6, pp. 7425–7441, Nov.-Dec. 2019, doi: 10.1109/TIA.2019.2920807.
21. Mousavi MJ, Butler-Purry KL. A novel condition assessment system for underground distribution applications. *IEEE Trans on Power Syst.*, 2009; *24*(3): 1115–1125.
22. Dhar S, Dash PK. Differential current-based fault protection with adaptive threshold for multiple PV-based DC microgrid. *IET Renew. Power Gener.*, 2017; *11*(6): 778-790.
23. Mohanty R, Pradhan, AK. A superimposed current based unit protection scheme for DC microgrid. *IEEE Trans on Smart Grid*, 2018; *9*(4): 3917–3919.
24. Som S, Samantaray SR. Efficient protection scheme for low-voltage DC micro-grid. *IET Gener, Trans & Distri*, 2018; *12*(13): 3322–3329.
25. Monadi M., Gavriluta C, Luna A, Candela JI, Rodriguez P. Centralized protection strategy for medium voltage DC microgrids. *IEEE Trans on Power Deliv.*, 2016; *32*(1): 430–440.

26. Li C, Rakhra P, Norman PJ, Burt GM and Clarkson P, Multi-Sample Differential Protection Scheme in DC Microgrids, in *IEEE Journal of Emerging and Selected Topics in Power Electronics*, vol. 9, no. 3, pp. 2560–2573, June 2021, doi: 10.1109/JESTPE.2020.3005588.
27. Mohanty R, Balaji USM, Pradhan AK. An accurate noniterative fault-location technique for low-voltage DC microgrid. *IEEE Trans on Power Deliv.*, 2015;*31*(2): 475–481.
28. Yu M, Wang Y, Zhang L, Zhang Z. DC short circuit fault analysis and protection of ring type DC microgrid. *IEEE 8th International Power Electr. and Motion Control Conference (IPEMC-ECCE Asia)* 2016; 1694–1700.
29. Fletcher SD, Norman PJ, Fong K, Galloway SJ, Burt GM. High-speed differential protection for smart DC distribution systems. *IEEE Trans on Smart Grid*, 2014; *5*(5): 2610–2617.
30. Mohammadi J, Ajaei FB. Adaptive Voltage-Based Load Shedding Scheme for the DC Microgrid. *IEEE Access*, 2019; *7*, 106002–106010.
31. Salehi M, Taher SA, Sadeghkhani I, Shahidehpour M.. A Poverty Severity Index-Based Protection Strategy for Ring-Bus Low-Voltage DC Microgrids *IEEE Trans on Smart Grid*, 2019; *10*(6), 6860–6869.
32. Farhadi M and Mohammed OA, A New Protection Scheme for Multi-Bus DC Power Systems Using an Event Classification Approach, in *IEEE Transactions on Industry Applications*, vol. 52, no. 4, pp. 2834–2842, July-Aug. 2016, doi: 10.1109/TIA.2016.2545639.
33. Mohanty R, Pradhan, AK. Protection of smart DC microgrid with ring configuration using parameter estimation approach. *IEEE Trans on Smart Grid*, 2017; *9*(6): 6328–6337.
34. Barton JP, Infield DG. Energy storage and its use with intermittent renewable energy. *IEEE Trans on Energy Conversion*, 2004; *19*(2): 441–448.
35. Maqsood A, Corzine K. DC microgrid protection: using the coupled-inductor solid-state circuit breaker. *IEEE Electrification Magazine*, 2016; *4*(2): 58–64.
36. Zha Y, Yang L, Lehman B, de Palma JF, Mosesian J, Lyons, R. Decision tree-based fault detection and classification in solar photovoltaic arrays. In *Twenty-Seventh Annual IEEE Applied Power Electr. Conference and Expos. (APEC)* 2012; 93–99.
37. Upendar J, Gupta CP, Singh GK. Statistical decision-tree based fault classification scheme for protection of power transmission lines. *International Journal of Electric Power & Energy Syst.*, 2012; *36*(1): 1–12.
38. Kotsiantis SB., Zaharakis I, Pintelas P. Supervised machine learning: A review of classification techniques. *Emerging Artificial Intelligence Applications in Computer Engineering*, 2007; *160*(1): 3–24.
39. Sun K, Likhate S, Vittal V, Kolluri VS, Mandal S. An online dynamic security assessment scheme using phasor measurements and decision trees. *IEEE Trans on Power Systems*, 2007; *22*(4):1935–1943.

40. Wehenkel L, Pavella M, Euxibie E, Heilbronn B. Decision tree based transient stability method a case study. *IEEE Trans on Power Systems* 1994; *9*(1): 459–469.
41. El Arroudi K, Joos G, Kamwa I, & McGillis, DT. Intelligent-based approach to islanding detection in distributed generation. *IEEE Trans on Power Systems*, 2007; *22*(2): 828–835.
42. S. K. Shukla, E. Koley and S. Ghosh, "DC offset estimation-based fault detection in transmission line during power swing using ensemble of decision tree," in IET Science, Measurement & Technology, vol. 13, no. 2, pp. 212–222, 3 2019.
43. Samantaray, SR. Decision tree-based fault zone identification and fault classification in flexible AC transmissions-based transmission line. *IET Generation, Transmission & Distribution,* 2009; *3*(5): 425–436.
44. Manohar M, Koley E, Ghosh S. Enhancing the reliability of protection scheme for PV integrated microgrid by discriminating between array faults and symmetrical line faults using sparse auto encoder. *IET Renew. Power Gener.*, 2018;*13*(2): 308–317.
45. Gautam S, Brahma SM. Detection of high impedance fault in power distribution systems using mathematical morphology. *IEEE Trans on Power Systems*, 2012; *28*(2):1226–1234.

14

Passive Islanding Detection Method Using Static Transfer Switch for Multi-DGs Microgrid

Rahul S. Somalwar[1]* **and S. G. Kadwane**[2]

[1]*Department of Electrical Engineering, Bajaj Institute of Technology, Wardha, India*
[2]*Department of Electrical Engineering, Yeshwantrao Chavan College of Engineering, Hingna, Nagpur, India*

Abstract

The designing and controlling of the microgrid in both modes, stand-alone mode and utility-connected mode, is a central concern. After main grid interruption and resynchronization with the grid, the power sharing is an interesting task. Similarly, in a single-phase inverter-based microgrid, the islanding occurs during grid disconnection, which is one of the important aspects.

In this paper, appropriate relay management in a single-phase system is projected for islanding and active power allocation in between multi-distribution generations (DGs) units. Micro-switch acting as a relay is used here for disconnection and synchronization of DGs by using Static Transfer Switch (STS). There are several types of islanding detection methods (IDM) out of these passive islanding methods that are competent, financially cheaper and easily implemented; however, they have large non-detection zones (NDZ). The conventional passive islanding methods are unable to sense islanding conditions when the generating power is approximately equal to load power. In this paper, various passive and active islanding methods are discussed along with their merits and demerits, and individual lower-order harmonics are investigated along with STS, which, as a new effective passive islanding method, is proposed for the inverter centered utility connected system. This scheme can be properly useful along with the existing passive methods. The proposed method overcomes the limitation of the passive method and has a benefit that it is able to reduce the Non-Detection Zone (NDZ). The test result of simulation and experimental set up is analyzed, which fulfills the requirement of IEEE 1547 standards.

**Corresponding author*: rahulsomalwar@gmail.com

Nikita Gupta, Mahajan Sagar Bhaskar, Sanjeevikumar Padmanaban and Dhafer Almakhles (eds.) DC Microgrids: Advances, Challenges, and Applications, (421–448) © 2022 Scrivener Publishing LLC

***Keywords*:** Distribution generation, islanding detection methods, microgrid, static transfer switch, synchronization

14.1 Introduction

One of the biggest concerns in today's world is the energy crisis. As the assets of non-renewable sources of energy are decaying in nature attention needs to be placed on some replacements of these sources [1, 2]. The renewable sources of energy are solar, wind, small hydro, etc., but these types of sources are distributed; hence if there could be a way to syndicate all these sources then loads with higher power demand can be aided easily. The outline in which various distributed energy sources are mutually served altogether is simply a microgrid [3]. Technically, a microgrid is a contained grouping of energy sources and loads that usually operate connected to the main utility, but can separate and function independently as physical conditions dictate [4]. The microgrid is a sound evolution of simple distribution networks and can observe a high power of several DG sources. A classic microgrid system as shown in Figure 14.1 consists of generators, wind turbines, solar photovoltaic (PV) arrays, and other renewable technologies, along with batteries for long- and short-term storage. Load is associated to the distribution system at a node called as the point of common coupling (PCC) [5].

There are various research projects around the world focusing on the microgrid. The main objectives of this research is to attain correct power distribution during islanding by analysing amplitude of frequency and voltage. A special feature of a microgrid is it permits both islanding operation as well as synchronization of grid when a fault occurs and is cleared respectively in the utility [6].

An islanding process can occur under various circumstances such as pre-planned or unintentional causes. The unintentional islanding may take place due to a fault which affects the distribution system in the form of power quality issues or damage to equipment [26]. For islanding, interconnected code has been established for DGs which are linked with the utility [7]. As per IEEE 1547 standard, year 2003 DGs are disconnected from the load within 2 sec after the failure of grid synchronization [27]. There are various technical challenges associated with distributed generation which are discussed later.

During disconnection of a microgrid from the main grid, the power sharing is done by a micro-switch in the microgrid. Generally, interfacing is done between the grid and a switch which contains a logic gate called as

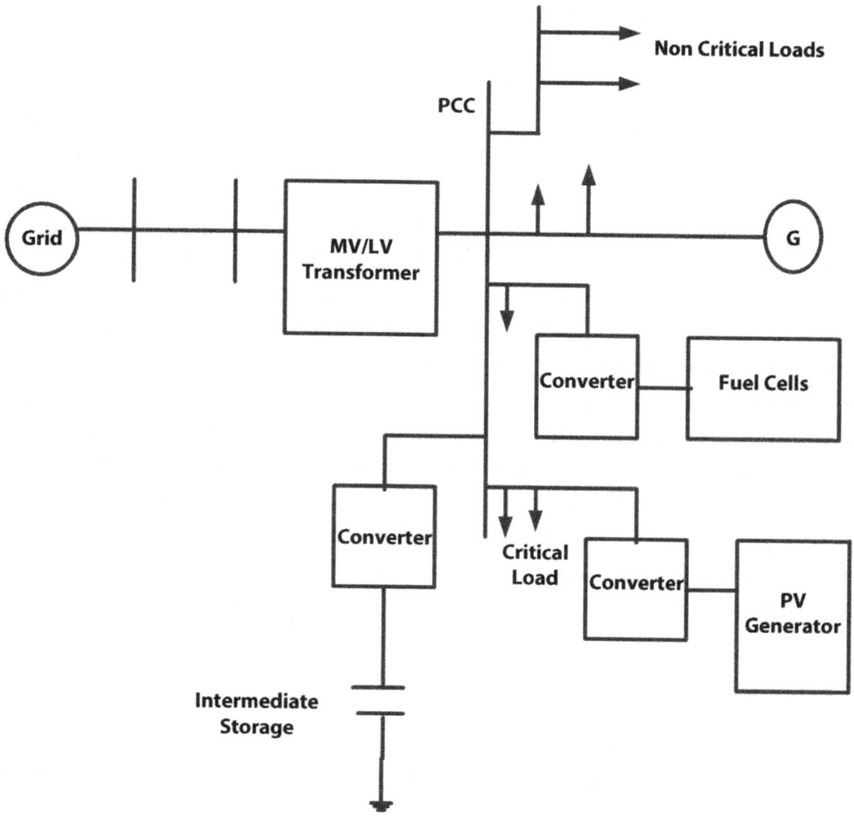

Figure 14.1 Typical microgrid structure.

a static transfer switch (STS) by micro-switch. This is a very important part for safety of the microgrid in both modes i.e., with utility and disconnected mode of grid operation in contrast to all faulty conditions [9].

In the section Islanding Detection Method (IDM), a detailed classification with its merits and demerits is reviewed. In brief, IDM are broadly classified as the remote and local method. The remote detection method is a reliable technique but its implementation is difficult and expensive. It is applicable for large-scale units with SCADA systems along with well-integrated control systems, and hence is not compatible with microgrid concepts. On the other hand, local methods are directly implemented at each generator, hence it is not expensive as a remote technique. Based on the information available on DGs side, the local method works, hence most of the researchers focus on these methods for detection of islanding. Local methods are again categorized into the active, passive, and hybrid method.

Local IDMs are applicable for synchronous generators, as well as inverter-based DGs or both.

In passive methods, detection of islanding is possible by monitoring the various system variables like phase difference, voltage amplitude, and frequency variation. Change in such parameters over its threshold values has been used to identify the condition of islanding. The passive islanding methods are unable to detect islanding conditions when the generating power is approximately the same as the load power. The islanding detection in a microgrid by using harmonic signatures is investigated in the literature for the fifth harmonic component [18]. Here in this paper there is a discussion about the power sharing at altered load circumstances and a static transfer switch which makes islanding and synchronization in the microgrid recommended. In line with above-mentioned work the individual lower-order harmonics are being investigated with STS in this paper with special emphasis on islanding detection when generated power is the same as load power requirement. The change in magnitude of the lower harmonic component at the power point coupling (PCC) between utility-connected and islanding modes of action is used for islanding detection.

14.1.1 Technical Challenges of Microgrid and Benefits

A microgrid is accomplished by functioning with the utility connected and a stand-alone mode and handling the evolutions among these stages. It delivers good solutions to provide power in case of a crisis and shortage of power during power disruption in the main grid. It allows an accelerated combination of renewable energy generation such as photovoltaic, wind and fuel cell generations without requiring redesign of the distribution system. Similarly, modern optimization can also be assimilated into the energy management system of a microgrid to enhance the efficiency, finances and resiliency. At the same time microgrids have some technical challenges regarding the protection systems and design of control. The most relevant challenges are:

a. Power flows in Bidirectional: Integration of DG units at low voltage levels can create an inverse power flow that may lead to generating problems, in coordination of protection systems, unwanted patterns in power flow, voltage control and fault current distribution.
b. Stability issues: In DG units, integration of control systems may cause local oscillations, requiring an in-depth small

disturbance stability analysis. Transient stabilities may form in both modes of operation in microgrid.

c. Modelling: The occurrence of a three-phase balanced condition, mainly inductive transmission lines and constant power loads, are not certainly valid for microgrid.
d. Low Inertia: The different large power systems, where a higher number of synchronous generators ensures a relatively large inertia, the microgrid might have low inertia characteristics, especially if there is a considerable share of power electronic interfaced DG units.
e. Uncertainty: In a bulk power system, the uncertainty is higher because of the lower number of loads and highly correlated variations of available energy resources. The operation of a microgrid contains very much uncertainty in which the economical and reliable operation of the microgrid is to be analyzed.

14.1.2 System with Multi-DGs

Figure 14.2 shows the block diagram of the multiple DGs distribution system having two DG sources and load connected at the PCC. In the figure, the two DGs which may be nonconventional power sources such as wind energy, photovoltaic or hydro power are connected with inverters. Due to any type of fault occurring in the system, the utility grid is detached by means of circuit breaker (CB) to avoid other damages and the DGs are adequate to complete the load requirement. As power generation of these DGs are such, an amount sufficient to fullfill the load demand.

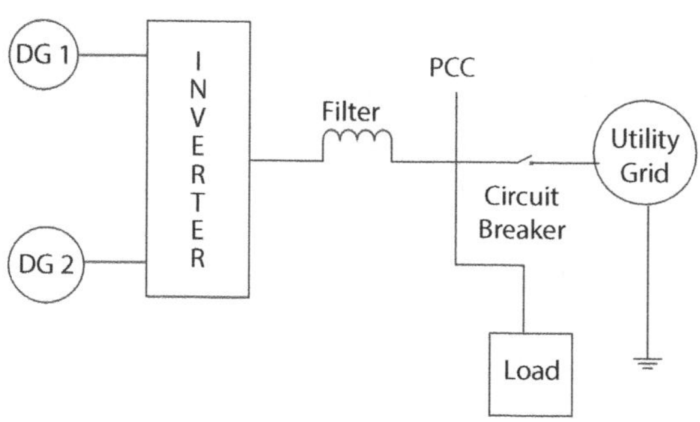

Figure 14.2 Block diagram of the multiple DGs distribution system.

14.1.3 Power Sharing Methods

The fast development of digital signal processors has brought an increase in control techniques for the parallel operation of inverters. These control schemes can be classified into two main groups, communication-based techniques and droop characteristics-based techniques. The first techniques are based on active load sharing, e.g., concentrated control, master/slave control, distributed control while the second techniques for the parallel operation of inverters are mainly based on the droop method. This technique consists of adjusting the frequency and voltage amplitude in terms of active and reactive power injected by the inverters. This novel method has been implemented for appropriate sharing of power and analysed the system stability at different condition. The droop control method is one of the important methods for appropriate sharing of active power without affecting the various variables at PCC. In the droop control concept, power sharing control strategies which operate without communications are considered. It can avoid complication and high costs as well as enhance redundancy and reliability of a supervisory system [25]. Also, this type of a system is simple to expand as the plug-and-play feature of the technique which permits to change an individual unit without closing the entire system.

14.1.3.1 Conventional Droop Control Method

The first method introduced for perfect sharing of power is the droop control method. It is implemented for perfect power sharing without changing its frequency and voltage at the PCC [8]. It is typically customary to provide reorganized control of power sharing without trusting on communications [9]. The sharing of active power is mostly accomplished by this method easily. Droop control is a control strategy commonly applied to creators for main frequency control (occasionally voltage control) to permit parallel operation of generator for load sharing. This is possible by categorization as active power sharing and reactive power sharing.

Disadvantages of existing droop control method: in islanding mode, the parameters like frequency and voltage are load dependent in microgrid. High frequency as well as voltage variations confirms by Steeper droop at the time of proper load sharing. The predictable droop method focuses on sharing of power but without considering harmonic in case of nonlinear loads. The impedances of the inverters at line and at PCC are different, due to which large circulating current flows in between inverters. Hence the method is not appropriate when such a system is connected in parallel and shares nonlinear load. There are many more techniques used in microgrid for power sharing.

14.2 Islanding

Islanding is a scenario or the condition in which the renewable energy sources (DGs) becomes electrically isolated from the utility, but still continue to provide power to the load by the DG unit connected to it. Islanding could be planned (intentional) or unplanned (unintentional). Intentional islanding is the resolute sectionalisation of the utility system in which the islanded portion is chosen to operate separately and DG is supplying adequate power to the loads, keeping voltage and frequency within essential limits. This is a harmless and common scenario especially planned for maintenance purposes. Unintentional islanding is hazardous and the major challenge in integrating DG with utility, as it has the perception that the system is unsynchronised due to frequency and voltage, and can destroy instruments; there is also the issue of worker safety. Reclosing under this condition can cause severe damage to the system, due to high current and large mechanical torque developed.

14.2.1 Challenges with Islanding

i. Safety concerns – electrician or workman safety can be endangered by DG sources serving a system after first/primary grid sources have been disconnected and marked out.
ii. Customer's equipment damage – end user equipment could be damaged, because the voltage and frequency may be varied compared with its standard limit.
iii. Power system may be islanded ineffectively earthed by the DG interconnection.
iv. Ending the failure – Immediate resynchronisation could result in 180-degree phase reclosing of DG, due to which huge mechanical torques are created and also high currents produced that can harm the prime movers or generators. At the same time, developed transients are possibly damaging to the grid and equipment of customers.

Due to all the above points, it is essential for DG to identify islanding accurately and immediately disconnect from the circuit; this is also referred to as anti-islanding. Hence standard rules like IEEE 1547, UL1741, and IEC 62116 are developed which order exact conditions while connecting DG with utility; these are discussed below. The key situation is that every DG system must recognize islanding within 2sec and must be capable to continue voltage and frequency within a specified standard limit [16].

14.2.2 Different Standards for Microgrid

Interconnecting Distributed Resources of Electric Power Systems having a standard of IEEE 1547 was permitted by the IEEE Standards Board in a June 2003 meeting and was available in July. This standard is the primary interconnection standard. For testing, installation, and communications some standards are developed by IEEE work groups. These standards were developed for uniform requirements and specifications globally needed to connect DG with the grid. The approval of this standard gives a guideline about business of the energy industry done and should impact the electrical distribution system to operate with DGs and two-way flow of electric energy. The main international standards are shown in Table 14.1 with their standard frequency and voltage range.

14.2.3 Islanding Detection Methods

In the Passive method, factors like magnitude of frequency and voltage, THD and phase displacement information at PCC are used for islanding detection. Selected modest methods are entrenched by using analysis of single amplitude. Traditional methods like over/under Voltage (OUV) and over/under Frequency (OUF) [10], Phase Jump in between voltage and current, Phase angle difference, are competent to identify islanding conditions by considering the variance between them at PCC [11]. By applying topical tools like wavelet transform [12], neural network or decision tree IDM have also been identified recently. System impedance [13], the rate of change of frequency (ROCOF) over power is presented which is represented by df/dp. Different methods having more than one quantity can also be found such as the change in frequency and active power. Similarly, variation in voltage imbalance and THD at PCC are used for islanding detection.

The harmonic distortion method is very useful for inverter-based DGs [17]. Passive methods are fast and do not inject any signal as a disturbance in the power system but having large Non-Detection Zone where it fails to notice the condition of islanding. In certain conditions where generated power and required load power is nearly equal, the variation in voltage and frequency is so small, which is within the permissible limit. In the

Table 14.1 Different anti-islanding standards. (* Normal)

Standard	Quality factor consider	Detection time of islanding	Frequency range f * (Nominal frequency f0)	Voltage range V * (% of nominal voltage)
IEC 6216	1	Time < 2sec	f between +/−1.5 of its Nominal frequency	between the range of 85% and 115%
IEEE 1547	1	Time < 2sec	f between 59.3 and 60.5 Hz	between the range of 88% and 110%
IEEE 929-2000	2.5	Time < 2sec	Same as IEEE 1547	between the range of 88% and 110%
Japanese Standard	0 rotating machinery	Passive: Time < 0.5s Active: 0.5s < Time < 1s	As per Setting value	As per Setting value
VDE 0126-1-1	2	Time < 2sec	f between 47.5 and 50.2 Hz	Same as IEC 6216, but lower side 80%
Korean standard	1	Time < 0.5s	Same as IEEE 1547	Same as IEEE 1547

active method, an additional external variable is to be introduced which is considered disturbance. It is inoculated in system voltage, frequency or phase angle to detect the islanding condition [14, 15]. It is more efficient than the passive method and less NDZ. However, it is not much faster as a passive method, because the system's inherent reaction time plays a vital role. On the basis of various parameters conventional passive, active and hybrid methods are matched in Table 14.2.

Table 14.2 Evaluation of IDM on different parameters

		Local islanding methods		
Sr no	**Parameters**	**Passive**	**Active**	**Hybrid**
1	Principle	considering local parameters like Voltage, current, frequency	Injection of small disturbances	Combination of Passive and Active both parameters
2	NDZ	Large	Less	Very Less
3	Detection Time	Small	Large	Large
4	Power Quality	No Degradation	High Degradation	Degradation
5	Cost	Low	Low	Low
6	Operation Failure	Depend upon threshold value	When Q factor high	Depends on the method select
7	Applicable for Multiple DGs	Possible	Difficult to implement	Possible
8	Effect on Power system	Nil	Yes, voltage fluctuation	Less than active method
9	Reliability	Low	High	High
10	Effectiveness	Depends on the condition	More effective	Very effective

14.3 Static Transfer Switch (STS)

Figure 14.3 shows the block diagram of a static transfer switch (STS). It is the only one way to connect the microgrid to utility grid that means the micro-switch is the elite construction in between the utility grid and microgrid [8].

The application of STS is to disconnect the microgrid to the utility grid under faulty condition as well as resynchronize the grid with the utility grid after the fault has been cleared. Switch S disconnected DGs from the utility grid and it resynchronized after receiving the signal from the STS control unit. The relay operation which is internally assembled in the micro-switch is discussed further.

Broadly, the STS consists of two units, Bridge of thyristor and Micro-switch unit control as shown in Figure 14.3. The task of a unit of micro-switch is to collect information of various parameters from both the grid and DGs sides and create the signal for islanding or resynchronization. A block diagram of micro-switch unit, and flow chart is discussed in [16] by considering various cases. For a proper synchronization, two quantities are essential; the first is the level of voltage and the second is the frequency of the grid and as well as at PCC also. In a single-phase system, phase sequence is not required. Micro-switch control block is a heart of the STS. For an under/over voltage and under/over frequency fault, itself micro-switch (STS) is used.

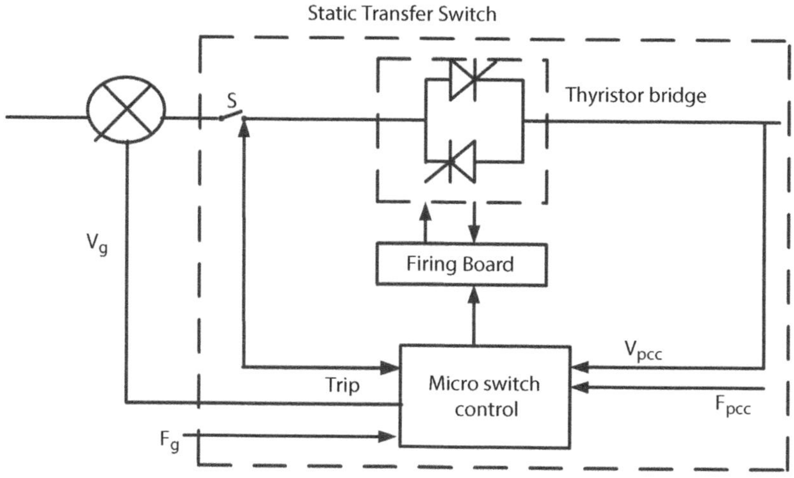

Figure 14.3 Block diagram of Static Transfer Switch (STS).

14.3.1 Simulation Results of STS

The grid connected with DGs, the first case is, if generating power is greater than Load Power ($P_L < P_{DG}$). In this case, when the grid disconnected at 0.1sec. Value of voltage suddenly goes high as shown in Figure 14.4 at PCC. The PCC voltage crosses its acceptable limit and hence the trip signal gives to DG_1. At 0.15, sec the load changes which increased the load power then generating power hence the DG_2 also tripped. The system is totally isolated from the grid and is in islanding condition. Now at 0.20 sec the load surges in an appropriate mode, the voltage limit is allowable, again synchronization condition tested and synchronized grid at 0.21sec.

In a next case the generating power is less than load power ($P_L > P_{DG}$) when the grid separated at 0.10sec. At present the PCC voltage is a smaller amount than the allowable limit hence UV relay trips the DG_1. At 0.15sec load decreases due to which DG_2 shares the needed power. Synchronization is done after 0.15 sec as the voltage at PCC within permissible limit, it will continue till 0.20 sec as shown in Figure 14.5. This relay organization specifies that sharing of power of the load is possible by considering the capacity of the DGs. Similarly, testing is done for DG_2 and finds satisfactory results.

In all these conditions the synchronization of a grid is manageable. Now synchronization is to be tested shown in Figure 14.6. The grid disengaged

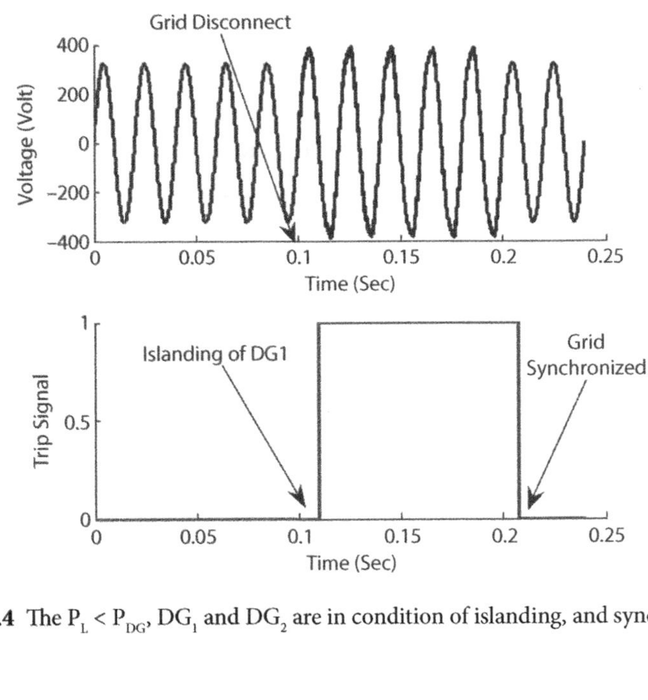

Figure 14.4 The $P_L < P_{DG}$, DG_1 and DG_2 are in condition of islanding, and synchronized grid.

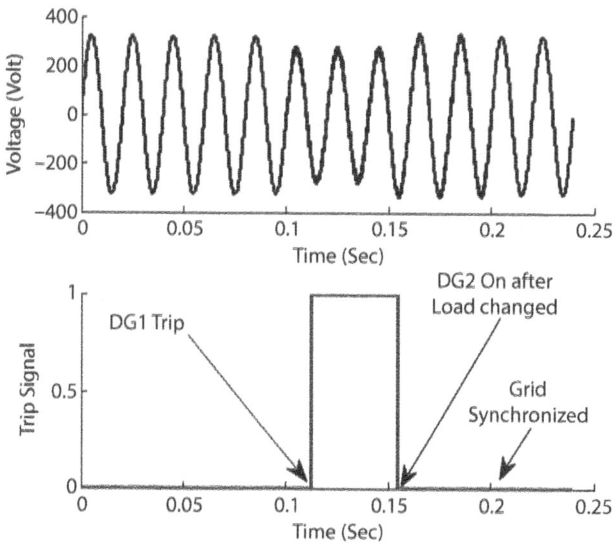

Figure 14.5 The $P_L > P_{DG}$ islanding condition of DG_1 power sharing of DG2 and synchronized grid.

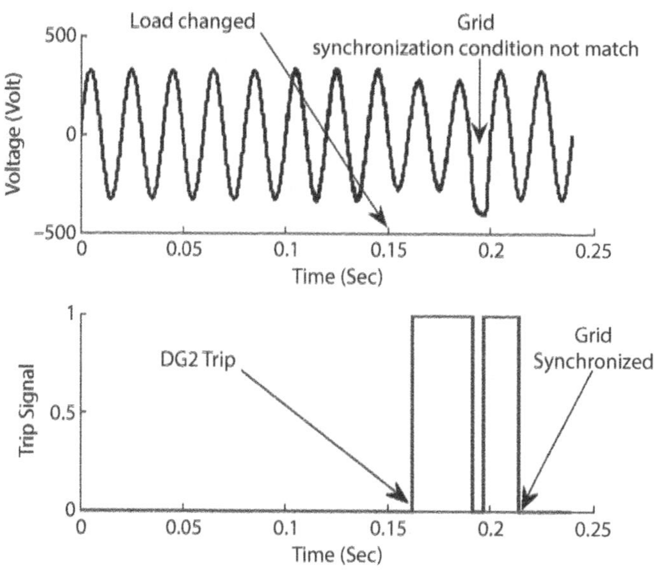

Figure 14.6 The $P_L > P_{DG}$ Power sharing of DG_1, islanding condition of DG_2, Grid not synchronized.

at 0.10 sec, DGs sharing the power as per the requirement. At 0.15sec the load again reformed which reflects on a voltage of PCC. The voltage of DG_2 crosses the allowable limit, and trips the DG_2 due to under voltage relay.

The load changes at 0.19, again testing the condition of synchronization. Here the condition of voltage does not fulfil synchronization condition hence relay operates and system is not resynchronized. After correcting the load, the grid is resynchronized as shown in Figure 14.6.

Equal Power Sharing is also tested by using the STS, details are available in [15]. Figure 14.7 shows that the voltage of DG_1, DG_2, and PCC are equal during grid disconnection at 0.10 sec, load changes at 0.15sec and at the grid synchronization. The relay coordination for this condition is also confirmed.

The voltage at DG_1, DG_2 and at PCC is nearly equal to grid voltage, when grid disconnected at 0.10 sec as well as at 0.15 after load changed. Grid resynchronized at 0.20 sec as the voltage value is equal even in this condition. The relays do not create the trip signal, the signal is high during all these conditions as shown in Figure 14.8.

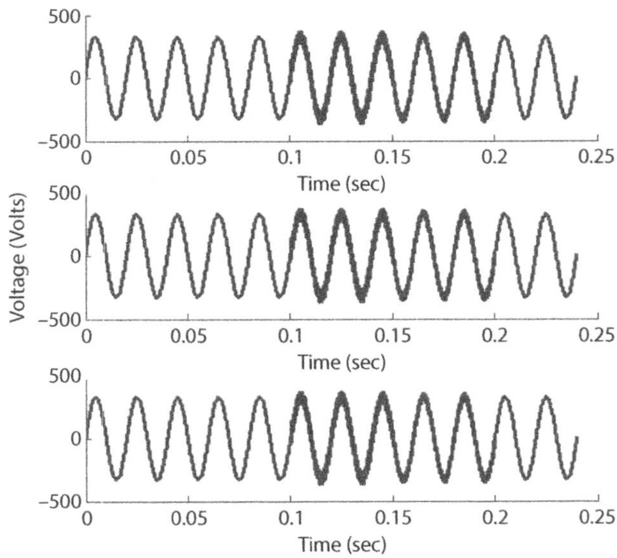

Figure 14.7 Voltage level of both DGs and PCC when $P_L = P_{DG}$.

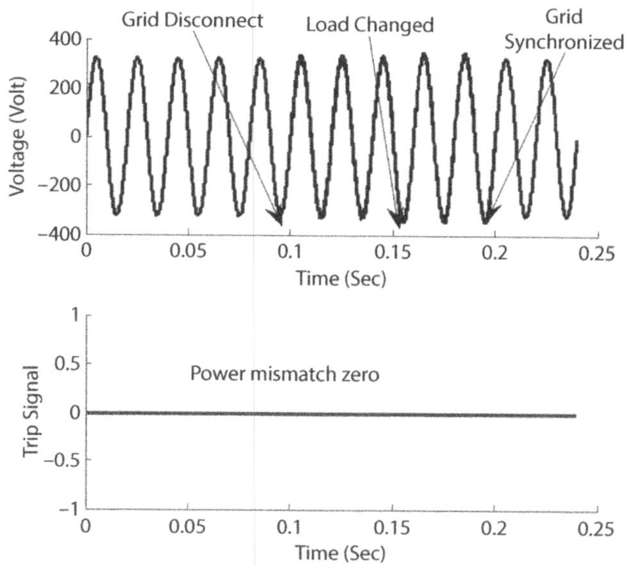

Figure 14.8 Voltage at PCC during three different condition and No tripping signal for the condition $P_L = P_{DG}$.

14.4 Proposed Scheme of Islanding

14.4.1 Proposed PV System

In the block diagram, two DGs are connected with an inverter, and inverters are associated with a grid as shown in Figure 14.9. The system contains a PV array, DC–DC boost converter, inverter, filter, grid supply, and the local load. In the controller unit, using the PI controller at PLL, current control is possible. The PCC is the point where all the devices are connected, and voltage and frequency are measured at this node for islanding. The outputs of the controller were used to generate triggering signals to the inverter by referring parameters of PCC and grid like V_{pcc}, I_{pcc}, V_g, and I_g. The triggering of the inverter is applied by the static transfer switch in an abnormal condition. The controller unit stops triggering the inverter and disconnects the load from the DGs. Structure of islanding unit which is inbuilt in the control unit and Grid interfacing are explained in [16, 17] with detailed architecture to enhance the voltage quality.

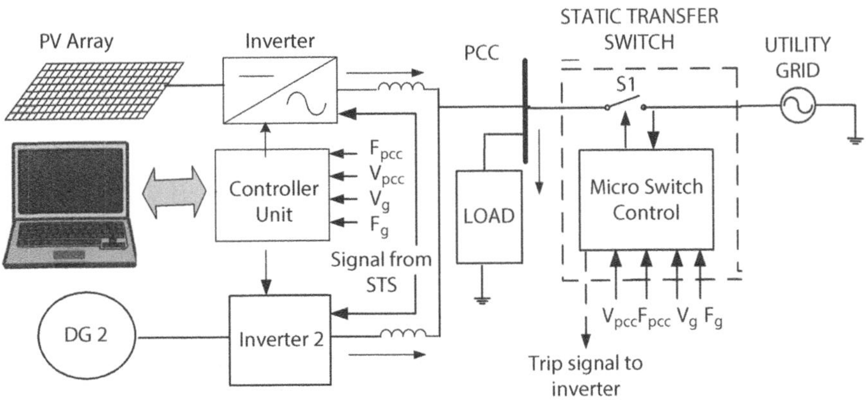

Figure 14.9 Proposed grid-connected PV system.

14.4.2 Mathematical Analysis of Harmonic Extraction

Using the individual harmonic extraction from the signal the equation has been developed.

$$U(t)=U_{dc}+\sum_{n=1}^{\infty}U_{(n)s}\sin(n\omega t)+U_{(n)c}\cos(n\omega t) \tag{14.1}$$

Any periodic waveform can be defined by a series of sine and cosine functions, which is also called Fourier series.

$$U_{(n)s}=\frac{1}{\pi}+\int_{0}^{2\pi}u(t)\sin(n\omega t)d\omega t \tag{14.2}$$

$$U_{(n)c}=\frac{1}{\pi}\int_{0}^{2\pi}u(t)\cos(n\omega t)d\omega t \tag{14.3}$$

Where n represents an integer and ω= 2π/T. Here the fundamental time period is T [19, 20].

In the Discrete Fourier Transform (DFT) for sample signal, the frequency spectrum obtained in [21] as follows:

$$X(k)=\sum_{n=0}^{N-1} x(n)e^{-i2\pi\frac{k}{N}} \quad (14.4)$$

Where k = 0, 1, 2 ...,N-1, x (n) is the amplitude at each sample and N is the number of samples over the period T.

In the case of DFT the 1st step is to calculate the fundamental frequency of the waveform to be investigated. A DFT permits the use of any integer number of trials in the analysis. In the practical application, where only a partial number of harmonics are essential, more precise DFT calculation can be approved in real time, hence it is not necessary to sacrifice performance for speed. DFT is the superior technology for investigating current/voltage waveforms [22–24].

14.5 Flow Chart

The working of the STS has already been discussed in the previous section. The various cases also consider for analysing the performance of the STS. The critical case is zero power mismatch, in which the islanding detection is not possible by conventional passive methods. At this stage the individual third harmonic is measured. Here the proposed system is explained in the flowchart as shown in Figure 14.10. The performance of the system was tested in three various conditions. All these cases are considered here in detail. The various parameters like V_{pcc}, V_{DG}, V_{g}, are measured. In the practical system, load may vary upward or downward, both the cases are reflected in the flowchart. The voltage affected if the power mismatched occurred, the islanding condition to be tested. The various three cases are discussed in detail as follows.

Case 1: If the required load power is less than the generated power of DG, voltage at PCC increases, and it exceeds the permissible limit, then DG should trip by STS. If value is in the permissible limit, check the value of 3rd order harmonics. The value exceeds the threshold limit islanding occurred, otherwise again the parameters should be checked.

Case 2: In this case, the load power is more than the generated power of DG, voltage reduced. If it crosses its edge limit then STS trip DG. If not then check the 3rd order harmonic.

Case 3: This is a critical condition, where passive islanding having large NDZ and STS is not able to identify Islanding. In this condition the generated power and load power are nearly equal. In this situation, if the

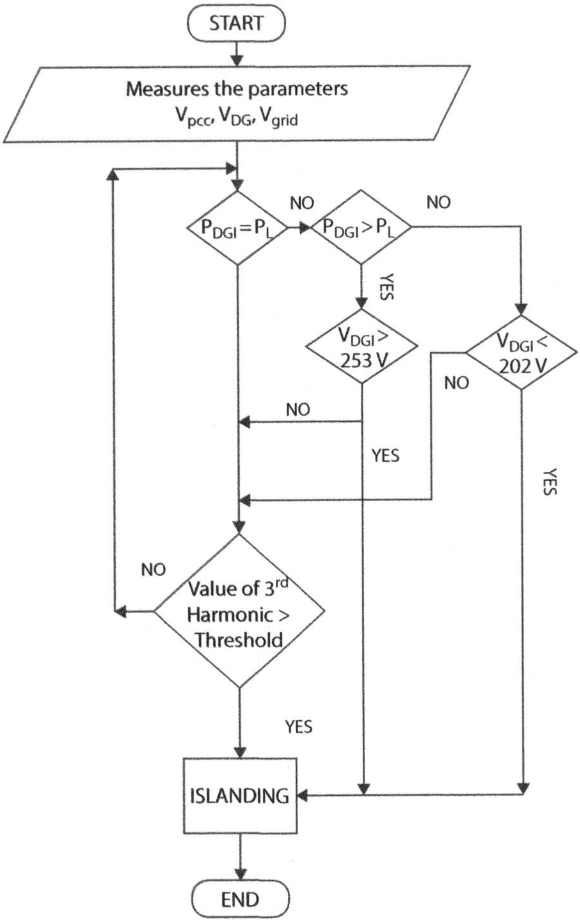

Figure 14.10 Flow chart.

individual harmonic crosses its threshold value (more than 5%) then islanding is done. If the value of 3rd harmonic is in the permissible limit then measure the various parameters again in the flowchart. When the system is in grid-connected mode the 3rd order harmonic signal is very small. The distortion in 3rd harmonics signal occurred when the grid disconnected.

14.6 Simulation Results

In the simulation system, the inverters are associated with the grid. The voltage of grid is 230V, the consumed power of the load 2.2kW, the value of the inductor used as a filter is 4 mH. Value of RL parallel load is depending upon the value of Q factor. A selection range of Q factor is from 0.5 to 2.5. The three operating conditions are considered in this paper.

In the first one, the load power (P_L) > the DG power (P_{DG}). The result of this condition is shown in Figure 14.11(a). The amplitude of the voltage is reduced which is below the threshold value, hence the under voltage relay trip the circuit. The lower order harmonics of voltage was analyzed, which is below 5%. This is shown in Figure 14.11 (b).

In the second condition, the $P_L < P_{DG}$, the magnitude of voltage increases above the threshold value shown in Figure 14.12 (a), over voltage relay trip in this condition. The individual harmonics of voltage was observed for this condition, which is also below 5% as shown in Figure 14.12(b). During this, after the grid disconnection, the DG power increases as the load increases.

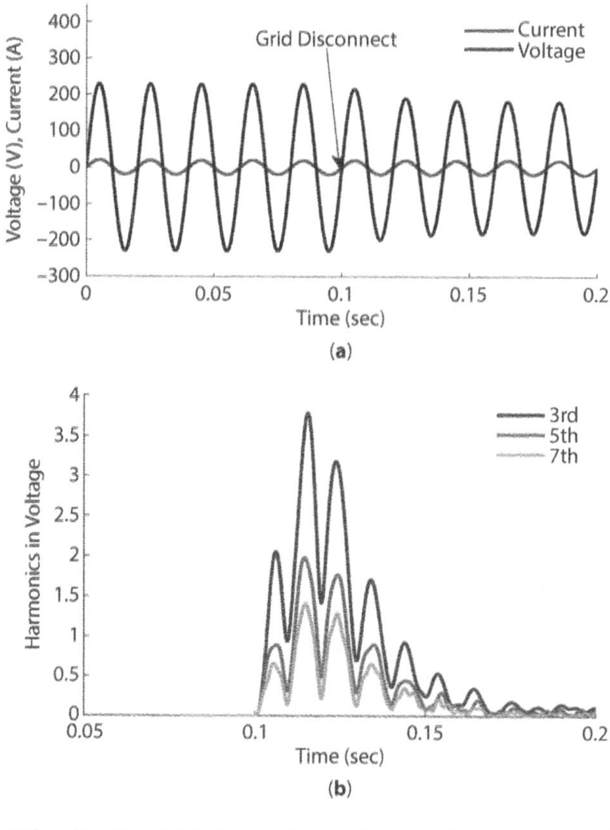

Figure 14.11 When $P_L > P_{DG}$. (a) Voltage reduces after grid disconnection. (b) Extraction of 3rd, 5th and 7th harmonics at PCC.

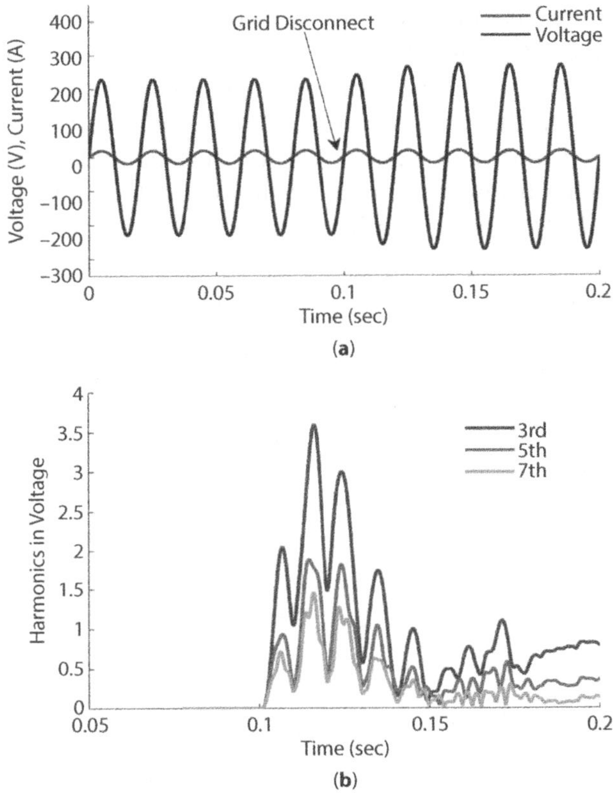

Figure 14.12 When $P_L < P_{DG}$. (a) Voltage increases after grid disconnection. (b) Extraction of 3rd, 5th and 7th harmonics at PCC.

In both of the above conditions relay trips the circuit, while in the third condition which creates the NDZ, the load power and DG power is nearly the same. In this case, the magnitude of voltage is within the permissible limit as shown in Figure 14.13(a). The individual harmonic of voltage is shown in Figure 14.13(b), the extraction of the third harmonic suddenly crosses the limit of 5% after the grid disconnected at 0.1 sec. During this condition the power drop is less. The lower order harmonic extraction is useful for islanding in this condition. The harmonic value crosses the permissible limit of harmonics as per IEEE standard 519-2014 for the power system.

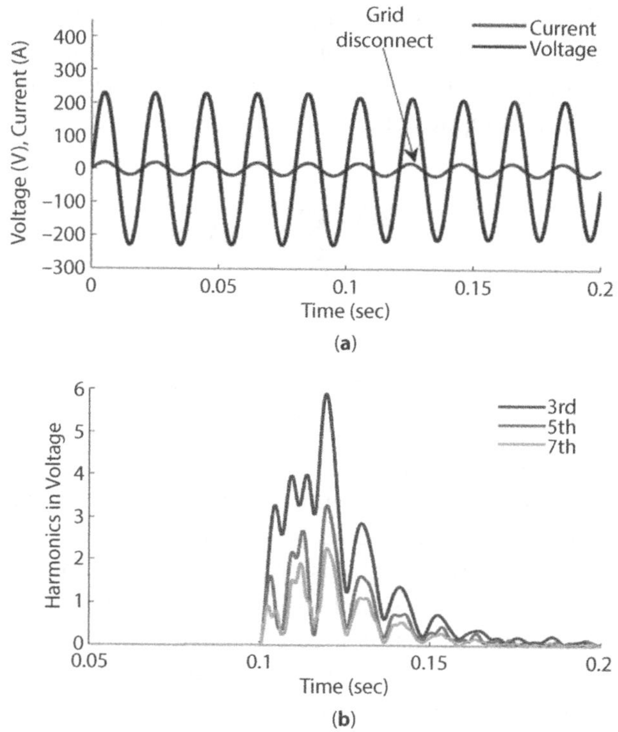

Figure 14.13 When PL = PDG. (a) Voltage same after grid disconnection. (b) Extraction of 3rd, 5th and 7th harmonics at PCC.

14.7 Experimental Results

Here decomposed the periodic waveform f (t) into the summation of a number of sinusoids waveform using the Equation of Discrete Fourier Transform from [10, 11]. The amplitude of DC components is A_0 and for AC voltage waveform A_0 is zero.

$$F(t) = A_0 + \sum_{V=1}^{\infty} (B_V \sin \omega t + C_V \cos \omega t) \tag{14.5}$$

For each harmonic, amplitude can be computed from equation 14.2.

$$A_v = \sqrt{B_v{}^2 + C_V{}^2} \tag{14.6}$$

By multiplying the corresponding sine wave and cosine wave, the coefficients B_v and C_v for every harmonic are calculated respectively for the input signal (equation 14.3 and 14.4). Where Δ_t is time between the sample, T is period, V_y is input signal.

$$B_V \approx \frac{2\Delta t}{T} \sum_{\gamma=1}^{\eta} V_\gamma \sin V \omega t_\gamma \tag{14.7}$$

$$C_V \approx \frac{2\Delta t}{T} \sum_{\gamma=1}^{\eta} V_\gamma \cos V \omega t_\gamma \tag{14.8}$$

Measuring circuit is going to connect to the incoming device of which the harmonics are to be calculated. This circuit has voltage divider circuit followed by the step-down transformer, which will convert 230/5V as shown in Figure 14.14. This signal fed to the microcontroller and microcontroller has a crystal oscillator circuit to provide the clock pulse for synchronizing purpose. One more circuit is connected to micro-controller which is in two parts used for indication purpose. The first one is displayed to show the values of THD, third harmonics, and fundamental voltage. The second is the trigger circuit which will work only when the value of THD goes beyond limit, i.e., 5%. The hardware of the STS is prepared by using the NAND and NOR gate circuit. Along with this, the comparator is used for the under/Over voltage and Under/Over frequency. The resynchronization is completed by comparing the magnitude and frequency.

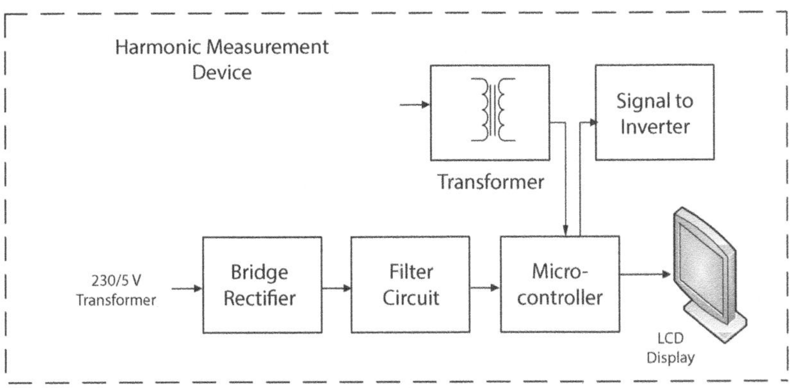

Figure 14.14 Block diagram of individual harmonic extraction.

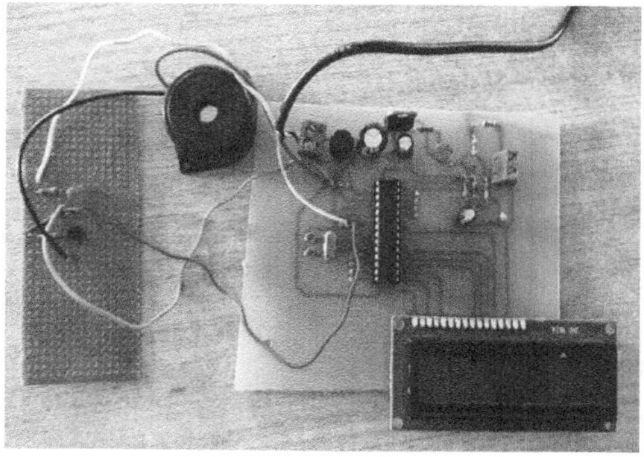

Figure 14.15 Harmonic measurement circuit.

Using the microcontroller with a resistor, capacitor, and crystal oscillator, a lower order harmonic measuring circuit developed which is shown in Figure 14.15. In the developed circuit, the individual 3rd order harmonic is measured with a different load.

A grid-connected prototype model was developed in the lab to measure an individual harmonics along with STS. The Inverter was developed by using the IRF 840 MOSFET switches and gate driver circuit designed using the TLP250. To generate the triggering signal digital signal processor disPIC33EP256MC202 was used. The autotransformer was used as the grid and coupling transformer isolate the circuit. 150 ohm resistance with small value of inductor is used as a load for the testing purpose. The results of hardware are shown in Figure 14.16. Figure 14.16 (a) shows the operation of STS when the load changes after grid disconnection. DG1 voltage is below its permissible range hence DG_1 trip, as observed in the simulation result.

Figure 14.16 (b) shows the result of islanding by operation at under voltage condition, Important case were power mismatch is zero, i,e., generated power and load power is the same, no voltage change is seen in Figure 14.16 (c), due to which the STF is unable to detect islanding. After connecting the device of harmonic detector, the islanding occurred which is shown in Figure 14.17 (a). Harmonic measurement circuit connected at PCC, measured 3rd harmonic before grid disconnection was 4.1%. After grid disconnection, the value was 5.51% which is more than the permissible limit due to which the circuit trips. The harmonic analysis is carried out by using the power analyser, a wave of third harmonic and fifth harmonic is shown in Figure 14.17 (b). As the simulation results shows that third harmonic is more than the fifth harmonic, similar result obtained in hardware, third harmonic waveform suddenly shoot up (blue in colour) as compared to the fifth harmonic.

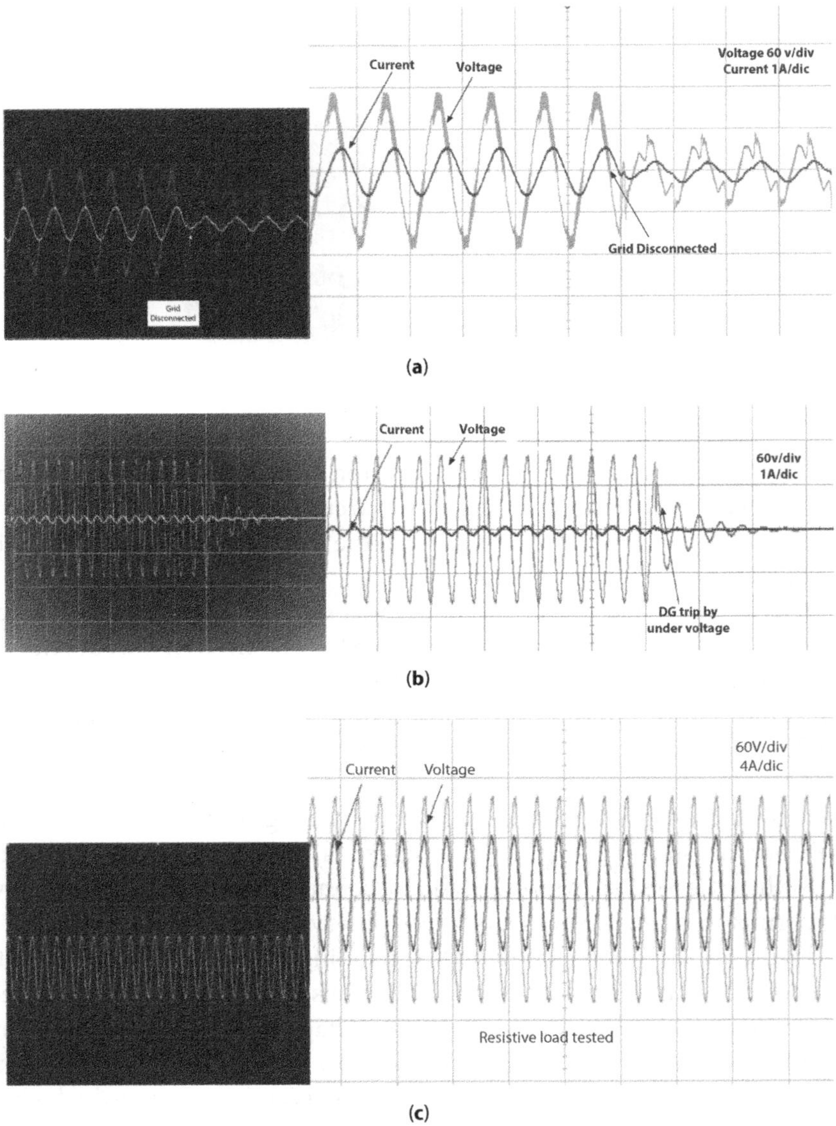

Figure 4.16 Hardware results. (a) Operation STS after load change. (b) Tripping at under voltage. (c) Power mismatch is zero (PG and PL).

The circuit which is developed during this work is very effectively used for measuring the lower order harmonic and its signal is used for the islanding purpose. For islanding, the harmonic measurement unit is developed at low cost. After doing some changes in the programming it is possible to measure the 5th, 7th order harmonics by this circuit. Limitation of this

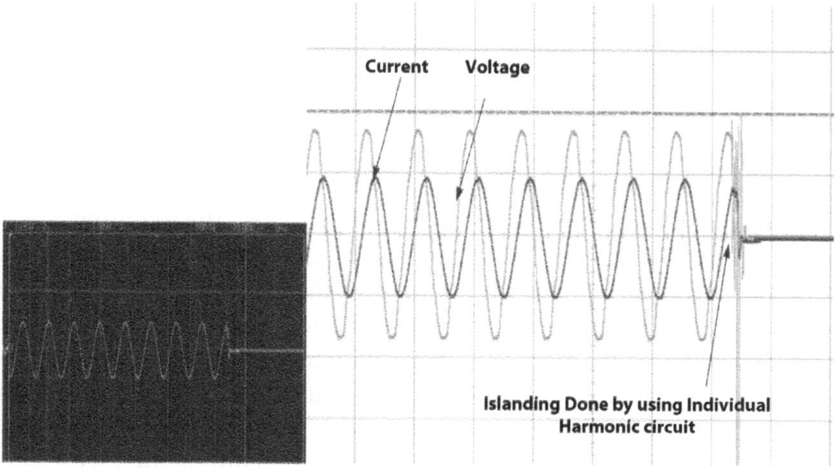

(a)

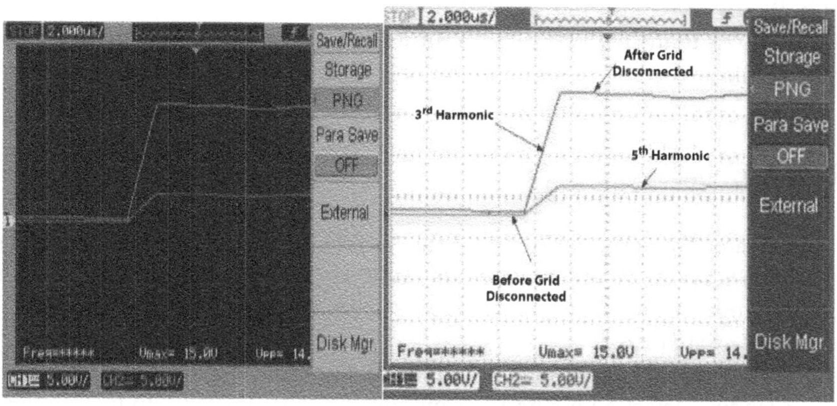

(b)

Figure 14.17 Voltage waveform at PCC with harmonic detection. (a) Islanding with harmonic detector. (b) Waveform of 3rd and 5th Harmonic.

proposed method is that even during the fault condition the variation in the harmonic may detect islanding. To avoid this drawback, the delay time was incorporated in the program.

14.8 Conclusion

A conventional passive method has limitations of a large non-detection zone in Multi-DGs microgrid. In this paper, the low-cost harmonic measurement circuit is developed to measure the value of individual harmonic and used for islanding detection along with STS. The working of STS has been tested and analyzed the various conditions. It is observed that if generating power is insufficient as compared to requirement of load then DGs trip. If this condition is momentary, then checking synchronization status and resynchronized microgrid using STS. STS is useful for islanding and resynchronization but not able to detect the islanding at zero mismatch condition. For this condition the magnitude of the voltage at PCC within the permissible limit after grid disconnection. The proposed method is useful for the condition where the active power of DG and load are nearly equal. Extraction of 3rd order individual harmonic analyzed for this condition and observed sudden changes at the moment when the grid disconnected. The variation in the value of 3rd harmonic before and after the grid disconnected is sufficiently high, where islanding identify. Hence the proposed method is used for islanding which overcomes the drawback of the conventional passive method. The simulation and hardware result shows that the proposed method is effectively used for the islanding detection.

References

1. Olivares, Daniel E., Ali Mehrizi-Sani, Amir H. Etemadi, Claudio A. Cañizares, Reza Iravani, Mehrdad Kazerani, Amir H. Hajimiragha *et al.* "Trends in microgrid control." *IEEE Transactions on Smart Grid*, Vol. 5, No. 4, pp. 1905–1919, March 2014.
2. Deshbhratar,P., Rahul Somalwar, and S. G. Kadwane. "Comparative analysis of islanding detection methods for multiple DG basedsystem." In *International Conference on Electrical, Electronics, and Optimization Techniques (ICEEOT)*, pp. 1525–1530. IEEE, 2016.
3. Ambrish Chandra, Singh B., Singh S. "Microgrid: Configurations, Control and Applications." *IEEE Transactions on Smart Grid*, Vol. 10, No. 2, pp. 1290–1302, March 2019.
4. Nagraju Pogaku, Milan Prodanvoic, Timothy C. Green, "Modeling, analysis and testing of Autonomous operation of an inverter based microgrid." IEEE Transactions on Power Electronics, Vol. 22, No. 2, pp. 917–926, March 2017.
5. Basak, Prasenjit, A. K. Saha, S. Chowdhury, and S. P. Chowdhury. "Microgrid: Control techniques and modeling." In *Universities Power Engineering Conference (UPEC), 2009 Proceedings of the 44th International*, pp. 1–5. IEEE, 2009.
6. Tang, Fen, Josep M. Guerrero, Juan C. Vasquez, Dan Wu, and Lexuan Meng. "Distributed active synchronization strategy for microgrid seamless

reconnection to the grid under unbalance and harmonic distortion." *IEEE Transactions on Smart Grid*, Vol. 6, No. 6 pp. 2757–2769, Aug. 2015.

7. Ng C.H., Busawon K., Putrus G.A. and Ran L. "Fast individual extraction technique", *IEEE Proc.-Gener. Transm. Distrib.*, Vol. 152, No. 4, July 2005.
8. Kroposki, Ben, Christopher Pink, Jonathan Lynch, Vinod John, S. Meor Daniel, E. Benedict, and I. Vihinen. "Development of a high- speed static switch for distributed energy and microgrid applications." In *Power Conversion Conference-Nagoya, 2007*. PCC'07, pp. 1418–1423. IEEE, 2007.
9. Farid Katiraei, and Mohammad Reza Iravani. "Power management strategies for a microgrid with multiple distributed generation units." *IEEE Transactions on Power Systems*, Vol. 21, No. 4, pp. 1821–1831, Dec. 2006.
10. Zeineldin, H. H., and Kirtley, J. L., "Performance of the OVP/UVP and OFP/UFP method with voltage and frequency dependent loads", *IEEE Trans. Power Del.*, Vol. 24, No. 2, pp. 772–778, Apr. 2009.
11. Kang S.W., Kim K.H, "Sliding mode harmonic compensation strategy for power quality improvement of a grid connected inverter under distorted grid condition", *IET Power Electron.*, Vol. 8, no. 8, pp. 1461–1472, Feb. 2015.
12. Pigazo A. Moreno V., Lissere M.,Aquila A., "Wavelet-based islanding detection algorithm for single-phase PV distributed generator systems," *IEEE, Trans on Power Del.* pp. 2409–2413, 2007.
13. Jia K., Wei H., Bi T., David W., and Tomas P., "An Islanding detection method for multi DG system based on High frequency impedance estimation" *IEEE Tran. on Sustainable Energy*, Vol. 8, no. 1, pp. 74–83, Jan. 2017.
14. Roop, M.E., Begovic M, and Rohatgi A., "Analysis and performance assessment of the active frequency drift method of islanding prevention", *IEEE Trans. Energy Conversion*, Vol. 14, No. 3, pp. 810–816, Sep. 1999.
15. Dash, P.K., Swain, D.P., Routray, A.,and Liew, A.C., "Harmonic estimation in a power system using adaptive perceptron", *IEEE Proceeding: Generation, Transmission and Distribution*, Vol. 143, No. 6, pp. 565–574, November 1996.
16. Rahul Somalwar, Kadwane S.G., S.Vaidya , Gawande S. "An analysis of active power sharing and synchronization in Multi DG Microgrid," INDICON 2017, 14th IEEE India Council International Conference, IIT Roorkee, Dec. 2017.
17. Rahul S. Somalwar, Sumant G. Kadwane, and Dusmanta Kumar Mohanta. "Harmonics-Based Enhanced Passive Islanding Method for Grid-Connected System." *Electric Power Components and Systems*, Vol. 45, No. 14, pp. 1554–1563. Dec. 2017.
18. Ng, C.H., Ran, L., Putrus, G.A., and Busawon, K.: 'A new approach to real time individual harmonic extraction'. *IEEE Power Electron. and Drive Systems Conference, Singapore, 2003.*
19. Hasan D., Azdemir M., "New extraction method for active reactive and individual harmonic components from distorted current signal." *IET Gener. Transm. Distrib.*, Vol. 8, Iss. 11, pp. 1767–1777, Mar. 2014.

20. Massoud A.M., Ahmed K.H., Finney,S.J.,and Williams,B.W.,"Harmonic distortion-based island detection technique for inverter-based distributed generation", *IET, Renewable Power Generation*, Vol. 3, No. 4, pp. 493–507, Dec. 2009.
21. Macias, J., Gomez Exposito, A., "Self-tuning of kalman filters for harmonic computation", *IEEE Trans. on Power Delivery*, Vol. 21, no. 1, pp. 501–503. Jan. 2006.
22. Cupertino, Francesco, *et al.* "Running DFT-based PLL algorithm for frequency, phase, and amplitude tracking in aircraft electrical systems." *IEEE Transactions on industrial Electronics*, Vol. 58, no. 3, 1027–1035, Jan. 2011.
23. Wang, Y. F., and Yun W. L.. "A grid fundamental and harmonic component detection method for single-phase systems," *IEEE Trans. on Power Electronics*, Vol. 28, no.5, pp. 2204–2213, July 2013.
24. Faqhruldin O. N., Saadany E. F., and Zeineldin H. H., "A universal islanding detection technique for distributed generation using pattern recognition," *IEEE Trans. Smart Grid*, Vol. 5, no. 4, pp. 1985–1992, Jul. 2014.
25. J. Zhou *et al.*, "Distributed Power Sharing Control for Islanded Single-/Three-Phase Microgrids with Admissible Voltage and Energy Storage Constraints," in *IEEE Transactions on Smart Grid*, Jan. 2021, doi: 10.1109/TSG.2021.3057899.
26. S. Murugesan and V. Murali, "Disturbance Injection Based Decentralized Identification of Accidental Islanding," in *IEEE Transactions on Industrial Electronics,* vol. 67, no. 5, pp. 3767–3775, May 2020, doi: 10.1109/TIE.2019.2917361.
27. A. Vukojevic and S. Lukic, "Microgrid Protection and Control Schemes for Seamless Transition to Island and Grid Synchronization," in *IEEE Transactions on Smart Grid*, vol. 11, no. 4, pp. 2845–2855, July 2020.

Index

AC Circuit Breakers (ACCBs), 16
Active Neutral-Point Clamped Multilevel Inverter (ANPC), 316
Adaptive Neuro-Fuzzy Interface System (ANFIS), 32
Arc Fault Current Interrupter, (AFCI), 45
Arc faults, 10
Arcing, 84
Artificial intelligence, 123

Bidirectional utility interface converter, 208
Boost converter, 77

Capacitor discharge stage, 14
Capacitor Clamped Multilevel Inverter (CCMLI), 299
Cascaded H-Bridge Multilevel Inverter (CHBMLI), 313
Centralized control strategy, 66
Clamp circuit, 288
Closed-loop control, 205
Common mode voltage, 18, 280, 299
Common-mode current, 280
Commutation, 280
Consensus-based algorithm, 203
Constant Power Loads (CPLs), 17
Constraints, 109
Continuous Conduction Mode (CCM), 329
Convergence of the power sharing correction, 200
Cuckoo search algorithm, 102
Current Injected Equivalent Circuit Approach (CIECA), 245
Current Source Converter (CSC), 17

Data-based protection scheme, 32
DC bipolar fault characteristics, 370
DC distribution network topology, 363
DC fault, 6
DC microgrid, 1,64
DC unipolar fault characteristics, 368
DC-DC converters, 243
Decentralized controller, 67
Decision tree, 401
Demand side management, 167
Denominator polynomial of lower-dimensional model, 259
Diesel generator, 106
Differential protection, 25
Diode freewheeling stage, 14
Diode-Clamped Multilevel Inverter (DCMLI), 305
Direct current, 63
Discontinuous Conduction Mode (DCM), 335
Distributed controller, 68
Distributed Energy Supplies, (DERs), 17
Distributed Generators (DGs), 2
Disturbance Detection (DI), 32
Droop Control Method, 426

Droop controller, 71
Dual active bridge topology, 208
Dynamic analysis for power converters, 243
Dynamics of the power sharing correction, 200

Economic dispatch, 101
Economics, 25
Energy management system (EMS), 122
Energy storage system, 78

Faddeev leverrier algorithm, 253
Fault clearing time, 84
Fault current capacity, 17
Fault Current Limiting, (FCL) 44
Feature Condensation, (FC) 32
Feature Extraction (FE), 32
Field-programmable gate array, 378
Flying capacitor multilevel inverter, 309
Fuel cell, 77
Full bridge modular multi-level converter, 362
Fuses, 47

Gaussian distribution fitting, 13
Genetic algorithm, 110
Geographical, 168
Green energy conversion applications, 243
Greenhouse gas, 161
Grey wolf optimizer, 133
Grid-connected mode, 414
Grounding, 3, 18

Half bridge modular multi-level converter, 362
Harmonic extraction, 436
Hessian matrix, 171
High Impedance Faults (HIFs), 25
High-frequency pulse width modulated converter, 280
Hybrid DC Circuit Breakers (HCB), 38
Hybrid Renewable Energy System (HRES), 161

Impedance-based protection, 29
Indian Metrological Department (IMD), 168
Input filter, 284
Intelligent Electronic Device (IED), 88, 376
Intermittent, 77
International Electrotechnical Commission (IEC) Standard, 5
Islanded mode, 414
Islanding, 95
Islanding detection method, 423

Jacobian matrix, 171

Kalman Filter (KF), 11

Leverrier's algorithm, 241
Line Commutated Converters (LCCs), 2
Line To Ground (L-G) fault, 3
Line To Line (L-L) fault, 3
Linear programing, 123
Local islanding methods, 430
Lock-In Amplifier (LIA), 10

MATLAB/Simulink, 115
Matrix converter, 275
Matrix converter fed drive system, 283
Maximum Power Point Tracking (MPPT), 354
Mean Biased Error (MBE), 165
Meta-Heuristic, 123
Metal Oxide Varistor (MOV), 35
Meteorological, 168
Micro hydro, 162
Microgrid, 63
Mixed Integer Linear Programming (MILP), 123
Model predictive, 123

Modeling techniques, 245
Modulation, 280
Modulation methods for matrix converter, 274
MPPT algorithm, 77
Multilevel control, 68
Multilevel inverters, 299

National Aeronautics and Space Administration (NASA), 168
National Electric Code (NEC), 46
Neutral grounding, 366
Nominal voltage, 209
Non-Detection Zone (NDZ), 421
Numerator polynomial of lower-dimensional model, 261

Optimization, 101
Order reduction of transfer function, 255
Over/Under Frequency (OUF), 427
Over/Under Voltage (OUV), 427
Overcurrent (O/C) relays, 7

Particle swarm optimization, 101
Pattern search algorithm, 111
Percentage differential protection (PDP), 10
Performance, 21
Perturb and observe algorithm, 355
Phasor measurement unit (PMU), 31
PI control loop, 74
Plug-In electric vehicles, 184
Point of common coupling, 422
Pole clustering method, 258
Pole to pole fault, 4
Power electronic interface, 241
PQ control, 31
Protection challenges, 3
Protective Devices (PDs), 30
Protection schemes, 10
Protection zone division, 373

Quasi-newton techniques, 171
Quasi-Z-source networks, 325

Rate of Change of Frequency (ROCOF), 427
Recovery of DC line, 378
Reduced gradient, 102
Reduced-order linear differential equation-based modeling method, 244
Reduce-order modeling 253
Reliability, 21
Renewable Energy Sources (RESs), 2
Resistor-Capacitor-Diode (RCD), 35
Reverse Blocking-Integrated Gate Commutated Thyristor (RB-IGCT), 45
Root Mean Square Error (RMSE), 165

Short circuit stable control stage, 14
Short-Circuit (SC) analysis, 13
Simplicity, 25
Solar generation, 106
Solar PV, 162
Solid State Circuit Breakers (SSCBs), 8, 42
Sparse matrix converter, 277
Speed, 21
Static transfer switch, 424
Stochastic & robust programming, 123
Switched inductor based quadratic boost converter, 254
Switching commutation of bidirectional switches, 280

Two-stage conversion, 276

Ultrafast Mechanical Switch (UFMS), 41
Ultra-Sparse Matrix Converter, 278
Unintentional islanding, 422

Venturini's direct transfer function approach, 280
Very sparse matrix converter, 277
Virtual impedance, 69
Voltage gain, 331
Voltage multiplier cells, 326
Voltage Source Converter (VSC), 17
Voltage Source Inverters (VSIs), 299
Voltage stress on semiconductors, 331
Voltage-Frequency (U-F) Control, 31
Voltage-mode control law for regulation, 244

Wind generation, 106

Zero-crossing, 84
Z-Source DC circuit breakers, (ZSB), 35

Also of Interest

Check out these other related titles from Scrivener Publishing

SMART GRIDS AND MICROGRIDS: Concepts and Applications, Edited by P. Prajof, S. Mohan Krishna, J. L. Febin Daya, Umashankar Subramaniam, and P. V. Brijesh, ISBN: 9781119760559. Written and edited by a team of experts in the field, this is the most comprehensive and up to date study of smart grids and microgrids for engineers, scientists, students, and other professionals. *NOW AVAILABLE!*

MICROGRID TECHNOLOGIES, Edited by C. Sharmeela, P. Sivaraman, P. Sanjeevikumar, and Jens Bo Holm-Nielsen, ISBN 9781119710790. Covering the concepts and fundamentals of microgrid technologies, this volume, written and edited by a global team of experts, also goes into the practical applications that can be utilized across multiple industries, for both the engineer and the student. *NOW AVAILABLE!*

INTEGRATION OF RENEWABLE ENERGY SOURCES WITH SMART GRIDS, Edited by A. Mahaboob Subahani, M. Kathiresh and G. R. Kanagachidambaresan, ISBN: 9781119750420. Provides comprehensive coverage of renewable energy and its integration with smart grid technologies. Encyclopedia of Renewable Energy, by James G. Speight, ISBN 9781119363675. Written by a highly respected engineer and prolific author in the energy sector, this is the single most comprehensive, thorough, and up to date reference work on renewable energy. *NOW AVAILABLE!*

Green Energy: Solar Energy, Photovoltaics, and Smart Cities, edited by Suman Lata Tripathi and Sanjeevikumar Padmanaban, ISBN 9781119760764. Covering the concepts and fundamentals of green energy, this volume, written and edited by a global team of experts, also goes into the practical applications that can be utilized across multiple industries, for both the engineer and the student. *NOW AVAILABLE!*

Energy Storage, edited by Umakanta Sahoo, ISBN 9781119555513. Written and edited by a team of well-known and respected experts in the field, this new volume on energy storage presents the state-of-the-art developments and challenges in the field of renewable energy systems for sustainability and scalability for engineers, researchers, academicians, industry.

Energy Storage 2nd Edition, by Ralph Zito and Haleh Ardibili, ISBN 9781119083597. A revision of the groundbreaking study of methods for storing energy on a massive scale to be used in wind, solar, and other renewable energy systems. *NOW AVAILABLE!*

Hybrid Renewable Energy Systems, edited by Umakanta Sahoo, ISBN 9781119555575. Edited and written by some of the world's top experts in renewable energy, this is the most comprehensive and in-depth volume on hybrid renewable energy systems available, a must-have for any engineer, scientist, or student. *NOW AVAILABLE!*

Progress in Solar Energy Technology and Applications, edited by Umakanta Sahoo, ISBN 9781119555605. This first volume in the new groundbreaking series, Advances in Renewable Energy, covers the latest concepts, trends, techniques, processes, and materials in solar energy, focusing on the state-of-the-art for the field and written by a group of world-renowned experts. *NOW AVAILABLE!*

A Polygeneration Process Concept for Hybrid Solar and Biomass Power Plants: Simulation, Modeling, and Optimization, by Umakanta Sahoo, ISBN 9781119536093. This is the most comprehensive and in-depth study of the theory and practical applications of a new and groundbreaking method for the energy industry to "go green" with renewable and alternative energy sources. *NOW AVAILABLE!*